Interdisciplinary Applied Mathematics

Volume 57

Problems in engineering, computational science, and the physical and biological sciences are using increasingly sophisticated mathematical techniques. Thus, the bridge between the mathematical sciences and other disciplines is heavily traveled. The correspondingly increased dialog between the disciplines has led to the establishment of the series: Interdisciplinary Applied Mathematics.

The purpose of this series is to meet the current and future needs for the interaction between various science and technology areas on the one hand and mathematics on the other. This is done, firstly, by encouraging the ways that mathematics may be applied in traditional areas, as well as point towards new and innovative areas of applications; and secondly, by encouraging other scientific disciplines to engage in a dialog with mathematicians outlining their problems to both access new methods as well as to suggest innovative developments within mathematics itself.

The series will consist of monographs and high-level texts from researchers working on the interplay between mathematics and other fields of science and technology.

Jim M. Cushing
Department of Mathematics
University of Arizona
Tucson, AZ, USA

Shandelle M. Henson
Department of Mathematics and Department of Biology
Andrews University
Berrien Springs, MI, USA

James L. Hayward
Department of Biology
Andrews University
Berrien Springs, MI, USA

ISSN 0939-6047 ISSN 2196-9973 (electronic)
Interdisciplinary Applied Mathematics
ISBN 978-3-031-34282-0 ISBN 978-3-031-34283-7 (eBook)
https://doi.org/10.1007/978-3-031-34283-7

Mathematics Subject Classification: 92D40, 92D50

This Springer imprint is published by the registered company Springer Nature Switzerland AG
The registered company address is: Gewerbestrasse 11, 6330 Cham, Switzerland

Paper in this product is recyclable.

Jim M. Cushing • Shandelle M. Henson • James L. Hayward

Modeling Behavior and Population Dynamics

Seabirds, Seals, and Marine Iguanas

Springer

Preface

We have entered the Anthropocene, an era of unprecedented, human-induced changes in the biosphere, geosphere, and atmosphere. An understanding and appreciation of the animals with which we share the planet is important for esthetic, ethical, and practical reasons. A significant part of that understanding concerns the behavior of animals and how behavior changes in response to environmental change.

In simplest terms, animal behavior can be defined as muscle movement. Muscle movement occurs in response to internal stimuli such as hormone levels or nervous system activity, and to external stimuli such as predator activity or changes in weather. The study of animal behavior is focused on patterns of animal movement, the internal and external stimuli that elicit these patterns, and the evolution of these patterns. Patterns of movement include things like shifts from one habitat to another, migration, visual signaling, vocalization, courtship, copulation, and predation.

Niko Tinbergen (1907–1988) shaped many of our contemporary concepts of animal behavior. He posited that every behavior should be examined within the context of four questions: How does it develop? How is it physiologically controlled? What is its function? What is its evolutionary origin? [19] Our studies on habitat selection and reproductive behavior in gulls, flightless cormorants, harbor seals, and marine iguanas address the third and fourth of these questions. We study how weather, food supplies, and other environmental variables influence behaviors, and we make predictions about how those behaviors will change in response to changes in environmental parameters and in evolutionary time.

Significant variation in behavior among individuals often thwarts attempts to make predictions about things like population dynamics, habitat occupancies, and the spread of behaviorally driven diseases. An understanding of the dynamics of behavior, however, can be achieved through the application of mathematical modeling, a powerful set of tools that allows not only the perception of how behavioral systems work in the present but also the prediction of how they will function in the future. Since the beginning of our collaboration in 2001, our Seabird Ecology Team has been among the first to use dynamical systems-based models to successfully describe and predict the behavior of vertebrate animals in their natural habitats in response to environmental change. This book summarizes our research and highlights the techniques we used to achieve our goals.

Our research involves three main procedures:

(1) The collection of large sets of temporally dense behavioral data on natural populations in the field
(2) The construction of realistic mathematical models rigorously tied to data, which describe, explain, and predict behavioral dynamics in relation to environmental variables
(3) The construction of simplified, proof-of-concept models to probe dynamic mechanisms, to suggest testable hypotheses, and to study the consequences of environmental change and evolving traits

Purpose and Structure of Book

Results from our work have been published in several dozen papers scattered among diverse mathematical and biological journals. This monograph assembles our main ideas, techniques, and data sets into a coherent whole. By assembling this information within a single volume, we hope to facilitate researchers who wish to apply our modeling approach to their own systems of study, whatever the species. As this volume reveals, we have successfully applied our approach to groups of non-avian reptiles, birds, and mammals, and it could be applied to any group of behaving organisms, from paramecia to humans.

This is not a textbook. Each chapter is self-contained, although we have endeavored to arrange the chapters within each section in a somewhat logical sequence of increasing complexity of techniques. This was not always possible because the same modeling technique might be applied to different organisms in successive chapters, with each chapter describing a unique data gathering process on which the model was predicated.

The book is organized into three parts. The first two parts are empirically based and connect models to data; the third part is theoretical but is motivated by the empirical studies in Part II. Topics are organized by the types of scientific questions being asked rather than by mathematical methods.

Part I deals only with the dynamics of aggregate animal behavior as it occurs locally in time and space, without following population dynamics across generations. We use various methods, including logistic regression, differential equations, and a hybrid of regression and differential equations, to model the behaviors of glaucous-winged gulls, harbor seals, marine iguanas, and flightless cormorants.

Part II focusses on two specific behaviors with major population and evolutionary import in gulls: egg cannibalism and the related adaptation consisting of egg-laying synchrony, which are documented with statistical methods involving logistic regression and discrete-time models of synchrony. These studies are presented as a package because they motivate Part III. Part II also contains a chapter on eagle-gull predator-prey dynamics, which employs a differential equation model of Lotka-Volterra type.

Part III is a step-wise theoretical investigation of the population and evolutionary consequences of cannibalism and reproductive synchrony from a bifurcation point of view using discrete-time matrix models for structured populations. We first consider a simple but general two-dimensional discrete-time model of adult-on-juvenile cannibalism and use a Darwinian dynamic extension to consider whether cannibalism can be an evolutionarily stable strategy. We then add a second adult class to the model in order to consider the interplay between cannibalism and reproductive synchrony. A crucial scientific problem is how to model the population dynamic consequences of these behaviors when they occur on a time scale shorter than the generational time scale. We conclude Part III by addressing this issue with a general framework based on nonlinear Leslie models that contain two time scales—daily and yearly—in which reproductive and feeding behaviors occur on a daily basis whereas generations are tracked from year to year.

Data Collection

To capture the dynamic patterns of animal behavior, time-series data must be collected on a finer temporal scale than behavioral and environmental fluctuations. For the marine organisms we study, behaviors often are driven by tidal and diurnal patterns, and we have found that data collected hourly over complete 2-week tidal cycles work best. Depending on the project, we typically collect data at the top of each hour for 16 or 17 hours per day for several weeks. This kind of dense data collection requires either assistants or continuous video or sound recordings from which data are later transcribed.

We work with large aggregations of animals at seabird colonies, seal haul-out sites, and marine iguana haul-out sites. If the number of individuals is not too large, we count every individual in the study area; if the group is very large, we count only individuals within representative sample areas. If we are evaluating fluctuations in behavior, we determine the number of individuals exhibiting the behavior in proportion to the total number of individuals present.

There is no substitute for thorough, temporally dense, long-term observation of the biological system to be modeled. This is time intensive and can be logistically difficult, but it yields two opportunities that cannot be realized in any other way. First, a thorough observation of a system, day in and day out, week after week, year after year, affords a deep understanding of the system that sometimes bears little resemblance to studies based on short-term samples. This is the kind of understanding that allows a scientist to pose realistic models that can be tied rigorously to data. Second, long and temporally dense time series are exactly what are needed to parameterize continuous-time models.

Construction of Realistic Models

Historically, behavior dynamics have been modeled using Markov chain analysis, game theory, and individual-based models [4, 5, 15, 17]. In addition, ordinary differential equations (ODEs) with "motivational" dependent variables were developed in the 1970s as qualitative behavior models of individual animals [8, 14]. These qualitative models, however, were not tied rigorously to empirical data because motivational variables are not measurable [18].

Our research relies on an interdisciplinary paradigm derived from the interface between population ecology and dynamical systems theory [3]. Models designed to quantitatively describe and predict animal behavior in field populations serve as testable scientific hypotheses, with measurable dependent and independent variables operating on scales at which deterministic trends emerge from variability among individuals. We use compartmental models (difference equations and ODEs) in which each compartment represents a specific behavioral state at a specific spatial location, and in which the state variables track the numbers of individuals in each compartment [9]. Applying these models to any particular behavioral system requires specifying the flow rates between compartments by means of modeling assumptions. In general, these flow rates depend on environmental factors. They also can be functions of population densities and/or time, in which case the models are nonlinear and/or nonautonomous. If the system recovers rapidly after disturbance, ODEs can be reduced to algebraic models on two time scales: one for disturbance dynamics and one for steady state dynamics [12].

Dynamic patterns depend, of course, on scale [13]. A seabird colony, for instance, is a complicated place at the scale of the individual. Single birds appear quite autonomous and exhibit a wide variety of complicated behavioral choices; yet we have shown that some behaviors at the aggregate level are highly deterministic and can be predicted as a function of environmental variables. For example, gulls leave colony and loafing areas in large numbers to feed when the tide is going out and the sun is high, and they return with an incoming tide in the evening [7, 10]; harbor seals leave the beach and move into the water with rising tides because the incoming current brings in food [2, 6]; Galápagos marine iguanas move from land to feeding sites in the sea primarily in response to changing patterns of solar radiation [16]. Some behaviors are more deterministic than others, and it is possible to rank them according to the degree of determinism.

Some researchers feel that methods of modeling animal behavior at the aggregate level are too coarse and that modeling behavior must involve individual-based models. Although our approach is not the only way or always the best way to model animal behavior, our results show that ODEs and difference equations can indeed accurately predict the behavior dynamics of animal groups [9, 10, 11].

Model Parameterization

Estimating model parameters requires a stochastic version of the model that accounts for the noise structure [3]. For example, in many of the systems we have studied, stochastic perturbations are largely uncorrelated in the hourly sample times, and the stochastic model can be written

$$\varphi\left(N\left(\tau+1\right)\right)=\varphi\left(G\left(\tau,N\left(\tau\right)\right)\right)+E\left(\tau\right),$$

where $N=(N_1,N_2,\ldots,N_m)$ is the vector of state variables, E is a vector from a multivariate normal random distribution with variance-covariance matrix $\Sigma=(\sigma_{ij})$, and $G(\tau,N(\tau))$ is the deterministic model prediction at hour $\tau+1$ based on the state of the system at hour τ. Here φ is a variance-stabilizing transformation; the transformations $\varphi(x) = \ln x$ and $\varphi(x) = \sqrt{x}$ render environmental and demographic stochasticity, respectively, approximately additive [3].

The one-step conditioned residual error vectors are

$$\rho(\tau + 1) = \varphi(n(\tau + 1)) - \varphi(G(\tau, n(\tau))),$$

where n is the vector of observations. The likelihood function, which measures the likelihood that the residuals arose from a joint normal distribution, is a function of the model parameters, and its maximizer is the vector of parameter estimates [3].

Alternative Models and Model Selection

If models serve as testable hypotheses, then we can pose alternative models as a means to test alternative hypotheses. Information theoretic methods of model selection, such as the Akaike Information Criterion (AIC), take into account both the value of the likelihood function and the number of parameters, so that models with more parameters are penalized for over-fitting. This criterion allows one to select the best model from a suite of alternatives [1].

Model Validation

A good model not only describes and explains, but also predicts. Validating a model means testing its predictability on an independent data set that was not used to estimate its parameters. One can validate a model by estimating parameters from a "calibration" data set and computing the goodness-of-fit of the fitted model on that data set, and then comparing that to the goodness-of-fit on an independent "validation" data set without re-estimating parameters. Goodness-of-fit can be computed with a generalized R^2 [3]. For example, when we modeled numbers of gulls loafing on a pier (Chap. 1), we obtained $R^2 = 0.58$ for the calibration data set and $R^2 = 0.61$ for the validation data set [10], which supported model validation. For our model of Galápagos marine iguana haulout (Chap. 3), $R^2 = 0.77$ for the calibration data set and $R^2 = 0.80$ for the validation data set [16], again supporting model validation. Close correspondence in goodness-of-fit between calibration and validation data sets, as in these cases, suggests that a model captures the major dynamics of a system.

The most convincing models are those that make unexpected, a priori predictions that are borne out by new experiments. Most seabird biologists, for example, would have predicted that during high tides gulls, which are intertidal feeders, should be loafing near the breeding colony and not away feeding. Our model predictions, however, counterintuitively suggested that during high tides occurring at midday, gulls should be away feeding—which is exactly what we observed [10].

Proof-of-Concept Models

In addition to models tied rigorously to data, we also construct and analyze "proof-of-concept" models. These are simplified models which focus on a small number of key mechanisms of interest. They are typically of low dimension (involve a small number of state variables) and are amenable to rigorous mathematical analysis. Our goal is to use such models to probe the dynamic consequences of the mechanisms under consideration, thereby not only suggesting hypotheses that can explain the observations we (and others) have made but also guide future observational protocols. In the process, these models also sharpen ideas and clarify definitions.

Given the nature of the phenomena we investigate, we often use discrete-time models (difference equations or maps) of structured populations. This methodology distinguishes different categories of individuals and focuses on interactions among them. The methodology is most appropriate when the stages have strongly discrete characteristics, such as juvenile and adult or larva-pupa-adult stages, and life history events occur at well-defined time intervals. A mathematical advantage of these discrete models is that unlike continuous-time structured models, they pose no difficult problems with regard to basic questions concerning the definition and existence of solutions and their numerical simulation, and the validity of basic analytic methods (such as linearization, bifurcations, etc.). The unit of time in our discrete models is generally determined by either a basic time scale in the modeled population (e.g., a maturation period) or the census time interval.

The predictions of a model, for example its equilibrium states and stability properties, are determined by the model coefficients or parameters which represent various vital rates (birth, survival, resource consumption, etc.). A significant change in a model prediction caused by a change in a model parameter is called a bifurcation. By "significant change" one typically means a change in the stability properties of an equilibrium (or some other attractor, such as a periodic state). A destabilization of equilibrium generally signals the creation of a new attractor. Such bifurcations can be of significant interest because the old and new attractors generally predict very different dynamic and long-term consequences for a population. For example, the most basic difference concerns survival versus extinction. Others include changes from equilibrium to cyclic fluctuations (crash and boom outbreaks) or other dramatic states such as chaos. The parameter values at which a bifurcation occurs is called a bifurcation point (or sometimes a tipping point).

One compelling reason we are interested in bifurcation points in our models is that the location of most of our field studies, Protection Island, Washington in the Salish Sea, is experiencing significant climate change (specifically, increased mean sea surface temperature in surrounding waters) which, in turn, effects certain vital rates of the populations we are modeling. Thus, climate change is represented in our models by a change in model parameters, which can cause bifurcations to occur. A basic question to ask of a model is, then, where are the bifurcation points that indicate a threat to the survival of the population? Which vital rates are involved and are most important? We can also inquire as to what counter changes in other parameters might mitigate the threat of extinction. We can use a model to investigate whether observed behavioral changes in the population that highly correlate with climate change are likely to be adaptive in the long run. Using Darwinian dynamic (evolutionary game theory) modeling methodology, we also investigate conditions under which these adaptive changes can be attained by means of natural selection.

Conclusion

A burgeoning human population, propped up by synthetic fertilizers and energy from fossil fuels, is taking a decided toll on resource availability, climate, and biodiversity. Each of these ecosystem components plays a role in how animals and humans behave and thrive. There has never been a more opportune time for mathematicians and biologists to collaborate to create a better understanding of the behavior of animals—from the honeybees that pollinate our crops, to the birds that bring us pleasure but interfere with jet engines, to the vast herds of wildlife that provide esthetic enjoyment and crucial ecosystem services. We believe the techniques highlighted in the following chapters will play a positive role in the understanding and health of the animal components of our planet.

Jim M. Cushing, Tucson, AZ, USA
Shandelle M. Henson, Berrien Springs, MI, USA
James L. Hayward, Berrien Springs, MI, USA

References

1. Burnham KP, Anderson DR (2002) Model selection and multi-model inference: a practical information-theoretic approach, 2nd edn. Springer, New York
2. Cowles JD, Henson SM, Hayward JL, Chacko MW (2013) A method for predicting harbor seal (*Phoca vitulina*) haulout and monitoring long-term population trends without telemetry. Nat Resour Model 26:605–627
3. Cushing JM, Costantino RF, Dennis B, Desharnais RA, Henson SM (2003) Chaos in ecology: experimental nonlinear dynamics. Academic Press, San Diego, CA
4. Dugatkin LE, Reeve HK (1998) Game theory and animal behavior. Oxford University Press, Oxford, UK
5. Gottman JM, Roy AK (1990) Sequential analysis: a guide for behavioral researchers. Cambridge University Press, New York
6. Hayward JL, Henson SM, Logan CJ, Parris CR, Meyer MW, Dennis B (2005) Predicting numbers of hauled-out harbour seals: a mathematical model. J Appl Ecol 42:108–117

7. Hayward JL, Henson SM, Tkachuck RD, Tkachuck CM, Payne BG, Boothby CK (2009) Predicting gull/human conflicts with mathematical models: a tool for management. Nat Resour Model 22:544–563

8. Hazlett BA, Bach CE (1977) Predicting behavioral relationships. In: Hazlett BA (ed) Quantitative methods in the study of animal behavior. Academic Press, New York, pp 121–144

9. Henson SM, Dennis B, Hayward JL, Cushing JM, Galusha JG (2007) Predicting the dynamics of animal behavior in field populations. Anim Behav 74:103–110

10. Henson SM, Hayward JL, Burden CM, Logan CJ, Galusha JG (2004) Predicting dynamics of aggregate loafing behavior in glaucous-winged gulls (*Larus glaucescens*) at a Washington colony. Auk 121:380–390

11. Henson SM, Hayward JL, Cushing JM, Galusha JG (2010) Socially induced synchronization of every-other-day egg laying in a seabird colony. Auk 127:571–580

12. Henson SM, Hayward JL, Damania SP (2006) Identifying environmental determinants of diurnal distribution in marine birds and mammals. Bull Math Biol 68:467–482

13. Levin SA (1992) The problem of pattern and scale in ecology. Ecology 73:1943–1947

14. McFarland DJ (1971) Feedback mechanisms in animal behaviour. Academic Press, London, UK

15. Mangel M, Clark CW (1988) Dynamic modeling in behavioral ecology. Princeton University Press, Princeton NJ

16. Payne BG, Henson SM, Hayward JL, Megna LC, Velastegui Chá vez SR (2015) Environmental constraints on haul-out and foraging dynamics in Galápagos marine iguanas. Journal of Coupled Systems and Multiscale Dynamics 3:208–218

17. Railsback SF (2001) Getting 'results': the pattern-oriented approach to analyzing natural systems with individual-based models. Nat Resour Model 14:465–475

18. Slater PJB (1978) Data collection. In: Colgan PW (ed) Quantitative ethology. Wiley, New York, pp 8–24

19. Tinbergen N (1963) On aims and methods of ethology. Zeitschrift Für Tierpsychologie 20:410–433

Acknowledgments

We thank the US National Science Foundation for funding most of the research reported in this volume under grants DMS 0314512, DMS 0613899, DMS 1022494, and DMS 1407040 to SMH and JLH; and DMS 0847777, DMS 0917435, and DMS 1407564 to JMC. We also thank the Andrews University Office of Research and Creative Scholarship for the Faculty Grant support of our Galápagos Island research; director John F. Stout and his successor, Gary Burdick, facilitated the funding and financial management of our research and encouraged our efforts.

Refuge managers Robert Edens, Jean Takekawa, Kevin Ryan, and Jennifer Brown-Scott, Washington Maritime National Wildlife Refuge Complex, US Department of the Interior, permitted our work on Protection Island, and their associates Chris Columbus, Pamela Sanguinetti, Lorenz Sollmann, Sue Thomas, and Ulrich Wilson provided valuable support. USFW volunteer caretakers also aided our research efforts. Galápagos National Park issued a permit for our research on Isla Fernandina, Galápagos, in 2011.

We thank Ross Anderson, a journalist with a love for the sea and a passion for supporting researchers, for many generosities, including ferrying us to and from Protection Island when it was impossible or impractical to use our own boat. Behavioral physiologist Gordon Atkins joined our Seabird Ecology Team in 2011. Since then, he has made crucial intellectual contributions to our work. His energy and creativity added significantly to the quality of our research. Joe Galusha, a behavioral biologist on our team, opened his research site and facilities on Protection Island to us, provided valuable insights into gull behavior, shared important tips on data collection, and supported our efforts logistically. We have benefited from the statistical expertise of our colleagues Brian Dennis, Robert Desharnais, and Lynelle Weldon in several projects. James Watson, Washington Department of Fish and Wildlife, provided crucial data on bald eagle populations for Chap. 10. We also thank the 11 graduate students and more than 50 undergraduates at Andrews University who aided in the collection of data in the field, analyzed data in the lab, and helped to write up the initial reports of projects. Two former graduate students, Libby Megna and Amanda Sandler, continued to support our research efforts with their expertise long after completing their degrees. Amy Veprauskas contributed significant mathematical analysis to some of our models in her PhD dissertation (Interdisciplinary Program in Applied Mathematics at the University of Arizona, Tucson) and subsequent publications. Susana Velastegui Chávez provided important assistance with data collection during our work in the Galápagos Islands.

Rosario Beach Marine Laboratory (RBML) served as the base for our research on Protection Island; we are particularly indebted to David Habenicht, RBML manager, who cheerfully oversaw our boating and storage needs each field season. Cape George Colony Board generously allowed us to utilize their private Cape George Marina during our work on Protection Island; we are grateful to Ross Anderson for facilitating this arrangement. Mike Heckinger provided boat transport for us several times as well. Marty Bluewater, the only permanent resident on Protection Island, did the same and provided friendship over the decades.

In 2011, personnel of the Charles Darwin Research Station, Puerto Ayora, Galápagos, arranged for transportation to and from Isla Fernandina and provided safety training before we embarked for that remote island; we particularly appreciate the efforts of Sonia Cisneros and Roberto Pépolas in this regard. We benefited from the seamanship and hospitality of Captain Lenin Cruz, who safely piloted us to and from Isla Fernandina, Galápagos, in *La Pirata*.

Contents

Part I

Modeling Behavior in Space and Time

The timing, spacing, and location of animal behaviors, along with transitions from one behavioral state to another, occupy the attention of biologists for both theoretical and practical reasons. Theoretically, information regarding these phenomena allows for interpretations concerning origin and function. Practically, this information can lead to better resource management, as well as to improved health and safety for humans with whom animals share Earth's resources.

The following six chapters describe how we applied principles of mathematical modeling and statistics toward a better understanding of the three groups of marine animals: birds, mammals, and non-avian reptiles. Previous attempts to relate mathematics to changes in behavior states in animals often have been purely theoretical, without direct connections to actual data. In contrast, we tie mathematical models of behavior directly to data obtained from free-living animals in their natural habitats.

Chapters 1–3 use differential equation models to test predictions concerning specific behaviors engaged in by groups of glaucous-winged gulls, harbor seals, and marine iguanas, respectively. Chapter 4 illustrates how techniques developed in the previous three chapters can be used to make predictions concerning systems of multiple behaviors. In Chap. 5 we use generalized linear models and multi-model inference to examine the responses of flightless cormorants to environmental variables. Chapter 6 applies logistic regression and Darwinian dynamics to explore how a behavior used for one function might evolve to perform an additional function.

Chapter 1

Predicting the Dynamics of Aggregate Loafing Behavior in Gulls

Research reported in this chapter was the initial project carried out by the Seabird Ecology Team. It provided one of the first rigorous demonstrations that a differential equation model using environmental variables can be used to describe, explain, and accurately predict the aggregate behavior of marine vertebrates. It also served as the impetus and conceptual basis for much of the work in the remaining chapters of this volume. Hypothesis formation, data collection, and analysis were carried out by senior researchers Shandelle Henson, James Hayward, and Joseph Galusha, with the assistance of undergraduate students Christina Burden and Clara Logan. The initial report of this work was published in The Auk *in 2004 [20].*

1.1 Introduction

Many animals enhance their fitness by moving selectively from habitat to habitat [5, 9, 22, 38], and habitat selection is a central concern of ecologists, evolutionary biologists, and resource managers [36]. Ecologists examine relations between habitat and an animal's morphology, physiological requirements, and behavior in an effort to understand the role of the animal in an ecological community; evolutionary biologists are concerned with these same variables, except that they are viewed in the context of fitness and adaptive processes; and resource managers attempt to identify and create optimal habitat features to provide attractive settings within which animals of interest can thrive and multiply.

J. M. Cushing et al., *Modeling Behavior and Population Dynamics*, Interdisciplinary Applied Mathematics 57, https://doi.org/10.1007/978-3-031-34283-7_1

In ecology, the analysis of spatial distribution patterns of organisms is an important subdiscipline [34]. Many studies of spatial distribution involve statistical or stochastic analyses, as well as regression models. Moreover, theoretical studies commonly incorporate deterministic dynamic models (e.g., continuous-time differential equations or discrete-time difference equations), although typically these models have not been tested in the field nor are they well connected to data. Ideally, deterministic mathematical models function as testable hypotheses. Statistical techniques connect models with data, and laboratory or field experiments test model predictions. This ideal set of circumstances is often difficult to achieve given a paucity of data, difficulty of experimentation, or mobility of research subjects [6, 7].

A major challenge in modeling ecological dynamics is the determination of scales at which the asynchronous behavior of individuals forms patterns and the identification of mechanisms leading to these patterns [22, 23, 30]. Quantum-level randomness in matter self-assembles into deterministic rules for the properties and behavior of larger objects. Similarly, although individual animals may move relatively independently due to individual differences and social interactions, patterns in habitat utilization may emerge for animal groups [30].

In [20] we constructed and tested a differential equation model for the prediction of numbers of animals occupying a single habitat in relation to environmental variables. We used glaucous-winged gulls (*Larus glaucescens*, Fig. 1.1) as our model organisms. Gulls are among the most widespread and thoroughly studied birds on the planet. They make excellent field study subjects because they are large, active during the day, and feed, nest, and "loaf" together in large aggregations [1, 11, 15, 28, 31, 33]. "Loaf" is a behavior category that refers to preening, defecating, standing, resting, sitting, or sleeping in nonfeeding and nonbreeding locations. Gulls typically favor loafing habitats with good visibility [5, 38], a propensity that often puts them at odds with human cleanliness, health, and safety [2, 32, 39].

Loafing behavior by gulls is correlated with tide height, most likely due to changes in food availability and/or time of day [10, 13, 26, 38]. Typically, the number of loafing gulls in a habitat rises with rising tides and decreases around midday. These correlations led us to pose four hypotheses upon which our mathematical model is based:

(H1) Numbers of loafing gulls fluctuate in direct response to an environmental variable $E(t)$, selected from the following set of alternative sub-hypotheses:

- (H1A) $E(t)$ = tide height
- (H1B) $E(t)$ = 1/solar elevation
- (H1C) $E(t)$ = tide height/solar elevation

(H2) Numbers of loafing gulls during daytime can be described with a two-compartment model, with one compartment being the loafing area and the other compartment encompassing everywhere else.

Figure 1.1: Male (background) and female (foreground) glaucous-winged gulls on their breeding territory at Protection Island National Wildlife Refuge, Washington. Note the different head shapes of the two birds and the larger body size of the male (Photo by J. L. Hayward.)

> (H3) Gulls enter the loafing site at a per capita rate proportional to $E(t)$ and exit the site at a per capita rate inversely proportional to $E(t)$.
> (H4) The total number of gulls in the system is proportional to the weekly maximum number occupying the loafing site, which varies during the course of the year.

Hypotheses H1–H4 are biological hypotheses. They are also modeling assumptions which we converted into mathematical equations in order to test them with empirical data. In the following sections, we describe how to rigorously connect this mathematical model to hourly time series data for numbers of gulls at a loafing site where individual birds enter and leave the site asynchronously throughout the day. We show how we validated the model on an independent data set and then predicted future numbers of loafing gulls during each daylight hour for 29 consecutive days during the next nesting season.

1.2 Methods

Our study was carried out with glaucous-winged gulls loafing on a pier (Fig. 1.2) adjacent to a breeding colony at Protection Island National Wildlife Refuge (48°7'N, 122°55'W) in the Salish Sea, Strait of Juan de Fuca, Washington, USA. During this study, the breeding colony contained more than 3,000 gulls.

Figure 1.2: Glaucous-winged gulls used this pier in a small marina at Protection Island National Wildlife Refuge, Washington, for "loafing". (**a**) Pier structure. (**b**) Close-up of gulls loafing on the pier (Photos by J. L. Hayward.)

For an initial application of mathematical modeling to this field system, we chose to restrict our study area to the pier, a well-defined site that was easy to census. The pier was located at the west side of a small marina closed to the public, with limited use by humans. It was used during most daylight hours for loafing by both breeding and nonbreeding gulls.

1.2.1 Historical Data

Counts of loafing gulls on the pier were made each hour, from 05:00 to 20:00 hours (LST, here and below), May to August 1997, 1998, 1999, and 2001. Hourly sampling allowed us to track tidal and diurnal periodicities [22]. Tides in the water surrounding Protection Island are semidiurnal with strong diurnal inequalities in the lows. Approximately every 14 days, tidal "nodes" with minimal amplitudes occur.

We partitioned the historical data using stratified random sampling by dividing each 14-day tidal period into approximately four quarters and, from each quarter, randomly selecting half the data for model calibration and the other half for independent model validation (Fig. 1.3).

1.2.2 A General Mathematical Model

At each point in time we classified individual gulls in the system into two compartments: those on the pier and those not on the pier. We used a rate balance equation,

$$\frac{dN}{dt} = [\text{inflow rate}] - [\text{outflow rate}],$$

to account for the net rate of change of the number of gulls on the pier as the inflow rate minus the outflow rate. Here $N(t)$ is the number of birds on the pier at time t, and t is the day of the year, including the decimal fraction of the specific time of day. The inflow rate is the per capita flow rate to the pier multiplied by the number of gulls elsewhere in the system. The outflow rate is the per capita flow rate away from the pier times the number of gulls loafing on the pier. The mathematical model incorporating hypotheses H1–H4 is

$$\frac{dN}{dt} = \alpha E(t)(\beta K_p(t) - N) - \frac{1}{\alpha E(t)} N \tag{1.1}$$

$$0 < \alpha < \infty, \; 1 \leq \beta \leq 80.$$

The expression $\beta K_p(t)$ is the total number of gulls in this two compartment system at time t, where $K_p(t)$ is an estimate of the maximal historical numbers of gulls at the pier throughout the year (hypothesis H4). The difference $\beta K_p(t) - N(t)$ is the number of birds elsewhere in the system at time t. The proportionality constants $\alpha, \beta > 0$ are parameters to be estimated from the historical data. The maximal number of gulls nesting in the adjacent breeding colony was approximately 60 times the maximal value of $K_p(t)$, so we chose 80 as an ample upper bound for β. Similar compartmental models have been used for waterfowl [29].

We estimated $K_p(t)$ from historical data sampled from January 1 to March 21 and May 23 to December 31, 1997–1999 and 2001 [19]. Using these data we estimated seasonal maxima for the fluctuating pier counts by fitting a modified lognormal curve through the averages, for each week of the year, of the maximal pier counts (Fig. 1.4):

$$K_p(t) = 76.36 \exp\left[\frac{[\ln(40.29 - t/7) - 2.504]^2}{-0.7225}\right].$$

After October 2 and before March 6, the average maximal pier counts were zeros or ones, so we set $K_p(t) = 0$ for those intervals (Fig. 1.4).

We numerically integrated the differential equation model (1.1) using tidal and solar data obtained from the National Oceanic and Atmospheric Administration (NOAA) as described below.

1.2.3 Alternative Models

We normalized hourly tide data as follows: We first subtracted the minimal tide value from each tide height in order to set the minimum equal to zero, then divided each by the resulting maximum to set the new maximum equal to one, and finally added one to the quotient so that the result was between one and two:

$$T(t) = \frac{\text{tide} - \min(\text{tide})}{\max(\text{tide} - \min(\text{tide}))} + 1.$$

This resulted in a nondimensional tide height function with $1 \leq T(t) \leq 2$. For solar elevation, we set negative elevations to zero and then normalized solar elevation data as

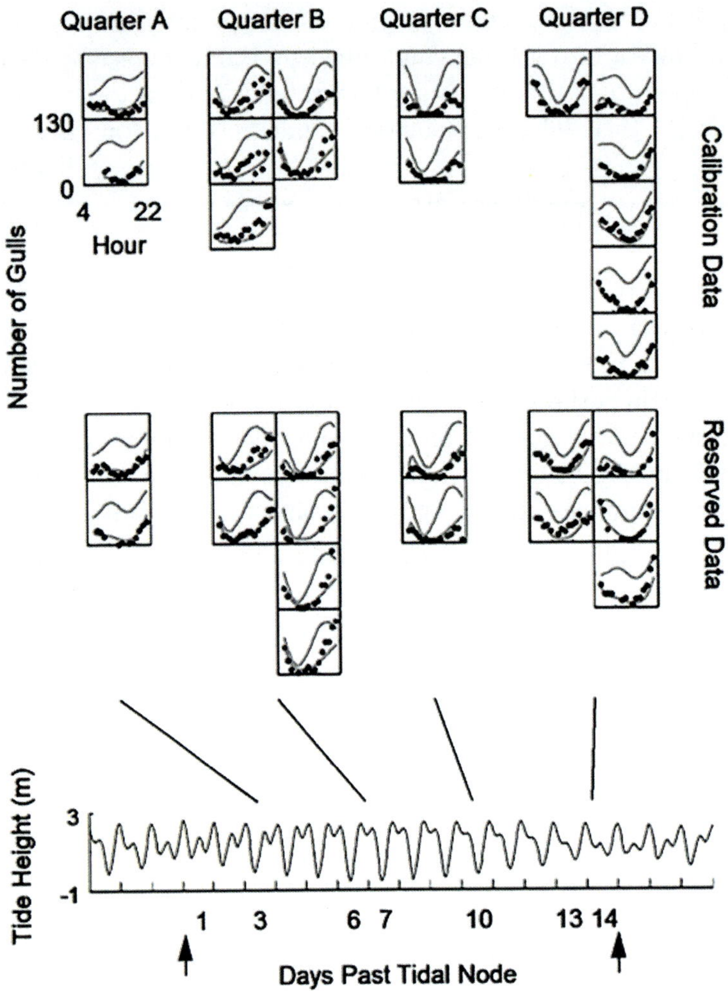

Figure 1.3: Tidal-solar model prediction (lower curve), historical data (circles), and tidal curve (upper curve). Each panel corresponds to 1 day. Tide height is graphed on a vertical scale of −1 to 3 m. Initial condition for each day's model prediction was taken to be the first data point collected that day. A typical tidal sequence for Protection Island is shown at the bottom; tidal nodes are indicated with arrows. Data from days occurring during the same quarter of the tidal sequence are stacked vertically. Data in a given column show similar diurnal patterns. Conditioned least-squares (CLS) parameters were estimated from the "calibration data." The "reserved data" were used to independently validate model performance without re-estimated parameters. The CLS parameter estimates appear in Table 1.1. Following are dates for days in each of the six columns, left to right, top to bottom: first column, July 21, 1999; August 18, 1999; July 7, 1999; August 4, 1999; second column, July 2, 1997; July 16, 1997; August 13, 1997; July 30, 1997; August 5, 1998;

for tide. For alternative hypothesis H1A, we let $E(t) = T(t)$; for H1B, we let $E(t) = 1/S(t)$; and for H1C, we let $E(t) = T(t)/S(t)$. When substituted into Eq. (1.1), the three alternative hypotheses produce three alternative models which we designate as the "tidal model," "solar model," and "tidal-solar model," respectively.

1.2.4 Model Parameterization

We parameterized each of the three alternative models on the historical calibration data using the conditioned least-squares (CLS) method. For a set of $n+1$ successive hourly observations $\{x_0, x_1, \ldots, x_n\}$, we used the model to produce n hourly one-step predictions $\{y_1, y_2, \ldots, y_n\}$, where y_{i+1} denotes the model prediction at time $i+1$, given the observation at time i as the initial condition for the model. The conditioned one-step residual errors are given by $x_i - y_i$. Unlike maximal likelihood methods, the CLS method relaxes many assumptions about the distribution of these residual errors [8]. We minimized the residual sum of squares

$$RSS(\theta) = \sum_{i=1}^{n} (x_i - y_i)^2$$

as a function of the vector θ of model parameters, where n is the number of residuals. The minimizer $\widehat{\theta}$ is the vector of CLS parameter estimates for the model. We produced the one-step predictions with the MATLAB ode45 integrator and minimized $RSS(\theta)$ with a Nelder-Mead downhill search algorithm [25, 27]. The CLS parameter estimates for each model appear in Table 1.1.

1.2.5 Model Selection

We compared the three parameterized models with the calibration data set using the residual sum of squares $RSS(\widehat{\theta})$ at the estimated parameter values. Because each alternative model had the same number of parameters, we did not need to use an information-theoretic model selection technique such as the Akaike information criterion (AIC) [3]. We selected the model with the smallest $RSS(\widehat{\theta})$ as the best.

Figure 1.3: third column, May 23, 2001; July 22, 1998; June 6, 2001; June 20, 2001; July 4, 2001; July 8, 1998; fourth column, July 14, 1999; July 28, 1999; June 30 1999; August 11, 1999; fifth column, July 23, 1997; July 9, 1997; August 6, 1997; sixth column, May 30, 2001; June 13, 2001; July 15, 1998; July 29, 1998; August 12, 1998; June 27, 2001; July 11, 2001; July 1, 1998 (Originally published in Henson et al., "Predicting dynamics of aggregate loafing behavior in glaucous-winged gulls (*Larus glaucescens*) at a Washington colony," *The Auk* 121:380–390. Copyright © 2004, Oxford University Press. Used with permission.)

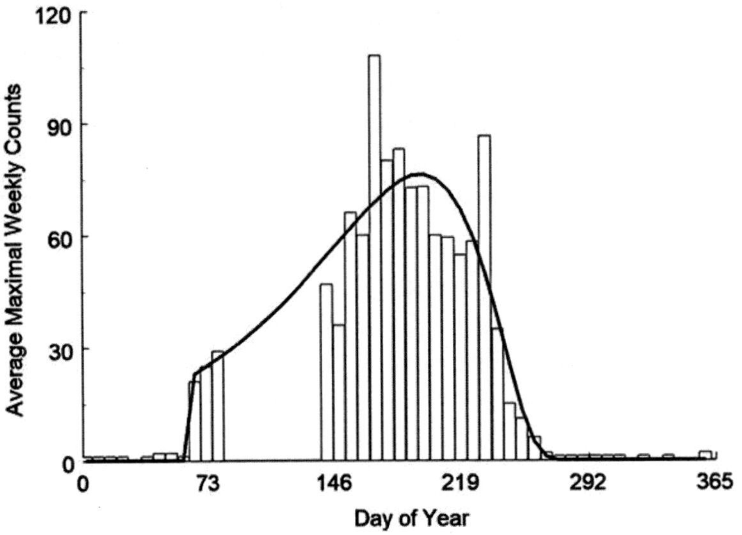

Figure 1.4: Maximal historical counts at the loafing area throughout the year. Each bar represents 1 week of the year. The height of each bar is the mean historical maximal count recorded at the loafing area for that week, averaged over the years 1997–1999 and 2001. A modified lognormal curve was fitted through the averages to estimate the function $K_p(t)$. After the 275th day of the year and before the 65th day of the year, the average maximal pier counts were zeros or ones, so we set $K_p(t) = 0$ for those intervals. (Originally published in Henson et al., "Predicting dynamics of aggregate loafing behavior in glaucous-winged gulls (*Larus glaucescens*) at a Washington colony," *The Auk* 121:380–390. Copyright © 2004, Oxford University Press. Used with permission.)

1.2.6 Model Validation

We validated the best model on the reserved historical data using the goodness-of-fit as measured by the generalized R^2

$$R^2 = 1 - \frac{RSS(\widehat{\theta})}{\sum_{i=1}^{n}(x_i - \overline{x})^2},$$

where $\overline{x}$ denotes the sample mean of the observations. The number R^2 estimates the proportion of the observed variability explained by the model, relative to using the mean as a predictor. We compared the R^2 for the estimation data to the R^2 for the validation data without refitting.

1.3 Results

The tidal-solar model was the best of the alternatives (Table 1.1). Fitted tidal-solar model simulations are shown with the calibration data in Fig. 1.3.

Table 1.1: Model selection on historical calibration data. Conditioned least-squares (CLS) parameters α and β were estimated for three alternative models using the historical calibration data; the variance parameter $\sigma^2 = RSS(\widehat{\theta})/n$ was estimated from the n residuals. The CLS parameters for the solar model occur on the boundary $\beta = 80$ of the feasible parameter range; with no restriction on β, the minimizer occurs at $\beta = \infty$. The tidal and solar models are discarded in favor of the tidal-solar model

Model	$\widehat{\alpha}$	$\widehat{\beta}$	$\widehat{\sigma}^2$	n	$\widehat{RSS}$	R^2
Tidal model	0.307	1.97	245	205	50,149	0.38
Solar model	0.0985	80.0	225	205	46,084	0.43
Tidal-solar model	0.349	2.85	167	205	34,195	0.58

The goodness-of-fit for the reserved data (without reparameterization) was $R^2 = 0.61$, similar to the goodness-of-fit of $R^2 = 0.58$ for the calibration data, supporting model validation. Model simulations with the reserved historical data appear in Fig. 1.3.

1.4 Test of Predicted Future Dynamics

In 2001, using the tidal-solar model parameterized on the historical calibration data set, we made a priori hourly predictions for counts of gulls on the pier during daylight hours for May and early June at the beginning of the future 2002 breeding season (Fig. 1.5). The model predicted periodicities at three temporal scales:

(1) high-frequency daily oscillations (Fig. 1.6c),
(2) the expected low-frequency seasonal oscillation (Fig. 1.6a)
(3) an unexpected medium-frequency oscillation (Fig. 1.6c) that coincided with a biweekly tidal oscillation (Fig. 1.6e, arrows)

On some days model predictions fluctuated out of phase with the tide (Fig. 1.5, days 142 and 155), which was surprising in the context of previous studies.

Data collection to test model predictions began at 09:00 hours on May 9 2002 and continued at hourly intervals from dawn to dusk (05:00–20:00 hours) until 20:00 hours on June 6, 2002. Our counts traversed two consecutive tidal nodes. If a human disturbance at the pier occurred within 30 minutes before or during a count, that count was eliminated. No reparameterization of the model was carried out with the new data.

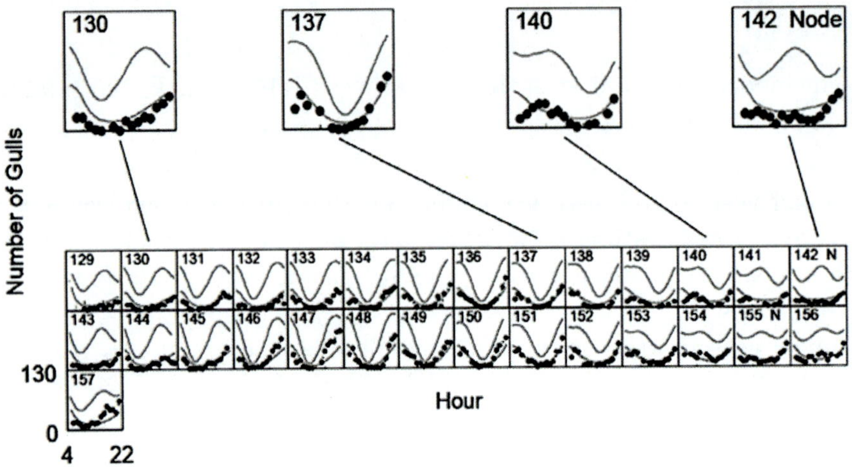

Figure 1.5: Tidal-solar model prediction (lower curve), data from spring 2002 (circles), and tidal curve (upper curve). Tide height is graphed on a vertical scale of −1 to 3 m. Each daily panel is identified with the day of the year. The model prediction was not reinitialized for each day but rather was run continuously for all 29 days. Each row of 14 panels corresponds to one 2-week tidal cycle. Tidal nodes (N) occur on or near days 142 and 155. Each column of panels contains similar patterns of data. The CLS parameter values appear in Table 1.1. (Originally published in Henson et al., "Predicting dynamics of aggregate loafing behavior in glaucous-winged gulls (*Larus glaucescens*) at a Washington colony," *The Auk* 121:380–390. Copyright © 2004, Oxford University Press. Used with permission.)

The 2002 data contained evidence of all three predicted periodicities (Fig. 1.6b,d). Model predictions closely approximated daily counts, both qualitatively (Fig. 1.5) and quantitatively (goodness-of-fit, $R^2 = 0.66$). Between nodes, daily minimal counts preceded low-low tides during the first several days (e.g., Fig. 1.5, day 130), moved to a point below mid-period midday low-low tides (e.g., Fig. 1.5, day 137), lagged afternoon low-low tides later in the period (e.g., Fig. 1.5, day 140), and coincided with high tide when reaching the node (e.g., Fig. 1.5, day 142). After the node, daily maximal counts increased, then decreased toward the next node in concert with the predicted biweekly oscillation (Fig. 1.6b,d). Moreover, average daily counts increased from the first to the second biweekly oscillation period (Fig. 1.6b,d), consistent with the expected rise in numbers during the spring (Fig. 1.6a,c).

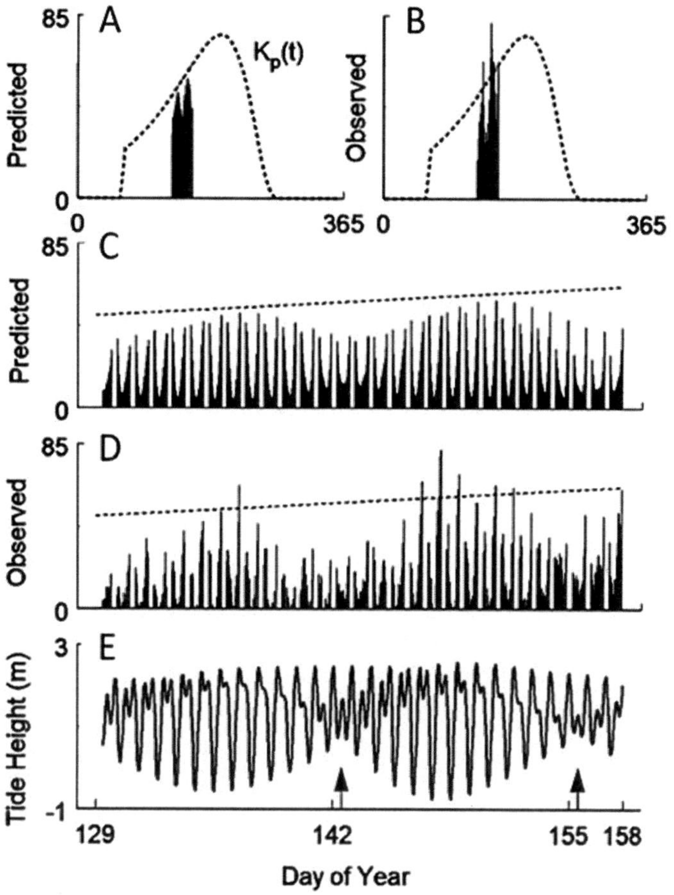

Figure 1.6: Tidal-solar model predicted oscillations in occupancy on three different time scales. The dotted curves show the seasonal oscillation $K_p(t)$. (**a**), (**c**) Model predictions for the spring of 2002. Oscillations are present on daily, biweekly, and yearly scales. (**b**), (**d**) Data observations corresponding to the predictions in (**a**) and (**c**). (**e**) Tidal oscillation for the data collection time period in 2002. Tidal nodes are indicated with arrows. (Originally published in Henson et al., "Predicting dynamics of aggregate loafing behavior in glaucous-winged gulls (*Larus glaucescens*) at a Washington colony," *The Auk* 121:380–390. Copyright © 2004, Oxford University Press. Used with permission.)

Our model does not consider dynamics during dawn and dusk, when it was light intensity that seemed to drive the system. Model predictions began to match the counts approximately 1 hour past sunrise. Moonlit nights extended attendance by gulls on the pier. Several factors may account for departures of model predictions from data, including

(1) Weather conditions (temperature, wind, rain)
(2) Count inaccuracies due to fog
(3) Low light intensity during early morning counts
(4) Number of boats moored on the pier (which changed visibility and area available for loafing)
(5) Human activity on the pier
(6) Bald eagle (*Haliaeetus leucocephalus*) disturbances [14, 18]

Higher-order effects of tide may be indicated by higher than expected counts on days 154, 156, and 157.

1.5 Discussion

Competing functional needs drive short-term fluctuations in animal numbers by habitat [4, 9, 35]. An individual bird, for example, may feed, bathe, incubate, and loaf, all during the course of a single day. In order to perform these behaviors, the bird moves multiple times from habitat to habitat [5, 38]. Physiological constraints and environmental variables play crucial roles in opening and closing opportunities for the performance of these behaviors [22, 35]. For example, because gulls are visually oriented foragers, they forage when food is exposed by low tides. When it is dark, foraging ceases. In view of this reasoning, accurate prediction of habitat occupancy in relation to tide and time of day might seem trivial and uninteresting. However, one needs to keep at least three considerations in mind.

First, gulls exhibit considerable individual variation. An individual gull decides when to move from one habitat to another within the context of what has preceded the move. A decision to enter or leave a habitat may be influenced by the behaviors of other gulls [12, 16, 37], or the decision may be independent of what other gulls are doing. We observed no indication that social facilitation or coaction was involved in decisions to arrive or depart the loafing area. Gulls appeared to arrive and depart independently throughout the day. (By contrast, gulls moved to and from feeding areas in groups.) Stochastic models of waterfowl movements between habitats have found that approximately 70% of the birds moved singly, whereas about 20% moved in pairs [30]. *Accurate mathematical prediction for the dynamics of an animal assemblage may be possible because deterministic forces at some group scales of behavior may be more important than stochastic individual variability* [23].

Second, *two unexpected outcomes occurred* that would not have been predicted on the basis of correlative studies alone:

(1) An increase in numbers of gulls loafing at the pier during low tides around tidal nodes
(2) Biweekly, medium-frequency oscillations in numbers of loafing gulls

These results suggest that the superposition of environmental oscillations, such as tide height, time of day, and day of the year can lead to predictable patterns of group behavior.

Third, we wish to emphasize the importance of accurate dynamical models for ecological field systems. The methodology outlined in this chapter should first be applied to relatively simple field systems before applying it to more complex systems [8]. To be successful, the modeler must apply this methodology rigorously to well-defined elements of the system.

Pressing world problems, such as increased food production, conservation of species diversity, resource management, and the spread of disease underscore the need to accurately predict the movement patterns of organisms [17]. *Dynamical models can provide testable hypotheses if data are rigorously tied to the models through parsimonious modeling assumptions, proper statistical techniques of model parameterization, use of model selection, and model validation.* Sufficiently dense sampling must be carried out to allow for proper model parameterization and validation. Sample intervals must be smaller than the pertinent environmental periodicities [22, 23]. *Moreover, whenever possible, parameterization and validation should occur on independent data sets, if the process is not to become a simple exercise in curve-fitting with little explanatory power.*

Finally, *the resultant models should not only describe and explain, but also predict.* The successful testing of a priori model predictions yields the most convincing evidence that the model captures the fundamental dynamics of the system; if the predictions are unexpected, all the better.

1.5.1 Manageable Modeling for Management: Multiple Time-Scale Analysis

We conclude this chapter with an introduction to multiple time-scale analysis, a technique which can vastly simplify the methods in this chapter for managers and which we use in the next two chapters.

Most nonlinear differential equations do not have analytic solutions and must be solved numerically. The prediction $N(T)$ of a differential equation model at time T is generated by numerically integrating over a continuum $[t_0, T]$. However, if the system of interest recovers quickly after disturbance, the steady state of the system depends approximately on the current time T rather than on the past history of the system. For example, in this current study both the parameterized differential equation model and our field observations suggest that recovery from disturbance happens in less than 1 hour.

For such systems with fast transient recovery times, it is possible to use a technique called multiple time-scale analysis, which produces two algebraic models: one for the transient dynamic and one for the steady-state dynamic [24]. Here we consider the steady-state model.

Returning to the differential equation

$$\frac{dN}{dt} = \alpha E(t)\left(\beta K_p(t) - N(t)\right) - \frac{N(t)}{\alpha E(t)},$$

we can imagine a fast time scale $\tau = t/\varepsilon$, where $\varepsilon > 0$ is small. The chain rule yields

$$\frac{dN}{dt}\frac{dt}{d\tau} = \frac{dN}{d\tau} = \alpha E(\tau)\left(\beta K_p(\tau) - N(\tau)\right) - \frac{N(\tau)}{\alpha E(\tau)},$$

and so

$$\varepsilon\frac{dN}{dt} = \alpha E(\tau)\left(\beta K_p(\tau) - N(\tau)\right) - \frac{N(\tau)}{\alpha E(\tau)}.$$

For small ε and bounded dN/dt, we have

$$0 \approx \alpha E(\tau)\left(\beta K_p(\tau) - N(\tau)\right) - \frac{N(\tau)}{\alpha E(\tau)}.$$

Solving for $N(\tau)$ and renaming τ as t, we obtain the fast time-scale (steady-state) solution $N_0(t)$, which is the algebraic model

$$N_0(t) = \frac{\beta K_p(t)}{1 + \left(\frac{1}{\alpha E(t)}\right)^2}.$$

For the tidal-solar model, the steady state is

$$N_0(t) = \frac{\beta K_p(t)}{1 + \frac{1}{\alpha^2}\left(\frac{S(t)}{T(t)}\right)^2}. \tag{1.2}$$

Clearly, the algebraic steady-state model (1.2) is much simpler to parameterize and use than the differential equation model (1.1). In this study, the steady-state model approximates the solution of the differential equation to within roughly one bird during 06:00–17:00 hours, to within two birds at the 05:00 hour, and to within five birds at the 20:00 hour, when the error is about 12%.

When appropriate, the time-scale approach can be useful in resource management [21], because it obviates the time-consuming necessity of parameterizing differential equations. Time-scale analysis will be important in the next two chapters.

References

1. Annett CA, Pierotti R (1999) Long-term reproductive output in western gulls: consequences of alternate tactics in diet choice. Ecology 80:288–297

2. Belant JL (1997) Gulls in urban environments: landscape-level management to reduce conflict. Landsc Urban Plan 38:245–258

3. Burnham KP, Anderson DR (2002) Model selection and multi-model inference: a practical information-theoretic approach, 2nd edn. Springer, New York

4. Cody ML (1985) An introduction to habitat selection in birds. In Cody ML (ed) Habitat selection in birds. Academic Press, Orlando, pp 3–56

5. Cooke F, Ross RK (1972) Diurnal and seasonal activities of a post-breeding population of gulls in southeastern Ontario. Wilson Bulletin 84:164–172

6. Cushing JM, Costantino RF, Dennis B, Desharnais RA, Henson SM (1998) Nonlinear population dynamics: models, experiments, and data. J Theor Biol 194:1–9

7. Cushing JM, Costantino RF, Dennis B, Desharnais RA, Henson SM (2003) Chaos in ecology: experimental nonlinear dynamics. Academic Press, San Diego, CA

8. Dennis B, Desharnais RA, Cushing JM, Henson SM, Costantino RF (2001) Estimating chaos and complex dynamics in an insect population. Ecol Monogr 71:277–303

9. DeWoskin R (1980) Heat exchange influence on foraging behavior in *Zonotrichia* flocks. Ecology 61:30–36

10. Drent RH (1967) Functional aspects of incubation in the herring gull (*Larus argentatus* Pont). Behaviour Supplement 17:1–32

11. Evans RM (1982a) Flock size formation in black-billed gulls. Can J Zool 60:1806–1811

12. Evans RM (1982b) Foraging-flock recruitment at a black-billed gull colony: implications for the Information Center hypothesis. Auk 99:24–30

13. Galusha JG, Amlaner CJ (1978) The effects of diurnal and tidal periodicities in the numbers and activities of herring gulls *Larus argentatus* in a colony. Ibis 120:322–328

14. Galusha JG, Hayward JL (2002) Bald eagle activity at a gull colony and seal rookery on Protection Island, Washington. Northwest Nat 83:23–25

15. Good TP, Ellis JC, Annett CA, Pierotti R (2000) Bounded hybrid superiority in an avian hybrid zone: effects of mate, diet, and habitat choice. Evolution 54:1774–1783

16. Götmark F, Winkler DW, Anderson M (1986) Flock-feeding of fish schools increases individual success in gulls. Nature 319:589–591

17. Hastings A, Palmer MA (2003) Mathematics and biology: a bright future for biologists and mathematicians? Science 299:2003–2004

18. Hayward JL, Galusha JG, Henson SM (2010) Foraging-related activity of bald eagles at a Washington seabird colony and seal rookery. J Raptor Res 44:19–29

19. Hayward JL, Henson SM, Bové J, Bové C, Gregory CJ (2017) Daily and annual habitat use and habitat-to-habitat movement by glaucous-winged gulls at Protection Island, Washington. Northwest Nat 98:180–189

20. Henson SM, Hayward JL, Burden CM, Logan CJ, Galusha JG (2004) Predicting dynamics of aggregate loafing behavior in glaucous-winged gulls (*Larus glaucescens*) at a Washington colony. Auk 121:380–390

21. Henson SM, Hayward JL, Damania SP (2006) Identifying environmental determinants of diurnal distribution in marine birds and mammals. Bull Math Biol 68:467–482

22. Hunt GL, Jr, Schneider DC (1987) Scale-dependent processes in the physical and biological environment of marine birds. In Croxall JP (ed) Seabirds: feeding, ecology and role in marine ecosystems. Cambridge University Press, Cambridge, pp 7–41

23. Levin SA (1992) The problem of pattern and scale in ecology. Ecology 73:1943–1967

24. Lin CC, Segel LA (1988) Mathematics applied to deterministic problems in the natural sciences. Society for Industrial and Applied Mathematics, Philadelphia, PA

25. Olsson DM, Nelson LS (1975) The Nelder-Mead simplex procedure for function minimization. Technometrics 17:45–51

26. Patterson IJ (1965) Timing and spacing of broods in the black-headed gull *Larus ridibundus* L. Ibis 107:433–459

27. Press WH, Flannery BP, Teukolsky SA, Vetterling WT (1986) Numerical recipes: the art of scientific computing. Cambridge University Press, Cambridge

28. Schreiber EA, Burger J (eds) (2001) Biology of marine birds. CRC Press, Boca Raton, FL

29. Silverman E, Kot M (2000) Rate estimation for a simple movement model. Bull Math Biol 62:351–375

30. Silverman ED, Kot M, Thompson E (2001) Testing a simple stochastic model for the dynamics of waterfowl aggregations. Oecologia 128:608–617

31. Spear LB, Pyle P, Nur N (1998) Natal dispersal in the western gull: proximal factors and fitness consequences. J Anim Ecol 67:165–179

32. Stout JF, Hayward JL, Gillett WH (1974) Aggregations of gulls (Laridae) on aerodromes and behavioral techniques for dispersal. In Croxall JP (ed) Proceedings of the conference on the biological aspects of the bird/aircraft collision problem. Clemson University, Clemson, NC, pp 125–148

33. Tinbergen N (1960) The herring gull's world. Harper and Row, New York

34. Turchin P (1998) Quantitative analysis of movement: measuring and modeling population redistribution in animals and plants. Sinauer Associates, Sunderland, MA.

35. Walsberg GE (1985) Physiological consequences of microhabitat selection. In Cody ML (ed) Habitat selection in birds. Academic Press, Orlando, FL, pp 389–413

36. Wiens JA (1976) Population responses to patchy environments. Annu Rev Ecol Syst 7:81–120

37. Wittenberger JF, Hunt GL, Jr (1985) The adaptive significance of coloniality in birds. In Farner DS, King JR, Parkes KC (eds) Avian biology, volume VIII. Academic Press, Orlando, FL, pp 1–78

38. Wondolowski L (2002) Diurnal activity patterns of wintering gulls at Jug Bay Wetlands Sanctuary, Maryland. MS thesis, Bard College, Annandale-on-Hudson, NY

39. Wright EN (1968) Modification of the habitat as a means of bird control. In Murton RK, Wright EN (eds) The problems of birds as pests. Academic Press, New York, pp 97–105

Chapter 2

Predicting Haul-Out Behavior by Harbor Seals

In this chapter we apply the differential equation-based modeling techniques developed in Chap. 1 to a marine mammal behavioral system. This research demonstrated that the same modeling techniques developed for bird behavior can be applied more broadly to other organisms. James Hayward, Shandelle Henson, and Brian Dennis, with undergraduate students Clara Logan, Catherine Parris, and Matthew Meyer, carried out this study which was initially reported in the Journal of Applied Ecology *in 2005 [10].*

2.1 Introduction

Harbor seals (*Phoca vitulina*, Fig. 2.1) are the world's most widespread pinnipeds. They live and breed within a broad north-south range along the North Pacific and North Atlantic coasts [6, 24]. From the early to mid-twentieth century, harbor seals were extensively persecuted under the false belief they consumed large quantities of salmon and other commercially important fish. Studies revealed, however, that the impact of seals on important fish was less extensive than believed. Moreover, during the mid-twentieth century, the public became increasingly alarmed about declining populations of marine mammals. As a result, the United Kingdom passed the Conservation of Seals Act (CSA) in 1970, and the United States passed the Marine Mammal Protection Act (MMPA) in 1972. In response, seal populations rebounded [3, 17, 18].

The CSA and MMPA mandate systematic monitoring of seal populations. Information obtained by these monitoring efforts inform decisions about seal conservation by government agencies. Monitoring of hauled-out seals (Fig. 2.2) is accomplished through radio and satellite telemetry of marked individuals and by aerial counts. Much effort has been

J. M. Cushing et al., *Modeling Behavior and Population Dynamics*, Interdisciplinary Applied Mathematics 57, https://doi.org/10.1007/978-3-031-34283-7_2

devoted to determining the best times to carry out maximal counts as a basis on which to make population estimates [1, 12, 14, 17, 19, 21, 22, 24, 25].

At pupping sites, harbor seals spend their time in coastal waters where they feed and along the shoreline where they interact with one another, rest, give birth, and nurse their young [15, 26]. Shoreline habitat chosen as haul-out sites vary from sandy or rocky beaches and tidal bars to drifting glacial ice. Shoreline occupation by these animals lowers their risk of predation by large sea predators, lowers stress associated with waves and currents, and raises the temperature of their peripheral tissues. Sex, locality, pelage dryness, and individual behavior all contribute to the timing of haulout [22, 24, 26], as do environmental factors such as time of day, time of year, wave action, shoreline topography, air temperature, solar radiation, wind speed, wind chill, and disturbance [20–22, 24, 26].

Figure 2.1: A female harbor seal and her nursing newborn pup along the north beach of Violet Point, Protection Island National Wildlife Refuge, Washington (Photo by J. L. Hayward)

Statistical methodologies such as linear regression, analysis of variance, and canonical correlation typically are used to characterize haul-out patterns [18, 20–22, 26], but these methodologies offer limited predictive capability. Mathematical modeling not only offers the advantage of predictability of future patterns but also allows the identification of driving forces for these patterns [9, 16]. Here we model and predict the timing of haulout by harbor seals at Protection Island, Washington.

2.2 Methods

2.2.1 Data Collection

Hourly counts of harbor seals hauled out along the north beach of Violet Point, Protection Island, Washington, were made from 05:00 to 20:00 hours (LST here and below) 1 day per week during July and August 1995, 1997, 1998, and 1999. Observers counted hauled-out seals from a 33-m-high bluff overlooking Violet Point. Counts made within 30 minutes following a major disturbance were not included in the study. Seals were considered hauled out if they rested on the substrate, even if parts of their bodies were partially submerged (Fig. 2.2). Hourly counts provided time series amenable for the detection of diurnal and tidal periodicities [13]. Tides around Protection Island are semi-diurnal characterized by strong diurnal inequalities in the lows. Minimal tidal amplitude (nodes) occur about every 14 days (Fig. 2.3, arrows).

We obtained tide and solar elevation data from the National Oceanic and Atmospheric Administration (NOAA) and current speed predictions from a University of South Carolina website (http://tbone.biol.sc.edu/tide/). From these we constructed non-dimensionalized tide height, solar elevation, and current functions $T(t)$, $S(t)$, and $C(t)$ using the method described in Chap. 1 so that $1 \leq T, S, C \leq 2$. We also considered the possibility that the animals might respond to tidal and/or current speed highs and lows rather than values. Thus, we constructed two more non-dimensionalized environmental variables $T_e(t)$ and $C_e(t)$ with equalized extrema, in which the local maxima of T and C were set to 2 and the local minima were set to 1 and the oscillations were splined between these points.

Figure 2.2: Harbor seals hauled out on Kanem Point, Zella Schultz Seabird Sanctuary, Protection Island, Washington (Photo by J. L. Hayward)

2.2.2 Modeling Assumptions

Our model in this study is based on the following five assumptions:

(H1) A two-compartment model involving the haul-out site and a remote location (everywhere else) can describe the numbers of seals that haul out in the study area during daylight hours.
(H2) Deterministic environmental variables drive the movement of seals back and forth between these two compartments in the following way: They leave the haul-out site for the remote compartment at a per capita rate that is proportional to a function $E_{21}(t)$ of deterministic environmental variables, and they return to the haul-out site at a per capita rate which is proportional to a function $E_{12}(t)$ of deterministic environmental variables. Neither of these per capita rates depend on the seal density in either compartment.
(H3) The maximum number $M(t)$ of seals that may haul out at the study area during pupping season can be approximated by

$$M(t) = \beta \exp\left(-\gamma\,(\text{day of year } + t/24 - \delta)^2\right),$$

where t is the hour of the day, and β, γ, and $\delta > 0$ are positive constants. The maximal weekly haul-out counts (Fig. 2.4) suggested the functional form of this assumption.
(H4) The values of $M(t)$, $E_{12}(t)$, and $E_{21}(t)$ remain approximately constant during the time it takes for the system to return to "steady-state" dynamics following a disturbance. In other words, the system recovers quickly following disturbance.
(H5) Demographic stochasticity, which can be modeled with a stochastic "arrival-and-departure" process, is the main source of noise in the census data. This assumption was made after a post hoc inspection of model residuals.

Note the following three points concerning assumption H3:

(1) $M(t)$ is a functional form assumed to be the upper number of seals that haul out at the study area; it is not the population size. We did not address or model population size in this study, although population size is often estimated from haul-out counts [12, 17–19, 23, 26].
(2) $M(t)$ is not the normal curve fitted to the data in Fig. 2.4. All the model parameters, including β, γ, and δ, were estimated from the census time series data described in the section on model parameterization below.
(3) Maximal counts are not normally distributed throughout the year; therefore, the functional form of $M(t)$ depends on the seasonal context.

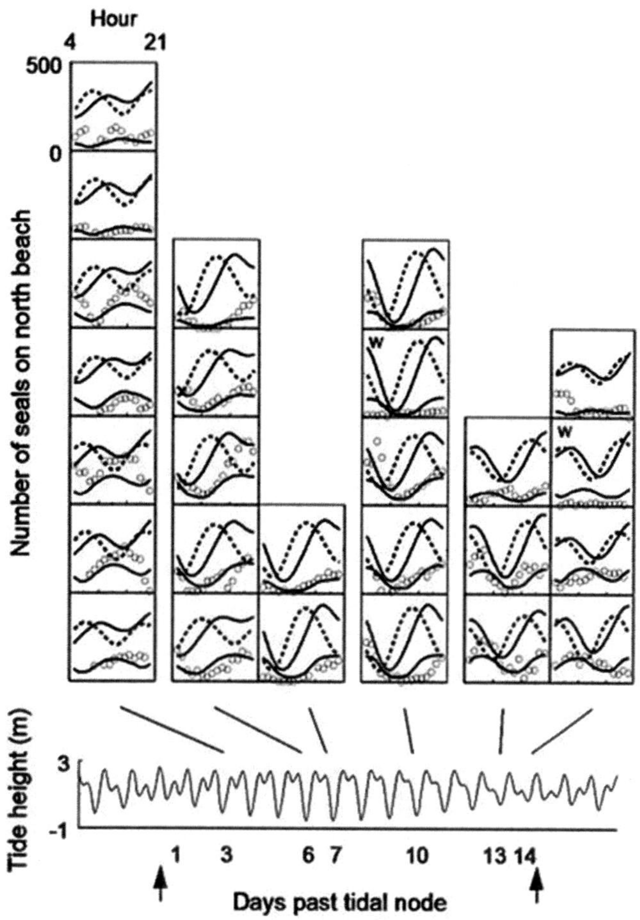

Figure 2.3: Model prediction (lower sold curve), seal haul-out data (circles), tidal curve (upper solid curve), and current velocity (dashed curve). Each panel corresponds to 1 day. Tide height is graphed on a vertical scale of -1 to 3 m, and current velocity on a scale of -2 to 1.5 knots. A typical 14-day tidal period for Protection Island is shown at the bottom; tidal nodes are indicated with arrows. Data from days occurring during the same time in the tidal period are stacked vertically. The data in a given column show similar diurnal patterns. Days preceded by high, sustained winds are designated W (see the Discussion). (Originally published in Hayward et al., "Predicting numbers of hauled-out harbour seals: a mathematical model," *Journal of Applied Ecology* 42:108–117. Copyright © 2005, British Ecological Society. Used with permission.)

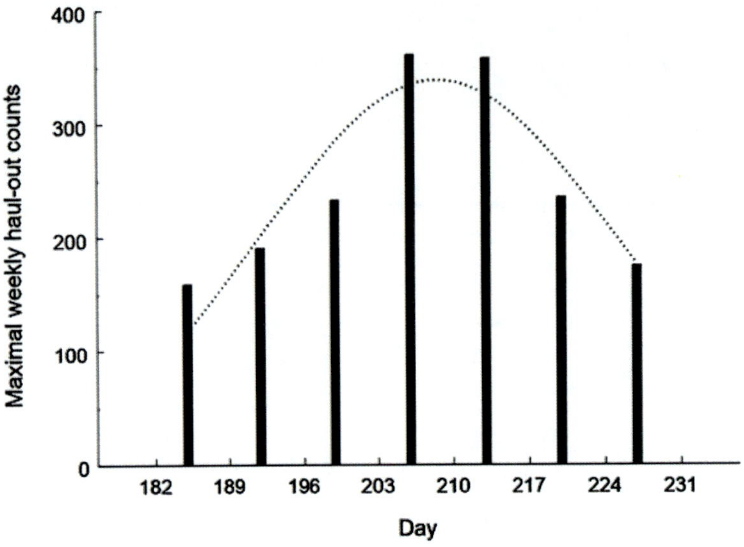

Figure 2.4: Maximal counts at the haul-out site during July 1 to August 18. Each bar represents 1 week. The height of each bar is the mean maximal count recorded at the haul-out site for that week, averaged over the years 1995 and 1997–1999. Based on this graph, the function $M(t)$ was assumed proportional to a normal curve. The dotted normal curve shown is not $M(t)$; see assumption 3 in the text. (Originally published in Hayward et al., "Predicting numbers of hauled-out harbour seals: a mathematical model," *Journal of Applied Ecology* 42:108–117. Copyright © 2005, British Ecological Society. Used with permission.)

2.2.3 Model

Assumptions H1–H3 above give rise to the differential equation model

$$\frac{dN}{dt} = aE_{12}(t)\left(M(t) - N\right) - bE_{21}(t)N, \tag{2.1}$$

where $N(t)$ is the number of seals hauled out at time t and $a, b > 0$ are constants of proportionality.

Assumption H4, along with the multiple time-scale method discussed in Chap. 1, Sect. 1.5.1, yields an algebraic model for the steady-state dynamics, namely,

$$N(t) = \frac{M(t)}{1 + \frac{bE_{21}(t)}{aE_{12}(t)}}.$$

Let $\alpha = b/a$ and $E(t) = E_{21}(t)/E_{12}(t)$. Then from assumption H3 we obtain the deterministic model

$$N(t) = \frac{\beta \exp\left(-\gamma\left(\text{day of year } + t/24 - \delta\right)^2\right)}{1 + \alpha E(t)}, \tag{2.2}$$

where $\alpha, \beta, \gamma, \delta > 0$ are parameters to be estimated from the data.

Demographic noise is approximately additive on the square-root scale [7]. Thus, from assumption H5, we obtain

$$\sqrt{N(t)} = \sqrt{\frac{\beta \exp\left(-\gamma\left(\text{day of year } + t/24 - \delta\right)^2\right)}{1 + \alpha E(t)}} + \sigma\varepsilon(t),$$

where $\sigma > 0$ is a constant parameter to be estimated from the data and the $\varepsilon(t)$ are standard normal random variables (mean 0 and standard deviation 1) uncorrelated in time. Thus, the stochastic model is

$$N(t) = \left(\sqrt{\frac{\beta \exp\left(-\gamma\left(\text{day of year } + t/24 - \delta\right)^2\right)}{1 + \alpha E(t)}} + \sigma\varepsilon(t)\right)^2,$$

where the right-hand side is taken to be zero if the quantity inside the square is negative.

2.2.4 Alternative Models, Parameterization, and Model Selection

A suite of 23 alternative choices for the function $E(t)$, as shown in Table 2.1, corresponds to a suite of alternative models of the form (2.2).

For each of the alternative $E(t)$ in Table 2.1, we estimated the parameters for model (2.2) using the method of nonlinear least squares (LS), which relaxes many assumptions about the distribution of the residuals [7]. That is, we minimized

$$RSS(\theta) = \sum_{data} \left(\sqrt{\text{observation}} - \sqrt{\text{model prediction}}\right)$$

as a function of the vector θ of parameters, thereby obtaining the LS estimates $\widehat{\theta}$.

We used the Akaike information criterion (AIC), which for LS is

$$AIC = n \ln \widehat{\sigma}^2 + 2\kappa,$$

to determine the best model in the suite of alternatives [5]. Here n is the number of residuals, $\widehat{\sigma}^2 = RSS(\widehat{\theta})/n$ is the estimated variance, and k is the number of estimated parameters, including σ^2. The AIC penalizes models having more parameters for overfitting, and the model with minimal AIC is considered the best model of the alternatives. The models are ranked according to the values of $\Delta_i = AIC_i - AIC_{\min}$, where AIC_i is the AIC of the ith model and $AIC_{\min}$ is the smallest AIC value. The best model is the one with $\Delta_i = 0$. Models with $\Delta_i > 10$ usually are considered inferior and are discarded [5].

Table 2.1: Model comparison. Least-squares (LS) parameters were estimated for each of the 23 alternative models grouped by the arrangement of variables in $E(t)$; the variance parameter σ^2 of the likelihood function was estimated from the residuals. AIC was computed from σ^2, the number of parameters (including σ^2), denoted here by κ, and the number of data points (389). *Best model in each group. The model with $E(t) = C_e^q(t)/T^r(t)$ was selected because it was the more parsimonious of the two best-fitting models. The LS parameters for this model are $\alpha = 3.735, \beta = 181.6, \gamma = 0.001158, \delta = 215.2, q = 3.774$, and $r = 7.255$

$E(t)$	σ^2	κ	AIC	Δ_i	R^2
1	13.5	5	1023	139	0.14
T^r	13.5	6	1025	141	0.14
T_e^r	13.5	6	1025	141	0.14
$1/T^r$	11.1	6	948	64	0.30*
$1/T_e^r$	13.5	6	1025	141	0.15
S^u	12.4	6	991	107	0.22*
$1/S^u$	13.5	6	1025	141	0.14
C^q	12.4	6	991	107	0.22*
C_e^q	12.6	6	998	114	0.20
$1/C^q$	13.5	6	1025	141	0.14
$1/C_e^q$	13.5	6	1025	141	0.14
C^qS^u	12.3	7	989	105	0.22
$C_e^qS^u$	11.6	7	968	84	0.27*
S^u/T^r	10.5	7	927	43	0.34*
S^u/T_e^r	13.3	7	1022	138	0.16
C^q/T^r	10.2	7	919	35	0.35
C_e^q/T^r	9.5	7	890	6	0.40*
C^q/T_e^r	13.0	7	1012	128	0.18
C_e^q/T_e^r	12.0	7	981	97	0.24
C^qS^u/T^r	10.2	8	921	37	0.35
$C_e^qS^u/T^r$	9.3	8	884	0	041*
C^qS^u/T_e^r	13.5	8	1028	144	0.15
$C_e^qS^u/T_e^r$	12.0	8	983	99	0.24

2.2.5 Goodness-of-Fit

We computed a generalized R^2 to measure the goodness-of-fit:

$$R^2 = 1 - \frac{RSS\left(\widehat{\theta}\right)}{\sum_{data}\left(\sqrt{\text{observation}} - \text{mean}\right)^2},$$

where "mean" denotes the mean of the square roots of the observations. The value of R^2 is a measure of the proportion of variability in the data that is explained by the model, relative to using the mean as a predictor. The higher the R^2, the better the fit, with $R^2 = 1$ denoting a perfect fit.

2.3 Results

The values of $\widehat{\sigma}^2$, κ, AIC, Δ_i, and R^2 for each model are reported in Table 2.1.

The highest R^2 (= 0.41) and lowest AIC ($\Delta_i = 0$) were associated with the model that included the environmental function $E(t) = C_e^q(t)S^u(t)/T^r(t)$ (Table 2.1). The next best model included the environmental function $E(t) = C_e^q(t)/T^r(t)$, with $R^2 = 0.40$ and $\Delta_i = 6$. The third best model was significantly less impressive with $R^2 = 0.35$ and $\Delta_i = 35 > 10$. All models except the best two received no further consideration. The model with $E(t) = C_e^q(t)/T^r(t)$ was selected as most parsimonious and led to the deterministic model:

$$N(t) = \frac{\beta \exp\left(-\gamma\left(\text{day of year } + t/24 - \delta\right)^2\right)}{1 + \alpha(C_e(t))^q/(T(t))^r}. \tag{2.3}$$

Simulations of model (2.3), using LS parameters $\alpha = 3.735$, $\beta = 181.6$, $\gamma = 0.001158$, $\delta = 215.2$, $q = 3.774$, and $r = 7.255$, are displayed in Fig. 2.3.

According to simulations of Eq. (2.3), seals start to leave the beach about midway between the high and low tides; they start to return to the beach between the low and high tides but prior to the midpoint. *Numbers of hauled-out individuals are predicted to be highest about midway between high and low tides and at a low point between low and high tides but prior to the midpoint.* The exact timings, however, depend in a nonintuitive way on the superposition of the tidal and current cycles. Consider, for instance, the first column in Fig. 2.3; the maximal predictions precede the midpoint by about 1 hour. However, the data are noisy enough that it is unlikely that deviations such as this are significant. *The model also predicts that daily haul-out patterns change depending on their temporal position with the biweekly tidal/current cycle.* For instance, the model predicts that over the biweekly cycle, haulout lows shift to later in the day in correspondence with shifts in tide/current fluctuations. Lastly, because of the seasonally dependent envelope of maximal counts $M(t)$ (Fig. 2.4), *maximum counts should occur toward the end of July, followed by a decrease, whereas counts are predicted to be low early in July.*

2.4 Discussion

Trends in the data generally align with model predictions, although with considerable variability. Therefore, interpretations of the results should be carried out with attention to the following caveats.

Environmental Stochasticity Given their high sensitivity to temperature, heat-stressed harbor seals enter the water to cool [26], and cold-stressed seals haul out on the beach where they absorb solar radiation. These animals also respond to additional environmental variables including wind speed and wave height. We did not include these variables in the deterministic model because they cannot be predicted in advance. The influence of environmental stochasticity can be strong. For example, on both July 15, 1998 and July 14, 1999, seal counts on the beach were low, contrary to model predictions. Both days followed 7 to 8 hours of strong, north-northwest winds (sustained wind speeds 38–56 km hour^{-1}, 64 km hour^{-1} gusts) during the afternoon and evening of the previous day. Wave action may have reduced food availability and discouraged haulout on the north beach. Omitting these 2 days from the analysis increased the R^2 value from 0.40 to 0.46.

Demographic Stochasticity Like most large mammals [27], harbor seals exhibit considerable individual behavior variation. Individual haul-out patterns may differ widely in groups of individual animals that use the same site repeatedly [4, 22].

Modeling Assumption Error Several additional haul-out sites on Violet Point were not visible from our observation point. Moreover, even though individual harbor seals exhibit haul-out site preferences [19], it is likely that disturbances, food availability, and/or pupping activity cause them to sometimes change sites [4]. The model does not account for these possibilities. Finally, the model does not take into account social facilitation, crowding, or other density-dependent effects.

Observational Error Aerial counts, such as are often carried out with harbor seals, would have been more accurate than our counts from a fixed observation post. Light conditions affected our counts and often it was difficult to distinguish individuals when clumped together on the beach. If the angle of vision from the horizontal was less than about 8°, counting was especially difficult.

Time-Scale Analysis It is easier and less time consuming to simulate and parameterize algebraic equations than differential equations, because differential equations must be integrated continuously using computer software. By contrast, algebraic equations are based only on the current state of the system. Two time scales are introduced when recovery time following disturbance is rapid (assumption H4), the recovery time scale and the time scale of the environmental variables. Under these conditions, the multiple time-scale theory allows the substitution of Eq. (2.2) in place of Eq. (2.1) if the system is not disturbed. See Chap. 1 for more details.

Environmental Cues for Movement Assumption H4 greatly simplifies model selection, parameterization, and simulation, but it also confounds the identification of the environmental functions $E_{12}(t)$ and $E_{21}(t)$, given that model (2.2) depends on the ratio of these functions instead of on the individual functions. Our model selection procedure showed that the ratio can be expressed as follows:

$$E = \frac{E_{21}}{E_{12}} = \frac{C_e^q}{T^r}.$$

This does not imply, however, that $E_{21} = C_e^q$ and $E_{12} = T^r$. Equation (2.3) could have been derived from an infinite number of equations in the form of model (2.1). Here are three representative possibilities:

$$E_{12} = \frac{1}{C_e^q} \quad \text{and} \quad E_{21} = \frac{1}{T^r} \tag{2.4}$$

$$E_{12} = T^r \quad \text{and} \quad E_{21} = C_e^q \tag{2.5}$$

$$E_{12} = \frac{T^{r/2}}{C_e^{q/2}} \quad \text{and} \quad E_{21} = \frac{C_e^{q/2}}{T^{r/2}}. \tag{2.6}$$

Equation (2.4) implies that hauled-out seals are influenced mainly by tide height or a direct correlate of tide height and that seals in the water are influenced mainly by current or a direct correlate thereof. This seems to be a more likely situation than what is implied by Eqs. (2.5) and (2.6). For example, Eq. (2.5) implies that hauled-out seals are influenced mainly by current and that seals in the water are influenced mainly by tide height. It is unlikely that hauled-out seals are responsive to current, although they might be indirectly responsive to current because current is approximately the rate of change of tide height with respect to time ($C = dT/dt$), depending on local coastal geometries, basin shapes, river discharges, and wind conditions [2, 8]. Based on our data set, it was impossible to choose among equations such as (2.4), (2.5), and (2.6). Counts of seal numbers following disturbance on a much finer temporal scale than 1 hour would be required to identify the individual functions $E_{12}(t)$ and $E_{21}(t)$ [11].

A functional hypothesis suggested by model (2.3) for seals using the north beach of Violet Point is as follows: Low haul-out numbers occur when food availability is high at flood current. Falling tides serve as a cue for seals to enter the water to feed. Movement to the water continues until few or no animals occur on the beach, about halfway between low and high tides. Flood current decline signals a paucity in food availability and a consequent return of the seals to the beach.

Although interactions between current and harbor seal movement have not been extensively explored, harbor seals near Eynhallow Sound, Orkney, UK, changed the pattern of hauling out depending on if the tide was falling or rising. In concert with our functional hypothesis above, it was hypothesized that the seals were responding to increased food brought in by incoming tides [22].

Interestingly, better predictions resulted from models using C_e than models incorporating C (Table 2.1), suggesting that current direction is more important than current strength for model predictions. This interpretation comports with our functional hypothesis. In addition, Eq. (2.3) performed similarly to the model that contained solar elevation: $E(t) = C_e^q(t)S^u(t)/T^r(t)$ (Table 2.1). Indeed, correlations between haul-out behavior and solar elevation or time of day have been noted in other studies [21, 24, 26]. Correlations with time of day may be a consequence of heat loading which often occurs about midday and later [26]. Alternately, they may be related to approximately similar tide patterns on consecutive days.

References

1. Adkinson MD, Quinn IITJ, Small RJ (2003) Evaluation of the Alaska harbor seal (*Phoca vitulina*) population survey: a simulation study. Mar Mamm Sci 19:764–790
2. Anonymous (1983) Current Atlas/Atlas des Courants: Juan de Fuca Strait to/à Strait of Georgia. Canadian Hydrographic Service, Fisheries and Oceans Canada, Ontario
3. Boveng P (1988) Status of the Pacific harbor seal population on the U.S. west coast. Administrative Report LJ-88-06, Southwest Fisheries Center, National Marine Fisheries Service, La Jolla, CA
4. Brown RF Mate BR (1983) Abundance, movements, and feeding habits of harbor seals, *Phoca vitulina*, at Netarts and Tillamook Bays, Oregon. Fish Bull 81:291–301
5. Burnham KP, Anderson DR (2002) Model selection and multi-model inference: a practical information-theoretic approach, 2nd edn. Springer, New York
6. Carretta JV, Muto MM, Barlow J, Baker J (2002) Draft U. S. Pacific marine mammal stock assessments: 2002. NOAA-TM-NMFS-SWFSC Technical Memorandum, U. S. Department of Commerce, National Oceanic and Atmospheric Administration, National Marine Fisheries Service, Southwest Fisheries Science Center, San Diego, CA
7. Dennis B, Desharnais RA, Cushing JM, Henson SM, Costantino RF (2001) Estimating chaos and complex dynamics in an insect population. Ecol Monogr 71:277–303
8. Duxbury AC, Duxbury AB, Sverdrup KA (2000) An Introduction to the World's Oceans, 6th edn. McGraw-Hill, Boston, MA
9. Hastings A (1997) Population modeling: concepts and models. Springer, New York
10. Hayward JL, Henson SM, Logan CJ, Parris CR, Meyer MW, Dennis B (2005) Predicting numbers of hauled-out harbour seals: a mathematical model. J Appl Ecol 42:108–117
11. Henson SM, Hayward JL, Damania SP (2006) Identifying environmental determinants of diurnal distribution in marine birds and mammals. Bull Math Biol 68:467–482

12. Huber HR (1995) The abundance of harbor seals (*Phoca vitulina richardsi*) in Washington, 1991–1993. Unpublished M.S. thesis, University of Washington, Seattle

13. Hunt GL, Jr, Schneider DC (1987) Scale-dependent processes in the physical and biological environment of marine birds. In Croxall JP (ed) Seabirds: feeding ecology and role in marine ecosystems. Cambridge University Press, Cambridge, UK, pp 7–41

14. Jeffries S, Huber H, Calambokidis J, Laake J (2003) Trends and status of harbor seals in Washington State: 1978–1999. J Wildl Manag 67:207–218

15. Kroll AM (1993) Haul out patterns and behavior of harbor seals, *Phoca vitulina*, during the breeding season at Protection Island, Washington. MS thesis, University of Washington, Seattle

16. Levin SA (1992) The problem of pattern and scale in ecology. Ecology 73:1943–1947

17. Matthiopoulos J, McConnell B, Duck C, Fedak M (2004) Using satellite telemetry and aerial counts to estimate space use by grey seals around the British Isles. J Appl Ecol 41:476–491

18. Moss J (1992) Environmental and biological factors that influence harbor seal (*Phoca vitulina richardsi*) haulout behavior in Washington and their consequences for the design of population surveys. MS thesis, University of Washington, Seattle

19. Pitcher KW, McAllister DC (1981) Movements and haulout behavior of radio-tagged harbor seals, *Phoca vitulina*. Can Field-Nat 95:292–297

20. Schneider DC, Payne PM (1983) Factors affecting haul-out of harbor seals at a site in southeastern Massachusetts. J Mammal 64:518–520

21. Stewart BS (1984) Diurnal hauling patterns of harbor seals at San Miguel Island, California. J Wildl Manag 48:1459–1461

22. Thompson PM, Fedak MA, McConnell BJ, Nicholas KS (1989) Seasonal and sex-related variation in the activity patterns of common seals (*Phoca vitulina*). J Appl Ecol 26:521–535

23. Thompson PM, Harwood J (1990) Methods for estimating the population size of common seals (*Phoca vitulina*). J Appl Ecol 27:924–938

24. Thompson, PM, Tollit DJ, Wood D, Corpe HM, Hammond PS, Mackay A (1997) Estimating harbour seal abundance and status in an estuarine habitat in north-east Scotland. J Appl Ecol 34:43–52

25. Thompson PM, Van Parijs S, Kovacs KT (2001) Local declines in the abundance of harbour seals: implications for the designation and monitoring of protected areas. J Appl Ecol 38:117–125

26. Watts P (1992) Thermal constraints on hauling out by harbour seals (*Phoca vitulina*). Can J Zool 70:553–560

27. Wilson EO (1975) Sociobiology: the new synthesis. Belknap Press of Harvard University Press, Cambridge, Massachusetts

Chapter 3

Modeling Timing of Haulout and Foraging in Marine Iguanas

The study reported in this chapter was carried out in the Galápagos Islands in parallel with the work on flightless cormorants reported in Chap. 5. Here the application of differential equation modeling to a marine reptilian system further demonstrates the broad applicability of this technique. This work formed the basis of a Master of Science thesis completed by Brianna Payne at Andrews University. Fellow graduate student Libby Megna and senior researchers Shandelle Henson, James Hayward, and Susana Velastegui Chávez aided with data collection and analysis. The results of this research were initially published in the Journal of Coupled Systems and Multiscale Dynamics *in 2015* [16].

3.1 Introduction

Marine iguanas (*Amblyrhynchus cristatus*, Fig. 3.1) are the only lizards that feed almost exclusively underwater in marine habitats [17, 21]. These animals are restricted to the Galápagos Islands where they are abundant. Nonetheless, given their restricted range and sensitivity to environmental change, marine iguanas are listed as "vulnerable" on the Red List of Threatened Species published by the International Union for Conservation of Nature. As such, they are protected under the Convention on International Trade in Endangered Species of Wild Fauna and Flora (CITES), United Nations Environment Programme, and Galápagos National Park.

Given their unique feeding behavior, marine iguanas have attracted the interest of biologists since the time of Charles Darwin. These animals experience sudden and dramatic changes in their thermal environment, alternating between arid terrestrial haul-out sites and much cooler marine feeding areas. Surface temperatures of haul-out sites sometimes approach 60°C [18], whereas the cool waters of their feeding areas range from 15 to 25°C

J. M. Cushing et al., *Modeling Behavior and Population Dynamics*, Interdisciplinary Applied Mathematics 57, https://doi.org/10.1007/978-3-031-34283-7_3

[1, 8, 18]. Significantly, marine iguanas accommodate to these drastic temperature differences largely through behavioral, not physiological, responses [1, 2, 8, 17]. Indeed, the metabolic costs of foraging underwater are similar to those of terrestrial foraging lizards [2, 10], and the differences between basal and maximal metabolic rates (the aerobic and anaerobic scopes) of marine iguanas parallel those of terrestrial lizards [2]. Marine iguanas do exhibit an enhanced capacity to excrete salt from their nasal glands [8], one of their few physiological adaptations to marine foraging. Multiple behavioral adaptations, by contrast, are prevalent. These include a reluctance to enter the water except to feed, optimal timing of foraging and haulout in relation to environmental conditions, alterations in body orientation and posture in response to changes in solar radiation, retaining heat through nocturnal clumping behavior, and intertidal feeding and crevice-seeking behavior by smaller iguanas [1, 4, 7, 17, 18].

Optimal timing of foraging and haulout in relation to environmental conditions constitutes the most significant behavioral adaptation of marine iguanas. Studies have focused on the effects of time of day, tide height, and sea condition on the timing of foraging [17], numbers of hauled-out animals in relation to thermal conditions [4], effects of environmental variables on the duration of feeding bouts [20], and a comparison of foraging times in relation to season and location [9]. Only one of these studies, however, examined the influence of individual environmental variables on behavior in which a multiple regression analysis was used to identify variables most responsible for feeding bout duration [20].

Figure 3.1: Marine iguanas resting on a rock at Cabo Douglas, Isla Fernandina, Galápagos Islands. Marine iguanas are the world's only diving lizards. They are endemic to the Galápagos Islands (Photo by J. L. Hayward)

In the present study we used mathematical modeling to indicate how the probabilities of haulout and foraging are influenced by a variety of environmental factors. The environmental constraint function that accounted for up to 80% of the variability in the system included solar elevation, tide height, heat index, and relative humidity. In addition, a model suitable for management that depended only on solar elevation and tide height, which can be predicted in advance, accounted for up to 72% of the variability in the system.

3.2 Methods

3.2.1 Data Collection

Data were collected from April 30 to May 16, 2011, at a haul-out site along a light-colored, sandy beach which opened north to shallow Sea Lion Bay, Cabo Douglas, Isla Fernandina, Galápagos (Fig. 3.2). The east and west sides of the site were bordered by lava rocks, and the upper part of the beach was bordered by a dense stand of salt bush (*Cryptocarpus pyriformis*). Hundreds of marine iguanas hauled out along this beach each day.

A Davis VantageVue^{TM} weather station located 1.8 m above site elevation on an adjacent white sandy beach and 350 m southwest of the study site logged meteorological data at 30-minute intervals. Data logged included temperature, humidity, dew point, wind speed, wind direction, wind chill, heat index, THW index (a measure of ambient conditions based on temperature, humidity, and wind speed), and barometric pressure (Table 3.1). A handheld TM-206 Solar Power Meter (Ambient Weather Corp., Chandler, Arizona, USA), recalibrated before each sampling event, was used to measure solar radiation. Solar elevation (in degrees above the horizon) was obtained from an online calculator provided by the National Oceanic and Atmospheric Administration (NOAA). Seawater temperature was obtained from a HOBO Temp Pro located at the point of Cabo Douglas at 10-m depth relative to the sea surface when deployed. A vertically positioned, 4-m surveyor pole mounted with its base just below the low-tide mark in a small cove SW of the study site was used to measure tide height. The cove was shielded from most surf by bordering rock outcrops.

Wind direction was converted from cardinal compass subdivisions to ° azimuth with 0° and 360° at true north and then transformed to a linear scale according to the following calculation:

$$W_d = \begin{cases} W & \text{if } W \leq 180 \\ 360 - W & \text{if } W > 180 \end{cases}.$$

The tide pole was shielded from major wave action and the observed water levels showed a clear tidal dynamic, but the tide height data were noisy. To increase accuracy, we used the Lowess smoothing function in Axum 5.0 (Mathsoft Inc.; [10 iterations, 25% points for each local regression, 100 output points]) to smooth the tide height data, and then we splined the 100-point output using MATLAB (Mathworks) so as to interpolate values at the observation times. Solar elevation values at 30- and 10-minute time intervals were interpolated using the same methods as for tide height. For the other environmental variables, we interpolated 10-minute time interval values linearly using the surrounding values.

Figure 3.2: Marine iguanas hauled out on a sandy beach at Cabo Douglas, Isla Fernandina, Galápagos Islands. The animals are exhausted, having recently come ashore after feeding in the cool sea. Once onshore, they depend on solar radiation to regain body heat (Photo by J. L. Hayward)

Each environmental variable $x(t)$ was nondimensionalized and scaled using the methods of Chap. 2 so that $1 \leq x(t) \leq 2$ [6, 12, 13].

Hauled-out iguanas tend to pile on top of one another, so it was not feasible to conduct accurate, repeated censuses at temporal intervals small enough to capture the main dynamics of the system. Thus, we conducted a single census each day close to noon when few animals were present on the beach. From an observation point 50 m from the beach, we counted influx and outflux across the beach-water interface during the remainder of the day. From April 30 to May 7, fluxes were tallied from 6:00 to 18:00 hours (LST here and below); from May 8 to May 16, fluxes were tallied during 7:00–17:30 hours. Iguanas moved little before and after these times. From April 30 to May 4, we summed the tallies at 30-minute intervals. However, these intervals were too long to capture system dynamics. So from May 5 to May 16, when most activity occurred (usually 8:30–15:30 hours), we summed the tallies at 10-minute intervals; before and after that time we summed at 30-minute intervals. The 10-minute time intervals began at variable times each day: before the first iguana entered the water and after most of the iguanas had oriented parallel to incident radiation and assumed elevated basking postures. Regardless of daily start time or time interval length, all collected data were included in the analyses. We placed size

Table 3.1: Environmental variables at Cabo Douglas, Isla Fernandina, Galápagos, April 30 to May 16, 2011, prior to nondimensionalization and scaling

Environmental factor (unit)	Variable	Minimum	Maximum	Mean
Solar radiation (Wm^{-2})	$S_r(t)$	1	1230	537.4
Solar elevation (°)	$S_e(t)$	–0.9	68.1	46.14
Tide height (m)	$T(t)$	0.3	1.83	1.12
Seawater temperature (°C)	$S_p(t)$	20.5	24.7	22.1
Ambient temperature (°C)	$A(t)$	22.7	31.3	28.1
Relative humidity (%)	$H_u(t)$	55	99	76.8
Dew point (°C)	$D(t)$	20	25.8	23.5
Wind speed (ms^{-1})	$W_s(t)$	0	25.7	14.1
Wind direction (°)	$W_d(t)$	15	360	190.1
Wind chill (°C)	$W_c(t)$	22.7	31.2	27.6
Heat index (°C)	$H_x(t)$	24.2	36.6	32.0
THW index (°C)	$T_x(t)$	24.2	36.2	31.5
Barometric pressure (mmHg)	$B(t)$	753.7	759.9	756.9
Hour of day (1–24)	$H_r(t)$	6	18	N/A

standards on the beach and found that only about 5% of animals using this haul-out site had snout-to-vent lengths of less than 25 cm, designated as "small" [20]. Because so few animals were small, we combined small and large animals in our counts.

We estimated the numbers of iguanas occupying the beach as follows: Let N_c be the census number for a particular day. Given that x iguanas entered the water and y iguanas hauled out of the water to the beach during the next time interval, the extrapolated occupancy at the beginning of the next time interval was estimated as

$$N_{c+1} = N_c - x + y. \tag{3.1}$$

Likewise, given that x iguanas entered the water and y iguanas hauled out of the water to the beach during the time interval preceding the census, the estimated occupancy at the beginning of the interval preceding the census was

$$N_{c-1} = N_c + x - y. \tag{3.2}$$

Inferred occupancies were thus extrapolated forward to the end of the day and backward to the beginning of the day, creating a time series of inferred occupancies for each day.

We never observed the animals exiting the beach via the salt bush area, but we were unable to count animals resting beneath the salt bushes. It is possible that some animals remained uncounted because of this. This limitation led to two problems:

(1) Inferred occupancies could yield negative "counts."
(2) Inferred occupancies for evenings and the following mornings did not necessarily correspond.

Thus, for each day having a negative inferred occupancy, we added the absolute value of the most negative inferred occupancy to each occupancy for that day. This had the effect of setting the minimum estimated occupancy to zero by adding in any iguanas beneath the salt bushes that day that exited the site by that route.

Inferred occupancies for the beach were divided using stratified random sampling into two bins: an estimation data set for model parameterization and a validation data set for model validation. The 14-day tidal cycle was divided into four segments of 3.5 days and the inferred occupancies were binned. A randomly assigned "tag" of 0 or 1 was applied to each inferred occupancy. Then for each bin either 0 or 1 tags were randomly designated as the subset for model validation.

3.2.2 Mathematical Model

We characterized the movement of iguanas to and from the beach in response to environmental fluctuations with the differential equation model

$$\frac{dN}{dt} = aE_{12}(t)(\beta M(t) - N(t)) - bE_{21}(t)N(t), \tag{3.3}$$

where $N(t)$ is the number of iguanas on the beach at time t, $\beta M(t)$ equals the total number of iguanas eligible to haul out along the beach, and $aE_{12}(t)$ and $bE_{21}(t)$ are the per capita rates at which iguanas entered and exited the beach, respectively. The constants a, b, $\beta > 0$ are parameters. We assumed the per capita rates $aE_{12}(t)$ and $bE_{21}(t)$ were products of powers of environmental variables [6, 11–13].

Using the multiple time-scale methods of Chap. 1, we derived an algebraic model for the steady-state dynamics of Eq. (3.3):

$$N(t) = \frac{\beta M(t)}{1 + \alpha E(t)}, \tag{3.4}$$

in which

$$E(t) = X_1^{\Omega_1} X_2^{\Omega_2} ... X_{14}^{\Omega_{14}} \tag{3.5}$$

was a product of powers of the 14 environmental factors X_i (Table 3.1) and in which α, $\beta > 0$ and $\Omega_i \in R$ were constant parameters.

We assumed that the noise in the system was largely demographic. As in Chap. 2, the associated stochastic model is

$$N(t) = \left(\sqrt{\frac{\beta M(t)}{1 + \alpha E(t)}} + \sigma\varepsilon(t) \right)^2, \tag{3.6}$$

where the $\varepsilon(t)$ are standard normal random variables (mean zero and standard deviation one) uncorrelated in time and $\sigma > 0$ is a constant parameter. The right-hand side was assumed to be 0 if the quantity within the parentheses was negative.

In Eq. (3.3), $\beta M(t)$ represents the total number of iguanas in the system that utilize the beach for haul-out. Thus, in Eq. (3.4) we can consider the expression

$$P(t) = \frac{1}{1 + \alpha E(t)} \tag{3.7}$$

to be an environmental constraint function in which $P(t)$ represents the probability that an iguana in the system is hauled out, given the current state of the environment $E(t)$. The probability that such an animal is not hauled out (is foraging) is $1 - P(t)$.

Tidal fluctuations seemed to circumscribe maximum haulout. Daily maximal haul-out was highest around neap tide (May 10, 2011, day 130, Fig. 3.3) and lowest at spring tide (May 16, 2011, day 136, Fig. 3.3). We utilized a sine wave as the functional form circumscribing maximum haul-out at this site:

$$M(t) = \eta + \delta \sin\left(\gamma\left(\text{day of year} + t/24\right)\right), \tag{3.8}$$

where t was the hour of day and η, δ, $\gamma > 0$ were constant parameters. Here η is the average value about which the sine wave oscillates, δ is the amplitude, and $\gamma/2\pi$ represents the frequency, that is, the number of oscillations per unit time.

3.2.3 Model Parameterization

The model obtained by incorporating Eqs. (3.5) and (3.8) into Eq. (3.4) contained 19 parameters. Direct estimation of this many parameters from the time series was intractable. Instead, we used the following procedure to reduce the number of parameters.

We fit Eq. (3.8) directly to the daily maximum inferred occupancies using the method of nonlinear least squares (LS), but under the assumption of environmental rather than demographic stochasticity, minimizing the sum of squared residuals on the log scale:

$$RSS = \sum_{data} \left(\ln(\text{observation}) - \ln(\text{prediction})\right)^2.$$

Of post hoc interest is the estimate of $\gamma = 0.4481$, which yields a period of just over 14 days, a complete tidal cycle.

Given the assumption of demographic stochasticity in haul-out numbers as in the stochastic model (3.6), we parameterized model (3.4) by minimizing RSS on the square-root scale:

$$RSS = \sum_{data} \left(\sqrt{\text{observation}} - \sqrt{\text{prediction}}\right)^2.$$

Given the three parameter estimates for Eq. (3.8), 16 parameters remained to be estimated in model (3.4): the coefficients α and β and the 14 exponents in Eq. (3.5). We used an information-theoretic model selection procedure to further reduce the number of parameters to just α and β. To do this, we posed a suite of alternative submodels using various combinations of the 14 environmental variables in Eq. (3.5). We limited the large number of alternatives in the following way. First, we tested for highly correlated ($r > 0.80$) variables so that we would not use them together in the same submodel. Then, for each single environmental variable X_i we parameterized model (3.4) with $E(t) = X_i^{\Omega_i}$ to see whether that variable had a positive or negative effect on haulout ($\Omega_i < 0$ or $\Omega_i > 0$). Finally, we created increasingly complex models through stepwise inclusion of environmental variables. At each step, we discarded variables whose inclusion did not further minimize the RSS after parameterization. In this way we obtained 155 biologically reasonable alternative models.

3.2.4 Model Selection

To determine the best model, we used the Akaike information criterion (AIC), which penalizes models with more parameters for overfitting. In the case of LS methodology, the criterion is equivalent to

$$AIC = n \ln \widehat{\sigma}^2 + 2\kappa,$$

where n is the number of residuals, $\widehat{\sigma}^2 = RSS/n$ is the variance of the (fitted) likelihood function, and κ is the number of model parameters, including σ. The best model is the one with the smallest AIC ($\Delta AIC = 0$); see Chap. 2 for the definition of ΔAIC. Models with $\Delta AIC > 10$ were omitted from consideration [3].

3.2.5 Model Validation

We validated the best model on the independent validation data set without reparameterizing by comparing goodness-of-fit on the estimation data to goodness-of-fit on the validation data, measuring goodness-of-fit as in Chap. 2:

$$R^2 = 1 - \frac{RSS}{\sum\limits_{data} \left(\sqrt{\text{observation}} - \text{mean} \right)^2},$$

where "mean" is the mean of the square roots of the observations and RSS is the fitted value.

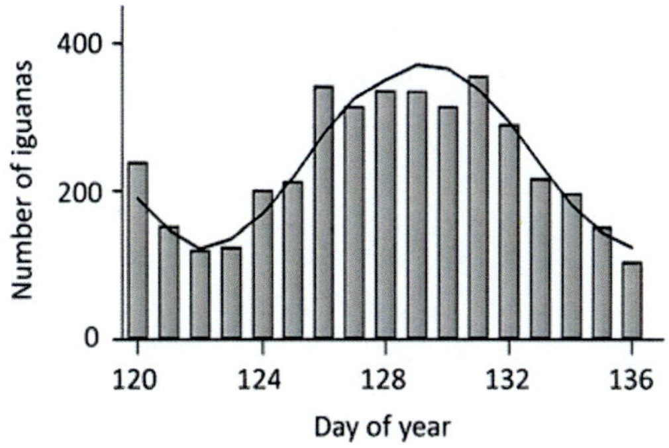

Figure 3.3: Maximal daily haul-out of marine iguanas at Cabo Douglas, Isla Fernandina, Galápagos, over the 17-day study period starting on April 30, 2011. The solid curve represents Eq. (3.9). (Originally published in Payne et al., "Environmental constraints on haul-out and foraging dynamics in Galápagos marine iguanas," *Journal of Coupled Systems and Multiscale Dynamics* 3:208–218. Copyright © 2015, American Scientific Publishers. Used by permission.)

3.3 Results

3.3.1 Environmental Trends

Descriptive statistics for the environmental variables are shown in Table 3.1. Solar radiation was highly variable over a short time scale due to intermittent cloud cover. But the overall daily trend matched that of solar elevation with which it was highly correlated ($r = 0.80$). Ambient temperature, wind chill, heat index, and THW index varied collinearly with one another ($r > 0.89$ for each correlation). Daily trends matched those of solar variation but were more variable. Dew point exhibited a daily trend similar to that of solar radiation as well, but with higher variability; moreover, dew point averaged about 2°C cooler during the second half of the study period. Relative humidity was lowest at midday and higher in the morning and evening, roughly the inverse of the trend in solar elevation. Variability over the trend in relative humidity was similar to that of ambient temperature, wind chill, heat index, and THW index.

Prevailing winds changed from easterly directions in the mornings to south to west-southwest directions by late morning to mid-afternoon. Brief gusts from other directions sometimes occurred. In contrast to those on other days, the winds on May 11 shifted back to easterly directions in the afternoon. Usually, wind speed exhibited a daily pattern similar pattern to that of solar elevation. Wind speed averaged over 10 ms^{-1} higher in the afternoons than in the mornings. High winds (35–44 ms^{-1}) occurred during nearly two-thirds of the afternoons and were predominantly from the southwest or south-southwest.

High tide shifted by approximately 50 minutes each day and was semidiurnal. Daily patterns in seawater temperature were dependent on rainfall, solar radiation, and wind patterns and were quite variable. They exhibited an overall decline of approximately 3°C during the study period. Barometric pressure rose briefly each morning and then mid-morning to later afternoon fell before increasing a little each evening.

3.3.2 The Best Models

The best haul-out models (those with $\Delta AIC < 10$; see Chap. 2) are listed in Table 3.2. The model with $\Delta AIC = 0$ and the highest R^2 (= 0.77) was

$$N(t) = \frac{\beta(247.7 + 125.2\sin(0.4481(\text{day of year } + t/24)))}{1 + \alpha S_e^{\varsigma} H_x^{\phi} / \left(H_u^{\psi} T^{\rho}\right)}, \tag{3.9}$$

where $\beta = 0.89595, \alpha = 5.0812 \times 10^{-6}, \varsigma = 24.057, \phi = 1.8040, \psi = 6.7644, \rho = 2.4393$, and each environmental variable is a function of time. Note that the period of the sine wave of model (3.9) is $2\pi/0.4481$, just over 14 days, a complete tidal cycle. From model (3.9) we obtain the environmental constraint function

$$P(t) = \frac{1}{1 + \alpha S_e^{\varsigma} H_x^{\phi} / \left(H_u^{\psi} T^{\rho}\right)}. \tag{3.10}$$

Thus, the probability P of haulout increases with increasing tide height and relative humidity and decreases with increasing solar elevation and heat index, all other factors held constant. The probability of foraging increases with increasing solar elevation and heat index and decreases with increasing tide height and relative humidity, all other factors held constant.

Table 3.2: Best marine iguana haul-out models ($\Delta AIC < 10$) based on data from Cabo Douglas, Isla Fernandina, Galápagos. Increasing variables in the denominator of $E(t)$ yields increased haulout, and increasing variables in the numerator of $E(t)$ yield decreased haulout

$E(t)$	σ^2	ΔAIC	R^2
$S_e^{\varsigma} H_x^{\phi} / \left(H_u^{\psi} T^{\rho}\right)$	5.89	0.000	0.7713
$S_e^{\varsigma} T_x^{\omega} / \left(H_u^{\psi} T^{\rho}\right)$	5.90	0.3478	0.7711
$S_e^{\varsigma} A^{\zeta} / \left(H_u^{\psi} T^{\rho}\right)$	5.90	0.7241	0.7709
$S_e^{\varsigma} W_c^{\varphi} / \left(H_u^{\psi} T^{\rho}\right)$	5.92	1.468	0.7705
$S_e^{\varsigma} / \left(H_u^{\psi} T^{\rho}\right)$	5.95	1.670	0.7691
$S_e^{\varsigma} W_s^{\lambda} / \left(H_u^{\psi} T^{\rho}\right)$	5.95	3.667	0.7691

Model predictions and inferred occupancies are shown in Fig. 3.4a, with corresponding solar elevations and tide heights shown in Fig. 3.4b. As earlier noted, daily maximal haulout was highest around neap tide and lowest at spring tide.

We computed the goodness-of-fit of the best model (3.9) on the validation data without re-estimating parameters, obtaining $R^2 = 0.80$, a strong support for successful model validation.

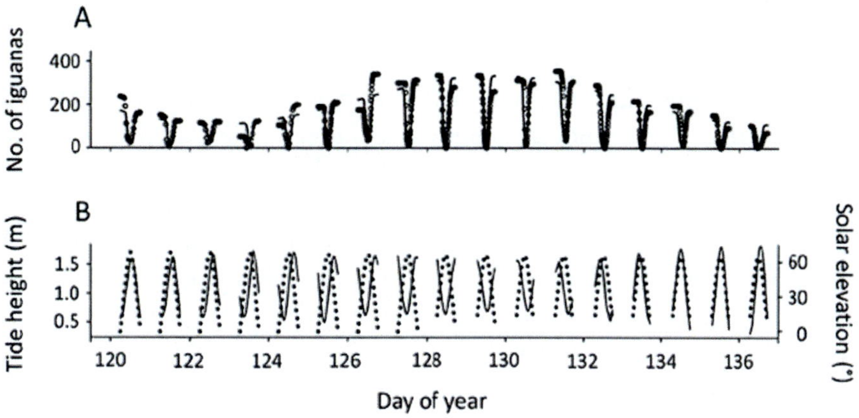

Figure 3.4: Best model of marine iguana haul-out. (**a**) Model predictions (solid line) are graphed with census and inferred occupancy data (circles). (**b**) Solar elevation (dashed line) and tide height (solid line), the most descriptive variables for the system, for reference. (Originally published in Payne et al., "Environmental constraints on haul-out and foraging dynamics in Galápagos marine iguanas," *Journal of Coupled Systems and Multiscale Dynamics* 3:208–218. Copyright © 2015, American Scientific Publishers. Used by permission.)

3.4 Discussion

The best model based on environmental variables predicted the haul-out/foraging dynamics with up to 80% accuracy. This value is high for an ecological system [14], particularly one for which data were combined for large and small iguanas which exhibit different foraging strategies [4, 20]. Note, however, that the study occurred during the start of the transition between the warm-wet and cool-dry seasons, a nonreproductive interval. Other dynamics may occur at other times of the year.

3.4.1 Environmental Constraints on Foraging and Haul-Out

Little or no influence on haul-out was exerted by wind speed and direction, dew point, and barometric pressure (Table 3.2). *All models with* $\Delta_i < 10$ *showed a direct relationship between haul-out and relative humidity* (Table 3.2), most likely because relative humidity was usually highest at the beginning and at the end of the day, before and after foraging, respectively.

Solar elevation and real or perceived temperatures were inversely related to haul-out (Table 3.2). Entering the water to forage once the body has warmed and returning to the beach before the body cools too much maximizes thermal efficiency [20].

Solar elevation accounted for 71% *of system variability* (Table 3.3). Solar elevation served as a proxy for, and was highly correlated with, solar radiation ($r = 0.80$). Indeed, our observations suggested that solar radiation influenced movement into the water more than any other variable. Only an occasional iguana entered the water before solar radiation reached 500 Wm^{-2}, typically on sunny mornings by 8:30 hours. This level of solar intensity was not reached until 10:30 hours or later on cloudy mornings, however, which resulted in a delay of entry to the water. On mornings in which the increase in solar radiation was rapid (350 Wm^{-2} to 950 Wm^{-2} in 10 minutes), iguanas moved rapidly to the water. Two studies [4, 20] have shown that for large iguanas, cloud cover and rock temperature were unrelated to the timing of water entry.

Table 3.3: Models useful for management of marine iguanas based on data from Cabo Douglas, Isla Fernandina, Galápagos. Increasing variables in the denominator of $E(t)$ yields increased haul-out, and increasing variables in the numerator of $E(t)$ yields decreased haul-out. Δ_i values are relative to the best model ($\Delta_i = 0$) given in Table 3.2

$E(t)$	σ^2	Δ_i	R^2
S_e^ς/T^ρ	7.10	68.0	0.7244
S_e^ς	7.55	89.0	0.7068

Haul-out to the beach varied directly with tide height (Table 3.2). This is surprising given that most of the iguanas in this study were large and fed subtidally at considerable distance from the shore, and subtidal feeders would be expected to be less influenced by tide than small iguanas which feed on algal beds exposed at low tides [4, 17, 20]. Marine iguanas are not efficient swimmers, however, and riding the incoming tide may be cost-effective especially during times of heavy surf [2, 8, 14].

Maximal daily haulout was highest at neap tide and lowest at spring tide (Fig. 3.4a,b). Maximal daily haulout exhibited a large range of animals (160 to >300), and this may have been partly due to intermittent use by non-colony individuals. Typically, marine iguanas exhibit terrestrial site fidelity [19, 20]. However, sometimes these animals swim as many as 400 m from their colony and temporarily exit to other haul-out sites [10, 17].

3.4.2 Inferred Occupancies

Inferred occupancies were strongly mismatched for some evenings and mornings that followed (Fig. 3.4a). Iguanas are quite inactive at night [4, 15], so this result was surprising. We never saw them leave the haul-out site by a route other than directly by water. Although it is possible that some animals left or entered via the brushy area at the top of the beach, this seems unlikely given the higher energetic cost of walking compared to swimming [10]. Moreover, nonbreeding individuals typically are found along the shoreline [5, 7]. This mismatch remains puzzling.

Inferred occupancies were sometimes artificially low, and adding the absolute value of the most negative inferred occupancy to all other occupancies for that day (see Methods) sometimes compounded mismatches, whereas in other cases the addition method ameliorated mismatches. Two factors may have exaggerated the mismatches. First, most animals at the study site were large and did not necessarily feed every day [17]. If on a given day x animals hauled out and crawled up to the salt bushes but failed to leave the site to forage on the following day, an overnight mismatch would result because our process of inferring occupancies back from the census of the next day would not account for these animals. Second, counting iguanas leaving the site required more vigilance than counting those entering the site from the water. This is because the hauling out process was more gradual than was the activity of leaving to forage. Consequently, counting error would tend to depress inferred occupancies before the census when the iguanas commonly entered the water but would inflate inferred occupancies following the census when iguanas commonly hauled out of the water.

3.4.3 Predictive Model for Managers

Two of the environmental variables in the best model (3.9), relative humidity and heat index, cannot be predicted far into the future. However, solar elevation and tide height alone account for 72% of the variability in haulout (Table 3.3), and both of these can be predicted quite accurately in advance. The associated parameterized model is

$$N(t) = \frac{\beta(247.7 + 125.2\sin(0.4481(\text{day of year } + t/24)))}{1 + \alpha S_e^{\varsigma}/T^{\rho}},$$

where $\beta = 0.90210$, $\alpha = 4.6447 \times 10^{-7}$, $\varsigma = 25.683$, and $\rho = 2.3556$. Managers can use a model of this form to predict haulout in advance, although the model should be constructed and parameterized on data from the site in question. Note that the model based on solar elevation alone had $R^2 = 0.71$ (Table 3.3); that model would not require data on tide height.

3.4.4 Conclusion

Given environmental data, the chance of haul-out and foraging by marine iguanas can be predicted by mathematical modeling with a high degree of accuracy. *Solar elevation and tide height were the two environmental factors that accounted for a large portion of the variability in count data, although other factors were also involved. This high level of predictability suggests these animals respond to an integrated set of environmental factors.* This capacity has evolved primarily behaviorally with a surprising paucity of physiological adaptation to a lifestyle dramatically different from that of other lizards.

References

1. Bartholomew GA (1966) A field study of temperature relations in the Galápagos marine iguana. Copeia 1966:241–250
2. Bennett AF, Dawson WR, Bartholomew GA (1975) Effects of activity and temperature on aerobic and anaerobic metabolism in the Galá pagos marine iguana. J Comp Physiol 100:317–329
3. Burnham KP, Anderson DR (2002) Model selection and multimodel inference: a practical information-theoretic approach. Springer, New York
4. Buttemer WA, Dawson WR (1993) Temporal pattern of foraging and microhabitat use by Galápagos marine iguanas, *Amblyrhynchus cristatus*. Oecologia 96:56–64
5. Carpenter CC (1966) The marine iguana of the Galápagos Islands, its behavior and ecology. Proc Calif Acad Sci 34:329–376
6. Damania SP, Phillips KW, Hayward JL, Henson SM (2005) Habitat patch occupancy dynamics of glaucous-winged gulls (*Larus glaucescens*) II: a continuous-time model. Nat Resour Model 18:469–499
7. Darwin C (1839) Narrative of the surveying voyages of His Majesty's ships Adventure and Beagle between the years 1826 and 1836, describing their examination of the southern shores of South America, and the Beagle's circumnavigation of the globe: journal and remarks 1832–1836. Henry Colburn, London, pp 466–469
8. Dawson WR, Bartholomew GA, Bennett AF (1977) A reappraisal of the aquatic specializations of the Galápagos marine iguana (*Amblyrhynchus cristatus*). Evolution 31:891–897
9. Drent J, Van Marken Lichtenbelt WD, Wikelski M (1999) Effects of foraging mode and season on the energetics of the marine iguana *Amblyrhynchus cristatus*. Funct Ecol 13:493–499

10. Gleeson TT (1979) Foraging and transport costs in the Galá pagos marine iguana *Amblyrhynchus cristatus*. Physiol Zool 52:549–557

11. Henson SM, Dennis B, Hayward JL, Cushing JM, Galusha JG (2007) Predicting the dynamics of animal behaviour in field populations. Anim Behav 74:103–110

12. Henson SM, Galusha JG, Hayward JL, Cushing JM (2007) Modeling territory attendance and preening behavior in a seabird colony as functions of environmental conditions. J Biol Dyn 1:95–107

13. Moore AL, Damania SP, Henson SM, Hayward JL. (2008) Modeling the daily activities of breeding colonial seabirds: dynamic occupancy patterns in multiple habitat patches. Math Biosci Eng 5:831–842

14. Møller AP, Jennions MD (2002) How much variance can be explained by ecologists and evolutionary biologists? Oecologia 132:492–500

15. Nagy KA, Shoemaker VH (1984) Field energetics and food consumption of the Galápagos marine iguana *Amblyrhynchus cristatus*. Physiol Zool 57:281–290

16. Payne BG, Henson SM, Hayward JL, Megna LC, Velastegui Chávez SR (2015) Environmental constraints on haul-out and foraging dynamics in Galápagos marine iguanas. Journal of Coupled Systems and Multiscale Dynamics 3:208–218

17. Trillmich KGK, Trillmich F (1986) Foraging strategies of the marine iguana, *Amblyrhynchus cristatus*. Behav Ecol Sociobiol 18:259–266

18. White EN (1973) Temperature and the Galápagos marine iguana—insights into reptilian thermoregulation. Comp Biochem Physiol A Physiol 45:503–504, IN9–IN10, 505–513

19. Wikelski M (1999) Influences of parasites and thermoregulation on grouping tendencies in marine iguanas. Behav Ecol 10:22–29

20. Wikelski M, Trillmich F (1994) Foraging strategies of the Gal ápagos marine iguana (*Amblyrhynchus cristatus*): adapting behavioral rules to ontogenetic size change. Behaviour 128:255–279

21. Wikelski M, Wrege PH (2000) Niche expansion, body size, and survival in Galápagos marine iguanas. Oecologia 124:107–115

Chapter 4

Simultaneous Modeling of Occupancy and Behavior Dynamics

Chapters 1–3 report studies that focus on single-state variables within behavioral systems. Behavioral systems, however, involve multiple-state variables. In this chapter we demonstrate the efficacy of differential equation-based modeling which focuses on two state variables, behavior and habitat occupancy. This research was initially published in the journal Animal Behaviour *in 2007 by Shandelle Henson, Brian Dennis, James Hayward, J. M. Cushing, and Joseph Galusha [16].*

4.1 Introduction

Behavior dynamics typically have been modeled using game theory, Markov chains, and individual-based models [9, 13, 21, 25]. Ordinary differential equation (ODE) models with "motivational" dependent variables have been developed in the context of control theory as qualitative models of individual behavior [15, 22]. Because "motivation" is not a measurable variable, however, such ODE models typically have not been tied to data [30].

In this chapter we propose a methodology for the quantitative description and prediction of animal behavior in natural habitats, and we use this methodology to describe and predict the joint dynamics of the occurrence of sleep and habitat occupancy in a glaucous-winged gull (*Larus glaucescens*) breeding colony. Our models are systems of compartmental ODEs, rigorously tied to empirical data from the field.

J. M. Cushing et al., *Modeling Behavior and Population Dynamics*, Interdisciplinary Applied Mathematics 57, https://doi.org/10.1007/978-3-031-34283-7_4

4.2 Modeling Approach

Individuals move between various habitats during the day while at the same time transitioning from one behavior state to another. We classify individuals as occurring in mutually exclusive compartments, each compartment representing a specific behavioral state at a specific habitat. We consider shifts from one behavior state to another conceptually the same as shifts between habitats, as discussed in Chaps. 1–3. The state variables (dependent variables) track numbers of gulls in each compartment and are thus measurable. Moreover, they track the behavior patterns of aggregated birds which allows for realistic estimates of demographic variability among individuals. Censuses of compartmentalized individuals are made at short enough time intervals to capture system dynamics.

Fluctuations in the data are considered to be from two sources:

(1) Deterministic fluctuations explained by the model
(2) Stochastic fluctuations which consist of the variability left unexplained by the model [5].

Deterministic and stochastic factors are considered "demographic" or "environmental." Demographic factors are experienced independently by single, or small subsets of, birds, whereas environmental factors are experienced collectively by all the birds in the population [5]. Demographic factors, environmental factors, or a combination of the two can drive both deterministic and stochastic fluctuations.

A system with b-specific behaviors occurring within h-specific habitats contains at most $m = bh$ compartments. A model with m compartments is a system of m equations with the following form:

$$\frac{dn_i}{dt} = \sum_{j=1}^{m} r_{ij} f_{ij} - \sum_{j=1}^{m} r_{ji} f_{ji}, \qquad i = 1, 2, ..., m, \tag{4.1}$$

where n_i represents the number of animals in compartment i, f_{ij} is the number of animals in compartment j that are eligible to move into compartment i, and r_{ij} is the per capita rate at which eligible animals move from compartment j to compartment i. Model (4.1) is a behavior model rather than a population model, and it does not incorporate birth and death rates. Consequently, the total population size $P = \sum_{i=1}^{m} n_i$ remains constant, and one of the state variables can be eliminated, for example, by writing $n_m = P - \sum_{i=1}^{m-1} n_i$.

To apply Eq. (4.1) to a specific system necessitates modeling the f_{ij} and the r_{ij}, which, in general, are functions of factors and/or compartment densities that vary over time.

4.2.1 Application to Colony Attendance and Sleep in Seabirds

Animals spend a significant amount of time in sleep, which has been defined as "a state of immobility with greatly reduced responsiveness and which can be distinguished from coma or anaesthesia by its rapid reversibility" ([29] page 1264). We have a limited understanding of the functional significance of sleep, and sleep may have different functions for different animals [19, 28, 29]. We do know, however, that sleep quality and duration exert a large influence on the efficiency of wakefulness [26].

Sleep in gulls consists of two types: "front sleep" in which the head and neck are retracted and the beak faces down and forward and the more common "back sleep" (Fig. 4.1) in which the head is turned back with the beak inserted beneath the scapulars [1, 2, 27]. Both types of sleep are accompanied by increased thresholds of arousal, but the duration of eye closure bouts is longer during back sleep suggesting that gulls in this posture are sleeping more deeply than those in front sleep posture [2]. Gulls in both sleep postures, however, engage in unihemispheral sleep during which one eye or the other monitors the environment [26]. For purposes of our modeling study, we use the term "sleep" to refer exclusively to "back sleep," even though this is an oversimplication.

Figure 4.1: A glaucous-winged gull engaged in back sleep at Protection Island National Wildlife Refuge, Washington (Photo by J. L. Hayward)

When raising chicks, adult gulls must divide their time on the colony between sleep and alertness, as well as feeding away from the colony. The identification of factors that lead to these various behavioral states through mathematical modeling helps clarify the functions of these behaviors.

Our model makes the following five assumptions:

(H1) Each gull counted in our study plots occupies one of three mutually exclusive compartments: "Wake" in the colony (W), "Sleep" in the colony (S), and "Away" from the colony (A). Each of these three variables represents the number of animals in the compartment.

(H2) The number of gulls $C = S + W$ present in the colony remains above a minimum threshold $\mu \geq 0$, because one member of each nesting pair remains on the territory during the chick-rearing season.

(H3) No gull moves from the S compartment into the A compartment, or vice versa, without moving into the W compartment first (Fig. 4.2); that is, $f_{AS} = f_{SA} = 0$.

(H4) The number of gulls eligible to leave the colony via the W compartment and enter the A compartment is the smaller of W or $C - \mu$; that is, $f_{AW} = \min\{W, C - \mu\}$. All gulls in the A and S compartments are eligible to move to the W compartment; that is, $f_{WA} = A$ and $f_{WS} = S$.

(H5) The number of individuals in the W compartment eligible to enter the S compartment is $\alpha C - S$, where $0 < \alpha \leq 1$, as long as $\alpha C - S$ is positive, and zero otherwise; that is, $f_{SW} = \max\{\alpha C - S, 0\}$. The coefficient α we interpret to be the proportion of gulls in the colony that are sleeping or in postures like rest that can transition directly to sleep. The quantity $1 - \alpha$, a fraction that changes over time, is the fraction of gulls in the colony engaged in behavior, such as preen, that rarely transition to sleep [23]. We assume that α is constant because sleep is the only behavior the model tracks.

With these assumptions, the system of differential equations (4.1) becomes

$$\begin{aligned} \frac{dW}{dt} &= r_{WS}S + r_{WA}A - r_{SW}\max\{\alpha C - S, 0\} - r_{AW}\min\{W, C - \mu\} \\ \frac{dS}{dt} &= r_{SW}\max\{\alpha C - S, 0\} - r_{WS}S \\ \frac{dA}{dt} &= r_{AW}\min\{W, C - \mu\} - r_{WA}A. \end{aligned} \tag{4.2}$$

From system (4.2) and the identities $W = C - S$ and $A = P - C$, we obtain a model for colony occupancy and the number of sleeping gulls in the colony:

$$\begin{aligned} \frac{dC}{dt} &= r_{WA}(P - C) - r_{AW}\min\{C - S, C - \mu\} \\ \frac{dS}{dt} &= r_{SW}\max\{\alpha C - S, 0\} - r_{WS}S. \end{aligned} \tag{4.3}$$

Solutions of Eqs. (4.3) satisfy the inequalities $\mu \leq C(t) \leq P$ and $0 \leq S(t) \leq C(t)$ for all time t, provided that the initial conditions do. Because there were 70 nesting pairs in our sample colony area (see Methods), $P = 140$ and $\mu = 70$. The coefficient α was an unknown parameter to be estimated from the data. We assumed that the four per capita transition rates r_{ij} were functions of time-varying abiotic environmental conditions.

4.3 Methods

4.3.1 Data Collection

Data were collected at Protection Island National Wildlife Refuge (48°08N, 122°55′W), Jefferson County, Washington, at the southeastern end of the Strait of Juan de Fuca. Violet Point, which contained a breeding colony of more than 2,400 pairs of glaucous-winged gulls during this study, extends southeast from the main part of Protection Island.

A 33 × 100-m area of the Violet Point colony that contained 70 nests was selected as a sample area. Hourly census counts and behavior scan counts were taken in that order from 5:00 to 20:00 hours (LST here and below) for 14 days on June 30 to July 2 and July 6–16, 2004, during the chick-rearing season. Counts were made using a 20–60 spotting scope from an observation point at the top of a 33-m bluff bordering the west end of Violet Point and 200 m from the proximal edge of the sample area. Behaviors were recorded by voice and then transcribed. Typically, there was a slight difference between the census counts and the number of gulls in the behavior scans. This difference was corrected by dividing the number of gulls exhibiting a particular behavior by the number of birds scanned and then multiplying the results by the census.

A weather station 2 m above site elevation at the northwest end of Violet Point automatically recorded temperature, humidity, wind speed and direction, barometric pressure, rainfall, solar radiation, and heat index (computed from temperature and relative humidity as a measure of how hot it feels [31]). Hourly measures of tide height, solar elevation, and wind speed over open water were obtained from the National Oceanographic and Atmospheric Administration (NOAA) website.

4.3.2 Submodels for r_{ij}

Full specification of system (4.3) necessitates modeling the per capita transition rates r_{ij} as functions of environmental variables. Correlations of occupancy and sleep with time of day, temperature, and tide height have been documented by earlier studies on gulls [7, 8, 11]. In our study, we partitioned the day into "Morning" (5:00–10:00 hours), "Midday" (10:00–14:00 hours), and "Evening" (14:00–20:00 hours). We assumed that the four per capita transition rates r_{WS}, r_{SW}, r_{WA},and r_{AW} in Eq. (4.3) could be modeled as functions of tide height (TIDE), time of day (HOUR), solar elevation (SOLAR), heat index (HEAT), wind speed over the colony (CWIND), and wind speed over the open water (WWIND):

$$\begin{aligned}
\text{Morning:}\quad r_{ij} &= m_{ij}\text{TIDE}^{b1_{ij}}\text{HOUR}^{c1_{ij}}\text{SOLAR}^{d1_{ij}}\text{HEAT}^{f1_{ij}}\text{CWIND}^{g1_{ij}}\text{WWIND}^{h1_{ij}} \\
\text{Midday:}\quad r_{ij} &= n_{ij}\text{TIDE}^{b2_{ij}}\text{HOUR}^{c2_{ij}}\text{SOLAR}^{d2_{ij}}\text{HEAT}^{f2_{ij}}\text{CWIND}^{g2_{ij}}\text{WWIND}^{h2_{ij}} \quad (4.4) \\
\text{Evening:}\quad r_{ij} &= e_{ij}\text{TIDE}^{b3_{ij}}\text{HOUR}^{c3_{ij}}\text{SOLAR}^{d3_{ij}}\text{HEAT}^{f3_{ij}}\text{CWIND}^{g3_{ij}}\text{WWIND}^{h3_{ij}},
\end{aligned}$$

where $m_{ij}, n_{ij}, e_{ij} > 0$ are constant coefficients and $b_{ij}, c_{ij}, d_{ij}, f_{ij}, g_{ij}, h_{ij}$ are constant exponents. We scaled and nondimensionalized the six environmental variables x so that $1 \leq x \leq 2$ [6, 14, 17, 18], as in Chap. 2.

Thus, Eq. (4.3) contained 13 coefficients $\alpha, m_{ij}, n_{ij}, e_{ij} > 0$ plus 72 constant exponents. Numerical estimation of 85 parameters would require a very large data set, so we restricted the model parameters to the 13 coefficients and treated the 72 exponents in the following way. We assumed that each of the 72 exponents $b_{ij}, c_{ij}, d_{ij}, f_{ij}, g_{ij}, h_{ij}$ had a value of -1, 0, or 1, an assumption that created a large number of alternative models having the form of system (4.3). However, we chose a small subset of these models based on

(1) What was most biologically reasonable for the system
(2) A statistical comparison of the data and environmental variables
(3) Trial-and-error searches for models that could fit the data

We selected the best of the subset models as the one with the lowest (fitted) sum of squared residuals and discarded the others. Because each of the alternative models had 13 parameters, it was unnecessary to use model selection indices such as the Akaike information criterion (AIC) that penalize models with more parameters for overfitting [14].

Once we selected the best model, the exponents with value -1 (or 1) were decreased (or increased) by integer units until the best integer exponents were obtained.

4.3.3 Parameter Estimation

The 14 days of data were randomly divided into two 7-day sets, the first for parameter estimation and the other for independent validation. The method of maximum likelihood was used to estimate the parameters for system (4.3). Given the observation (c_t, s_t) at hour t as the initial observation, numerical integration of system (4.3) produces a one-step prediction (C_{t+1}, S_{t+1}) at hour $t+1$. We generated one-step model predictions with the MATLAB ODE integrator ode45 (MathWorks).

Prior to determination of the maximum likelihood parameter estimates, the data and predictions were transformed to stabilize the variances of the one-step departures of the data from the model. For each state variable, the likelihood estimation procedure assumes that the variances of the residual prediction errors are not dependent on the sizes of the state variables. Assuming the one-step, conditional variance $\mathrm{Var}(x_t)$ of a dependent variable x_t is a function $v(x_{t-1})$ of the size of the system at time $t-1$, a transformed dependent variable $w_t = \phi(x_t)$ that exhibits an approximately constant variance can be found as the solution of the equation $v(x)[\phi'(x)]^2 = c$, where c is an arbitrary constant [5]. Two principle forms of variance scaling in ecological systems are recognized by ecologists: "demographic" and "environmental" [10]. Frequently, demographic scaling is modeled by $v(x) = \gamma x$, whereas environmental scaling is modeled by $v(x) = \delta x^2$, where γ and δ are positive constants. A model that includes both demographic and environmental stochasticity is $v(x) = \gamma x + \delta x^2$.

The variance-stabilizing transformation for both is

$$\phi(x) = \frac{2ln\left(\frac{1}{2}\sqrt{\psi x} + \frac{1}{2}\sqrt{\psi x + 4(1-\psi)}\right)}{\sqrt{\psi}},$$

where $\psi = \delta/(\delta + (\gamma/4))$ is a measure of the relative amount of environmental noise in the data. The quantity $\psi = 1$ corresponds to environmental stochasticity with transformation $\phi(x) = \ln x$, and $\psi \to 0^+$ corresponds to demographic stochasticity with transformation $\phi(x) = \sqrt{x}$ [5].

The parameter estimation procedure for a given value of ψ assumes that the vectors $\langle \phi(c_t) - \phi(C_t), \phi(s_t) - \phi(S_t) \rangle$ of transformed one-step residual errors derive from a joint normal distribution with variance-covariance matrix $\sum$ and that they are uncorrelated in time. Under these assumptions, the parameters are estimated by maximizing the log likelihood as a function of the model parameters [5]. The log-likelihood function was maximized numerically using the Nelder-Mead algorithm [24] under three types of stochasticity: primarily demographic ($\psi = 0.01$), a mixture of demographic and environmental ($\psi = 0.5$), and purely environmental ($\psi = 1$).

4.3.4 Model Validation

The (fitted) goodness-of-fit on the calibration data was compared to the goodness-of-fit on the validation data without re-estimating parameters. The colony goodness-of-fit was calculated to be

$$R_C^2 = 1 - \frac{\sum_{t=1}^{T}(\phi(c_t) - \phi(C_t))^2}{\sum_{t=1}^{T}(\phi(c_t) - \overline{\phi(c)})^2},$$

where $\overline{\phi(c)}$ denotes the sample mean of the transformed observations. We computed the R_S^2 for sleep similarly.

Presently, good statistical hypothesis tests of goodness-of-fit for dynamic models are unavailable, and it is not clear whether such tests would be useful if they were available. We think it is more informative to evaluate the predictive performance of a model under as wide a variety of conditions as possible. Thus, we use this generalized R^2 to quantify goodness-of-fit. We also use model validation analysis to probe the predictions of the model under environmental conditions encountered post-estimation.

4.4 Results

4.4.1 Best Model

Figure 4.2 lists per capita transition rates r_{ij} for the best model. It is important to realize these rates are for the chick-rearing season and may be quite different at other times of the year.

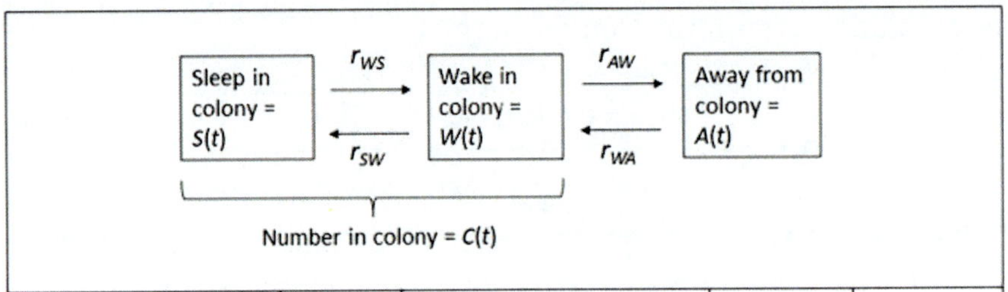

Time of day (PST)	r_{WS}	r_{SW}	r_{WA}	r_{AW}
Morning (0500 ≤ Hour ≤ 1000)	m_{WS}	m^*_{SW}*Hour2* (if hour ≤ 0800) m_{SW}/*Solar*^{2} (otherwise)	m_{WA}Tide^2	m_{AW}
Midday (1000 < Hour < 1400)	n_{WS} Heat^3	n_{SW}Solar x Cwind^2	n_{WA}Wwind^2	n_{AW}Solar^2
Evening (1400 ≤ Hour ≤ 2000)	e_{WS}	E_{SW}Hour^2/Solar^2	e_{WA}Hour^6	e_{AW}*Solar*^{2}

Figure 4.2: Per capita flow rate submodels as functions of six environmental factors: tide height (Tide), hour of day (Hour), solar elevation (Solar), heat index (Heat), wind speed on the colony (Cwind), and wind speed over the open water (Wwind). Assuming $\psi = 1$, the maximum likelihood parameters as estimated from the calibration data (all data) were $m_{WS} = 1.008$ (1.006), $m^*_{SW} = 0.08985$ (0.7328), $m_{SW} = 0.06299$ (0.07603), $m_{WA} = 0.2465$ (0.1046), $m_{AW} = 0.6115$ (0.3448), $n_{WS} = 4.493$ (4.591), $n_{SW} = 0.3184$ (0.2765), $n_{WA} = 0.1334$ (0.1546), $n_{AW} = 0.05996$ (0.07317), $e_{WS} = 8.759$ (11.77), $e_{SW} = 1.054$ (1.053), $e_{WA} = 0.02029$ (0.02822), $e_{AW} = 0.1198$ (0.1918), $\alpha = 0.6343$ (0.7640). The estimated entries of the variance-covariance matrix $\sum$ were $\sigma_{11} = 0.006279$ (0.006386), $\sigma_{22} = 0.3087$ (0.3955), $\sigma_{12} = \sigma_{21} = 0.009251$ (0.007457). (Originally published in Henson et al., "Predicting the dynamics of animal behaviour in field populations," *Animal Behaviour* 74:103–110. Copyright © 2007, The Association for the Study of Animal Behaviour. Used with permission.)

4.4.2 Parameter Estimates

Fixed values of $\psi = 0.01$, $\psi = 0.5$, and $\psi = 1$ for the best model yielded log-likelihood values of $-138, -3$, and 30, respectively, for the calibration data set and $-287, -22$, and 32, respectively, for the data set as a whole. Thus, we took $\psi = 1$ as the appropriate transformation. The maximum likelihood parameter estimates for system (4.3) with $\psi = 1$ are provided in the legend of Fig. 4.2.

4.4.3 Model Validation

The goodness-of-fits for the calibration data set were $R_C^2 = 0.60$ and $R_S^2 = 0.57$ and for the validation data set (without refitting) were $R_C^2 = 0.63$ and $R_S^2 = 0.46$. The goodness-of-fits obtained by fitting the model to the entire data set were $R_C^2 = 0.65$ and $R_S^2 = 0.52$.

A comparison of the data with one-step model predictions based on parameters estimated from the entire data set are shown in Fig. 4.3.

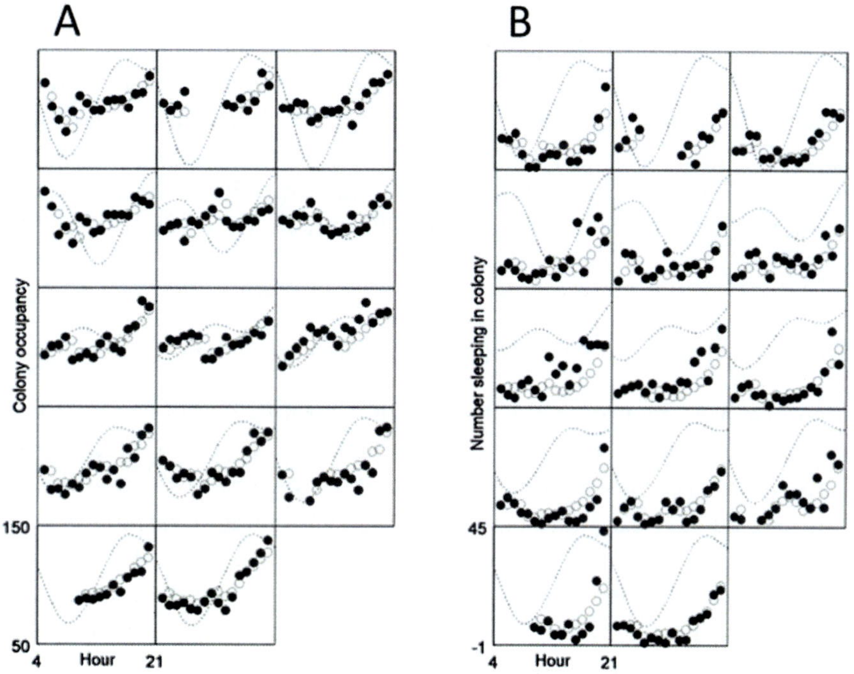

Figure 4.3: Hourly observations (solid circles), one-step model predictions (open circles), and tide height (dotted curve) as a function of time (LST). (**a**) Number of birds in the colony. (**b**) Number of birds asleep in the colony. In each of (**a**) and (**b**), 14 days of data are shown in chronological order from left to right, top to bottom. (Originally published in Henson et al., "Predicting the dynamics of animal behaviour in field populations," *Animal Behaviour* 74:103–110. Copyright © 2007, The Association for the Study of Animal Behaviour. Used with permission.)

4.5 Discussion

4.5.1 The Model

The integer-valued Markov chain model [4, 20] which accommodates the discrete nature of behavior compartment membership is an alternative modeling procedure for behavioral data. This model views animal behavior as transitions among discrete behavior units. The probabilities of the associated transitions are collected into a projection matrix. The mean compartment sizes in these models generally are solutions of linear differential equations of the same form as our model. In an effort to focus on the effect of the six environmental covariates, we build a "deterministic skeleton" for the means and add "noise" in a way that would potentially accommodate the ordinary variability of Markov models, along with the variability arising out of environmentally caused fluctuations in the transition probabilities. Instead of using discrete-time difference equations, we used continuous-time differential equations given that there was nothing inherently discrete in the time scale of the process except in our observations.

4.5.2 Demographic and Environmental Variability

Confusion can result from use of the terms "demographic" and "environmental" when applied to variability. These terms refer to types of variability in population growth models in ecology. In terms of mathematics, what ecologists call demographic variability is the variability in the sum of a fixed number of independent random variables: in other words, the net birth and death contributions of n individuals within a time interval. Environmental variability as used by ecologists refers to random changes through time in the rate constants of average per capita individual births and deaths themselves which induces a covariation of births and deaths among individuals.

The types of variability involved in our behavior model are directly described by these concepts. In this view, a "population" is the number of animals in a behavioral state at time t, and this population increases or decreases depending on the number of animals choosing this behavioral state and the number transitioning to an alternate behavior state. If the number of animals occupying a behavioral state is the sum of independent random binary choices (1 or 0) of n animals, then variability will be scaled in proportion to n, in the same way as in a population growth model with demographic variability. If the probability of choice varies randomly through time, the number of animals engaged in a behavior has an added variability component on a scale proportional to n^2, as in the scaling of environmental noise in an ecological model.

Demographic stochasticity is implicit in a Markov chain—individual animals occupying a specific behavior state independently face fixed and identical probabilities in transitioning to a different behavior. If large numbers of animals are governed by such a model, demographic stochasticity becomes essentially deterministic because the relative variability (coefficient of variation) of animal numbers engaged in a behavior becomes small.

4.5.3 Deterministic and Stochastic Fluctuations in the Colony-Sleep System

Using the methodology described in this chapter, we showed that *most of the temporal fluctuations in colony occupancy and sleep incidence were deterministic*. Moreover, we identified six abiotic environmental factors that determined the dynamics of these variables. It is important to note that the logical connection between "environmental determinants" and behavioral dynamics consists of mathematical implication, not scientific causation. In other words, environmental determinants are correlative but not necessarily causative [18]. However, the identification of environmental determinants narrows the search for causative behavior cues.

Our maximum likelihood analysis showed that *fluctuations not explained by our model were mostly due to environmental stochasticity instead of demographic stochasticity, or a mixture of the two*. We based this conclusion on the finding that $\psi = 1$ yielded higher log-likelihood values than did $\psi = 0.5$ or $\psi = 0.01$. In a Markov chain model, this would mean that the transition probabilities themselves were undergoing stochastic fluctuations. Using a deterministic system of differential equations with a stochastic noise component allowed for model fitting and evaluation under several types of stochastic variability. The conclusion of environmental variability comported with our data. We observed a variety of stochastic events including bald eagle (*Haliaeetus leucocephalus*) disturbances [12]. Demographic stochastic events included things such as individual differences in gull behavior and localized conspecific intrusions. But as our analysis demonstrated, demographic stochasticity was not a major influence on the data. Given the large number of animals in our system, demographic variability should have been relatively small.

4.5.4 Function

The functions of behavior can be clarified and new hypotheses can be generated by use of predictive mathematical models. Consider two examples from our system:

> (1) The Midday per capita tendency to transition to sleep increased with solar elevation and wind speed over the colony, and the tendency to transition out of sleep increased with heat index (Midday r_{SW} and r_{WS} in Fig. 4.2). This result is consistent with an earlier report showing increased colony sleep at midday in herring gulls (*L. argentatus*) [11] and also suggests that sleep posture has a thermoregulatory function.

(2) The Morning per capita rate of gulls leaving the colony was constant, whereas the per capita rate of returning gulls increased with tide height (Morning r_{AW} and r_{WA} in Fig. 4.2). This comports with reports of decreased/increased habitat occupancy in gulls at low/high tide, most likely due to increased/decreased food availability at low/high water levels [8, 11, 17].

A mathematical model in another study, however, suggests more specifically the gulls left loafing habitats near the Protection Island colony in response to low tide height, and they returned in response to low solar elevation [18]. We do not know if the contradiction with the present study is an artifact of insufficient data, or if it is due to behavior differences in gulls in the colony as compared with those of gulls away from the colony.

These examples demonstrate that each of the per capita flow rates in Fig. 4.2 generates new hypotheses and begs for further study.

4.5.5 Mathematical Models as Hypotheses

Mathematical models are precise formulations of modeling assumptions. In our study, assumptions included the three compartments W, S, and A, the specific form of the f_{ij} as given in system (4.2), and the general form of the r_{ij} as given in Eq. (4.4). Mathematical models and the assumptions on which they are based function as scientific hypotheses, hypotheses that can be tested with data, not in the sense of statistical hypothesis testing, but in the sense of applied dynamical systems theory through parameter estimation, comparison with alternative models, model selection, and model validation [5, 14]. Mathematical models must go beyond mere curve-fitting to be taken seriously as scientific hypotheses. In other words, mathematical models should be motivated by a mechanistic understanding of the system under study and be predictive. Of particular importance is model validation based on data not used to estimate parameters; this is what tests the predictive capability of a model.

4.6 Summary

Mathematical models are powerful tools for identifying factors that elicit behavior, clarifying functions of behavior, and generating hypotheses about behavior. These models explain and predict deterministic trends in behavior, and they distinguish between the contributions of environmental and demographic stochasticity. Despite the fact that the use of compartmental models is standard in fields such as epidemiology, population biology, pharmacology, physics, and chemistry [3], they have been considered too coarse to predict animal behavior because they group individuals into aggregates under simplifying assumptions [25]. *We have demonstrated, however, that some behaviors of gulls are determined largely by environmental factors and are predictable at the group level despite individual variability*. Compartmental models provide a powerful tool for the study of deterministic trends in animal, including human, behavior.

References

1. Amlaner CJ, Ball NJ (1983) A synthesis of sleep in wild birds. Behaviour 87:85–119

2. Amlaner CJ, McFarland DJ (1981) Sleep in the herring gull (*Larus argentatus*). Anim Behav 29:551–556

3. Anderson DH (1983) Compartmental modeling and tracer kinetics. In: Levin S (ed) Lecture notes in biomathematics. Springer, Berlin

4. Chen S, Lee AY, Bowens NM, Huber R, Kravitz EA (2002) Fighting fruit flies: a model system for the study of aggression. Proc Natl Acad Sci U. S. A. 99:5664–5668

5. Cushing, JM, Costantino RF, Dennis B, Desharnais RA, Henson SM (2003) Chaos in ecology: experimental nonlinear dynamics. Academic Press, San Diego, CA

6. Damania SP, Phillips KW, Henson SM, Hayward JL (2005) Habitat patch occupancy dynamics of glaucous-winged gulls (*Larus glaucescens*) II: a continuous time model. Nat Resour Model 18:469–499

7. Delius JD (1970) The effect of daytime, tides and other factors on some activities of lesser black-backed gulls, *Larus fuscus*. Revue de Comportement Animal 4:3–11

8. Drent RH (1967) Functional aspects of incubation in the herring gull (*Larus argentatus* Pont). Behaviour Supplement 17:1–32

9. Dugatkin LE, Reeve HK (1998) Game theory and animal behavior. Oxford University Press, Oxford

10. Engen S, Sæther BE (1998) Stochastic population models: some concepts, definitions, and results. Oikos 83:345–352

11. Galusha JG, Amlaner CJ (1978) The effects of diurnal and tidal periodicities in the numbers and activities of herring gulls *Larus argentatus*. Ibis 120:322–328

12. Galusha JG, Hayward JL (2002) Bald eagle activity at a gull colony and seal rookery on Protection Island, Washington. Northwest Nat 83:23–25

13. Gottman JM, Roy AK (1990) Sequential analysis: a guide for behavioral researchers. Cambridge University Press, New York

14. Hayward JL, Henson SM, Logan CJ, Parris CR, Meyer MW, Dennis B (2005) Predicting numbers of hauled-out harbour seals: a mathematical model. J Appl Ecol 42:108–117

15. Hazlett BA, Bach CE (1977) Predicting behavioral relationships: In Hazlett BA (ed) Quantitative methods in the study of animal behavior, pp 121–144. Academic Press, New York

16. Henson SM, Dennis B, Hayward JL, Cushing JM, Galusha JG (2007) Predicting the dynamics of animal behaviour in field populations. Anim Behav 74:103–110

17. Henson SM, Hayward JL, Burden CM, Logan CJ, Galusha JG (2004) Predicting dynamics of aggregate loafing behavior in glaucous-winged gulls (*Larus glaucescens*) at a Washington colony. Auk 121:380–390

18. Henson SM, Hayward JL, Damania SP (2005) Identifying environmental determinants of diurnal distribution in marine birds and mammals. Bull Math Biol 68:467–482

19. Hobson JA (2005) Sleep is of the brain, by the brain, and for the brain. Nature 437:1254–1256

20. Lusseau D (2003) Effects of tour boats on the behavior of bottlenose dolphins: using Markov chains to model anthropogenic impacts. Conserv Biol 17:1785–1793

21. Mangel M, Clark CW (1988) Dynamic modeling in behavioral ecology. Princeton University Press, Princeton, NJ

22. McFarland DJ (1971) Feedback mechanisms in animal behaviour. Academic Press, London

23. Murdoch M (1993) Factors affecting behavioral variation of individual glaucous-winged gulls (*Larus glaucescens*) while on territory. M.S. thesis, Loma Linda University, Loma Linda, CA

24. Press WH, Flannery BP, Teukolsky SA, Vetterling WT (1986) Numerical recipes: the art of scientific computing. Cambridge University Press, Cambridge

25. Railsback SF (2001) Getting "results": the pattern-oriented approach to analyzing natural systems with individual-based models. Nat Resour Model 14:465–475

26. Rattenborg NC, Amlaner CJ, Lima SL (2000) Behavioral, neurophysiology and evolutionary perspective on unihemispheric sleep. Neurosci Biobehav Rev 24:817–842

27. Shaffery JP, Ball NJ, Amlaner CJ, Jr (1985) Manipulating daytime sleep in herring gulls (*Larus argentatus*). Anim Behav 33:566–572

28. Siegel JM (1995) Phylogeny and the function of REM sleep. Behav Brain Res 69:29–34

29. Siegel JM (2005) Clues to the functions of mammalian sleep. Nature 437:1264–1271

30. Slater PJB (1978) Data collection. In: Colgan PW (ed) Quantitative Ethology. Wiley, New York, pp 8–24

31. Steadman RG (1979) The assessment of sultriness, Part I: a temperature-humidity index based on human physiology and clothing science. J Appl Meteorol 18:861–873

Chapter 5

Predicting Flightless Cormorant Behavior

This chapter is based on work carried out in the Galápagos Islands, Ecuador, in 2011. James Hayward and Susana Velastegui Chávez, with two graduate students, Libby Megna and Brianna Payne, collected the data. Shandelle Henson carried out the data analysis, which involved Poisson regression, logistic regression, and multi-model inference based on the Akaike Information Criterion (AIC). Results from the study were initially published in The Wilson Journal of Ornithology *in 2013 [18]. We present this work as an example of how regression models can be used to explain the aggregate behavior of marine organisms.*

5.1 Introduction

Flightless cormorants (*Nannopterum harrisi*, Fig. 5.1) were first encountered during the 1887–1898 Webster-Harris expedition to the Galápagos Islands [38]. The entire species of less than 1,000 breeding pairs [26, 29, 49] resides in small, scattered colonies on the islands of Fernandina and Isabela, Galápagos [15, 36]. Flightless cormorants weigh more than twice as much as other cormorants [51] and have been known to live for at least 17 years [25]. They are the only sequentially polyandrous and flightless members of their approximately 40-member family [27–29, 47]. Genetic analysis indicates that flightless cormorants are closely related to double-crested cormorants (*Nannopterum auritus*) and neotropic cormorants (*Nannopterum brasilianum*) [28] , but unlike these and other members of their family, the International Union for Conservation of Nature (IUCN) includes flightless cormorants on its red list of vulnerable species [24]. Consequently, information on their behavioral ecology is important as an aid toward conservation efforts.

Flightless cormorants were first described by Lionel Walter Rothschild [37], the wealthy banker, politician, and naturalist who financed the Webster-Harris expedition. He referred

J. M. Cushing et al., *Modeling Behavior and Population Dynamics*, Interdisciplinary Applied Mathematics 57, https://doi.org/10.1007/978-3-031-34283-7_5

to these birds as the "most remarkable" discovery of the expedition. Since Rothschild's initial description of these endemic forms, a modest number of reports have focused on flightless cormorant morphology, behavior, demography, response to El Niño-Southern Oscillation (ENSO) events , and breeding strategies [11, 15, 16, 19, 26, 38, 39, 42–45, 48, 49]. More recent studies have dealt with how environmental factors affect the survivability of these birds. One study demonstrated that flightless cormorants are seronegative to many avian pathogens, making them more vulnerable than most birds to foreign pathogens [46]. Genetic analysis has demonstrated that flightless cormorants exhibit significantly lower genetic diversity than other closely related species, a finding that supports the view that these birds are particularly vulnerable to environmental alterations associated with climate change [12].

Except for one study that reported thermoregulatory behavior in relation to temporal or environmental variables [19], prior to our study no systematic examination of the daily activity patterns of flightless cormorants in relation to environmental change had been made. In our research, habitat occupancy and behavior scans of flightless cormorants were made each half hour at a colony of approximately 35 individuals toward the beginning of heightened reproductive activity [16] and during a year when sea surface temperatures were cooler than average.

Generalized linear models and multi-model inference were used to test three hypotheses:

(1) The number of cormorants on land is related to environmental variables.
(2) The occurrences of preening, wing spreading (Fig. 5.1), rest posture, back sleep, and gular flutter are influenced by thermal stress.
(3) Thermal stress inhibits the incidence of courtship behavior.

Figure 5.1: Flightless cormorants, among the rarest of birds, are restricted to two islands in the Galápagos Archipelago. They are the only cormorants without the capacity to fly. This bird is drying its stubby wings in the late afternoon sun in the evening sun on Cabo Douglas, Isla Fernandina (Photo by J. L. Hayward)

5.2 Methods

5.2.1 Data Collection

The study was carried out from a vantage point adjacent to Sea Lion Bay, Cabo Douglas, Isla Fernandina, Galápagos ($0°18'13''S$, $91°39'8''W$). Scan counts of all visible flightless cormorants were made each half hour from 10:00–18:00 hours (LST here and below) on May 10, from 06:00–18:00 hours on May 11–16, and from 06:00–10:00 hours on May 17, 2011, for a total of 176 counts ranging from 11 to 35 birds. Each bird was counted as on land (Fig. 5.2) or in water (Fig. 5.3) and engaged in one of the following postures or behavioral states: preen (scratching, shaking, drawing feathers through the bill), wings spread, rest posture (standing or sitting without other activity), back sleep (head and neck turned back with bill tucked beneath the scapulars [2, 3, 41]), or courtship (snake-necking, nuptial dancing, copulation [43], Fig. 5.2). On the mornings of May 11, 12, and 15, 2011, we recorded the time at which the first cormorant entered the water.

At the same time as the scan counts, wind speed, barometric pressure, ambient temperature, and relative humidity were recorded by a Vantage Vue^{TM} wireless weather station (Davis Instruments Corp.) mounted 1.8 m above the white sandy beach. A 4-m surveyor leveling rod mounted in a small, quiet bay at the study site and protected from most wave action was used to measure local tide heights. Solar elevation was obtained from an online calculator at the National Oceanic and Atmospheric Administration website [10].

At half-hourly intervals from 10:00–17:30 hours on May 10 and from 07:00–17:30 hours on May 11–16, solar flux was measured with a handheld TM0206 Solar Power Meter. Thus, solar flux data were obtained for 148 of the 176 scan counts. Ranges and means of environmental variables for hours of data collection are presented in Table 5.1.

Table 5.1: Means (± SD) and ranges of environmental variables during 7 days of data collection on the behavior of flightless cormorants at Cabo Douglas, Isla Fernandina, Galápagos, May 10–17, 2011

Environmental variable	Mean ± SD	Range
Tide height (m)	1.0 ± 0.4	0.2–1.8
Ambient temperature (°C)	27.4 ± 2.0	21.6–31.1
Relative humidity (%)	75.1 ± 8.5	55–96
Wind speed (m/sec)	14.2 ± 7.0	0–25.7
Barometric pressure (mmHg)	757.7 ± 1.4	754.4–760.0
Solar elevation (°)	39.2 ± 23.2	−0.1–72.62
Solar flux (W/m^2)	486.0 ± 296.5	15–1,132

5.2.2 Smoothing of Tide-Height Observations

The water in the small bay with the tide pole experienced little wave action, but the tide-height data were noisy. Even so, a definite tidal dynamic was observed. We smoothed the noisy data by fitting a curve through the time series of tide heights using the Lowess smoothing function in the graphics program Axum with 10 iterations, 25% points for each local regression, and 100 output points [5]. Using Matlab [30], we splined the 100-point output from the Lowess smoothing function to obtain interpolated tide heights at 30-minute intervals.

Figure 5.2: A pair of flightless cormorants engage in snake-necking, a courtship behavior, at Cabo Douglas, Isla Fernandina, Galápagos Islands. Courtship in these birds involves extensive snake-necking, nuptial dancing, and unbirdlike vocalizations (Photo by J. L. Hayward)

5.2.3 Models

We used generalized linear modeling [31], a type of generalized linear regression, with intercept β_0 and no interaction terms to model count and behavior occurrence data as a function of environmental factors:

$$g(y) = \beta_0 + \sum_{i=1}^{n} \beta_i x_i, \tag{5.1}$$

where $g(y)$ is the appropriate link function. (Of course, the model could require interaction terms or be nonlinear in the predictors. Thus, it is important in the initial data summaries to inspect scatterplots to look for nonlinear trends.)

Count data typically are modeled with Poisson regression, which has link function

$$g(y) = \ln y.$$

Binary data (e.g., the behavior occurred or it did not) are modeled with logistic regression, which uses the log-odds link function

$$g(y) = \ln\left(\frac{P}{1-P}\right),$$

where P is the probability of the occurrence of the behavior and $P/(1-P)$ is the odds of the occurrence.

Interpretation of Regression Coefficients

Poisson and logistic regression slope coefficients lend themselves to convenient interpretation. If factor x_i increases by c units in Eq. (5.1) with all other factors held constant, the resulting change in the value of the link function is

$$\Delta g(y) = \beta_i c,$$

where β_i is the slope coefficient for factor x_i. For Poisson regression, if y_2 and y_1 are the counts after and before the c-unit increase in x_i, respectively, then

$$\ln y_2 - \ln y_1 = \beta_i c,$$

and so

$$\frac{y_2}{y_1} = e^{\beta_i c}.$$

Thus, for count data, $\exp(\beta_i c)$ is the count ratio, that is, the multiplier by which the count changes due to a c-unit increase in x_i if all other factors are held constant.

For logistic regression,

$$\ln\left(\frac{P_2}{1-P_2}\right) - \ln\left(\frac{P_1}{1-P_1}\right) = \ln\left(\frac{\text{odds}_2}{\text{odds}_1}\right) = \beta_i c,$$

and so

$$OR = e^{\beta_i c},$$

where OR is the odds ratio, that is, the multiplier by which the odds change due to a c-unit increase in x_i if all other factors are held constant.

The 95% confidence interval for the estimated coefficient $\hat{\beta}_i$ is

$$\left(\hat{\beta}_i - 1.96 \times SE,\ \hat{\beta}_i + 1.96 \times SE\right),$$

where SE is the standard error of the coefficient $\hat{\beta}_i$. The 95% confidence intervals for the associated count or odds ratios are

$$\left(\exp(c(\hat{\beta}_i - 1.96 \times SE)),\ \exp(c(\hat{\beta}_i + 1.96 \times SE))\right).$$

If the confidence interval for the coefficient $\hat{\beta}_i$ is wholly negative or wholly positive, then the confidence interval for the associated count or odds ratio will be completely below one or completely above one, indicating a significant negative or positive effect on the count or odds given a c-unit increase in x_i, all other factors held constant.

Parameter Estimation

Ideally, one would consider the global model (5.1) and all its submodels (regression models with all possible subsets of factors) as a suite of alternative models. One would obtain the maximum likelihood parameter estimates $\hat{\beta}_i$ for each model and would then use information-theoretic techniques such as the Akaike information criterion (AIC) to select the best model from the suite of alternatives [8]. The AIC for a given model in the suite is

$$AIC = -2\ln L + 2\kappa,$$

where L is the fitted value of the likelihood function for that model and κ is the number of estimated parameters for that model, including the variance parameters. The model with the least AIC value, denoted AIC_{min}, is considered the best of the alternatives. The AIC penalizes models for overfitting (having too many parameters). The jth model in the suite is ranked by the difference between its own AIC and the AIC of the best model:

$$\Delta^j = AIC^j - AIC_{\text{min}}.$$

The best model is the one with $\Delta^j = 0$, and that model would be used to draw inferences.

In this current study, however, the small data sets led to uncertainty in model selection, so we followed the multi-model inference techniques in [8]. We first fit each model to the data and computed the Akaike information criterion for small samples

$$AIC_c = AIC + \frac{2\kappa(\kappa+1)}{q-\kappa-1},$$

where q is the sample size. We then computed the Akaike weight w for each model. The Akaike weight w^j for the jth model is

$$w^j = \frac{\exp\left(-\frac{1}{2}\Delta^j\right)}{\sum_{k=1}^{N} \exp\left(-\frac{1}{2}\Delta^k\right)},$$

where the denominator is summed over all N candidate models. The Akaike weights, which sum to one, generally indicate how likely it is that a given model is actually the best.

Next, we calculated the relative importance of each factor, which is the sum of the Akaike weights over all models which contain that factor. If a factor appears in all of the models with large Akaike weights, then its relative factor importance will be close to 1.

Finally, we computed model-averaged parameter estimates. For a given parameter β_i, the model-averaged estimate is the average of the estimates $\hat{\beta}_i$ over all models in the suite containing that parameter, weighted by the models' Akaike weights [8]:

$$\overline{\hat{\beta}_i} = \frac{\sum w^j \hat{\beta}_i}{\sum w^j},$$

where the sums are over the models containing factor x_i. We did these computations with the package glmulti [9] in the statistical programming environment R [32].

For each type of scan count data for which we had solar flux measurements ($N = 148$), we used the global Poisson regression model

$$\begin{aligned} \ln y &= \beta_0 + \beta_1 TIME + \beta_2 TIDE + \beta_3 TEMP + \beta_4 HUM \\ &\quad + \beta_5 WIND + \beta_6 BAR + \beta_7 SOLEL + \beta_8 SOLF, \end{aligned}$$

where y is the count, $TIME$ is hour of day, $TIDE$ is tide height in meters, $TEMP$ is ambient temperature in degrees Celsius, HUM is percent relative humidity, $WIND$ is wind speed in meters/second, BAR is barometric pressure in mmHg, $SOLEL$ is solar elevation in degrees above the horizon, and $SOLF$ is solar flux in W/m^2. For those dependent variables (scan count variables) for which $SOLF$ was discarded as a factor, we repeated the analysis using the global model

$$\begin{aligned} \ln y &= \beta_0 + \beta_1 TIME + \beta_2 TIDE + \beta_3 TEMP + \beta_4 HUM \\ &\quad + \beta_5 WIND + \beta_6 BAR + \beta_7 SOLEL \end{aligned}$$

with the complete data set ($N = 176$).

We used the same approach for the occurrence of behavior but used instead a logistic regression model with occupancy as the number of trials:

$$\begin{aligned} \ln\left(\frac{P}{1-P}\right) &= \beta_0 + \beta_1 TIME + \beta_2 TIDE + \beta_3 TEMP + \beta_4 HUM \\ &\quad + \beta_5 WIND + \beta_6 BAR + \beta_7 SOLEL + \beta_8 SOLF. \end{aligned}$$

For every model-averaged coefficient $\overline{\hat{\beta}_i}$ with a confidence interval that was entirely positive or entirely negative, we determined the count ratio (for occupancy) or odds ratio (for behaviors) as $\exp(c\overline{\hat{\beta}_i})$ [23].

We did not include numbers of cormorants in the water in our regression analyses because the cormorants were difficult to count in that habitat. This was because birds underwater could not be seen, waves sometimes inhibited the view of swimming birds, and our view was sometimes blocked by a large rock outcrop (see also [19]). Counts of cormorants in the water summed by count period over the 7-day study are presented, however.

5.3 Results

Only four of the eight measured environmental variables influenced the behavior states and land occupancy of flightless cormorants: time of day, solar flux, wind speed, and barometric pressure (see Tables A5.1 and A5.2 for details).

> *Hypothesis 1: The number of cormorants on land is related to environmental variables.*

If all other factors were held constant, regression indicated that occupancy on land increased by 11% with each 3-hour increase in time of day (Table A5.1 and Fig. 5.4a). Consequently, Hypothesis 1 is supported for time of day but not for other variables.

Figure 5.3: A pair of adult flightless cormorants swim in water off Cabo Douglas, Isla Fernandina, Galápagos Islands. Flightless cormorants float low in the water due to their high body density. They are skilled divers. (Photo by J. L. Hayward)

Cormorants that were not on land were most likely in the water. The first cormorant to enter the water on May 11, 12, and 15 occurred just before dawn at 05:51, 05:57, and

05:55 hours, respectively. From 08:00–09:00 hours, the average number of cormorants in the water increased. As the afternoon progressed, however, a decline occurred toward dusk, and during and following our final 18:00-hour count, we observed no cormorants in the water. Due to the visibility problems noted above, counts of cormorants in the water were small and highly variable (Fig. 5.4b). When we could observe birds on moonlit nights and 2 hours before sunrise, virtually all observed individuals were in a back-sleep posture on land.

Hypothesis 2: The occurrences of preening, wing spreading, rest posture, back sleep, and gular flutter are influenced by thermal stress.

Our data partially supported this hypothesis. With other factors held constant, the odds of gular flutter increased 51% with a 200 W/m^2 increase in solar flux, increased 105% for each 3-hour increase in time of day, and increased 106% for each 3-mmHg rise in barometric pressure.

Gular flutter is an indicator of thermal stress. Obviously, cormorants engaged in this behavior could not at the same time engage in back sleep or preen. For each 200 W/m^2 rise in solar flux, back sleep decreased by 28%. Rises in solar flux did not affect the incidence of preen. Each 10 m/sec rise of wind speed raised the odds of back sleep by 48%, lowered the odds of preening by 23% (Table A5.1), and lowered the odds of gular flutter by 48% (Table A5.2).

Gular flutter occurred simultaneously with wings spread and rest posture. For each 200 W/m^2 increase in solar flux, the odds of rest posture increased by 12%. No environmental variable influenced the odds of wings spread.

Hypothesis 3: Thermal stress inhibits the incidence of courtship behavior.

Each 3-hr increase in time of day decreased the odds of courtship behavior by 46%. However, thermal stress, along with other environmental factors, failed to influence the odds of courtship. Consequently, our hypothesis remained unsupported.

5.4 Discussion

Two major constraints, flightlessness and prey abundance, constrain the behavior of flightless cormorants. Flightlessness, along with behavioral philopatry, limits the home range of these birds [56]. Restricted mobility has resulted in a high degree of genetic differentiation among colonies [12], although individuals may swim for considerable distances along the coastline [12, 48]. The second major constraint, prey abundance, fluctuates erratically in response to variation in local upwellings [7, 47]. El Niño events lead to high levels of mor-

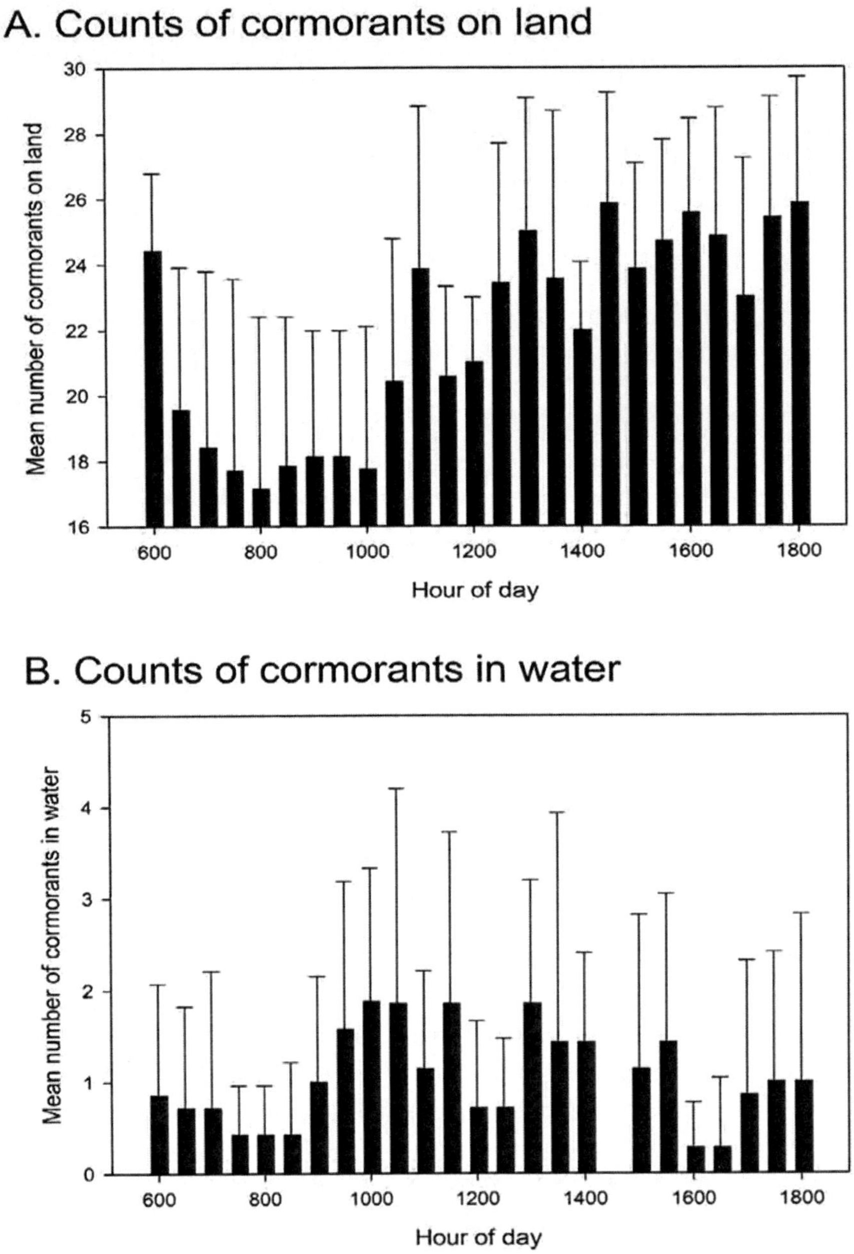

Figure 5.4: Mean number of flightless cormorants (**a**) on land and (**b**) in the water during half-hourly counts, 6:00–18:00 hours LST, May 10–17, 2011, Cabo Douglas, Isla Fernandina, Galápagos. Error bars are standard deviations. (Originally published in Hayward et al. "Temporal and environmental effects on the behavior of flightless cormorants," *Wilson Journal of Ornithology* 125:790–799. Copyright © 2013, Wilson Ornithological Society. Used with permission.)

tality and severe reproductive output in flightless cormorants [7, 49]. The way in which individual birds divide their time in relation to these constraints and to environmental variables determines their fitness.

Figure 5.5: A flightless cormorant incubates eggs at Punta Espinosa, Isla Fernandina, Galápagos Islands. (Photo by J. L. Hayward)

Although breeding may occur anytime during the year, our research took place toward the beginning of the most active period of reproductive activity (April to October; [16]). Particularly in the morning, long periods of courtship were frequent in water and on land (Table A5.1). Often in groups of four, cormorants moved from one place to another, walking, snake-necking, and swimming together (see also [16, 43, 44, 47]).

We found five active nests (Fig. 5.5) in the colony located on lava outcrops along the west side of Sea Lion Bay. One nest contained three eggs on May 4, but as a result of unknown causes, all three eggs disappeared after several days. The other four nests were in early stages of construction. Males often brought algae to females at the nest sites; the females arranged this material to form the nests [43]. Nonbreeding cormorants fed, rested, preened, and slept on land.

Preening was one of the most frequent behaviors of cormorants on land. An earlier study [19] similarly reported that cormorants on land but off the nest spent 49.9% of their time preening. High winds can knock birds off their feet, and we found that the odds of preening decreased as wind speed increased, findings similar to those for glaucous-winged gulls [20, 22]. During windy conditions, the cormorants frequently engaged in back-sleep posture which minimizes the surface-to volume ratio [2, 34] and provides stability under windy conditions.

Solar flux did not affect the frequency of wings spread (Table A5.1). Previous studies with little cormorants (*Microcarbo niger*, [52]), double-crested cormorants [19], and great cormorants (*Phalacrocorax carbo*, [40]) concluded that wing spreading dries feathers but is unrelated to thermoregulation.

The odds of back sleep increased with increasing wind speed and decreasing solar flux. The odds of rest increased and the odds of back sleep decreased with increasing solar flux (Table A5.1). Birds in rest posture can engage in gular flutter, an important means of unloading heat under high thermal loads [6]. The odds of gular flutter increased with solar flux, hour of day, and barometric pressure, but decreased with wind speed. Gular flux cannot occur simultaneously with preen, sleep, or courtship, all of which enhance reproductive fitness. Presumably, global warming could increase the incidence of gular flutter to the point that it interferes with physiological requirements and inhibits reproductive output. El Niño events precipitate population declines in flightless cormorants, but the high reproductive capacity of these birds promotes rapid population recovery [49]. Long-term global warming associated with climate change, however, may present a more significant threat. Unlike many species of birds which escape increasingly harsh conditions by range shifts, the inability to fly by flightless cormorants and the small size of the Galápagos Archipelago prevent effective range shifts by this species [14].

Only one or two environmental factors influenced occupancy on land, as well as the incidence of all behavior except of gular flutter. Gular flutter was influenced by wind speed, solar flux, time of day, and barometric pressure, implying that these factors are involved in an undefined heat index function. Heat index functions are species-specific and usually are based on the results of laboratory experiments. The upper critical temperature of the thermoneutral zone is most important in this regard because at this temperature the animal begins to expend energy to off-load heat [33]. The radiation of stored heat by dark lava rocks of the colony undoubtedly increases the incidence of gular flutter. Although our data are limited, Fig. 5.4 suggests that the cormorants may enter the water in the afternoon when thermal conditions on land are most intense. Although we did not see birds drinking seawater when thermally stressed, most likely they drink when swimming. Given the threats to the long-term survival of this species in the face of global warming, further study of the impact of heat stress on flightless cormorants is warranted [1, 13].

Cormorants in the water fed, courted, gathered nest material, and fed juveniles which likely were hatched during the preceding year. Foraging by members of this species is reported to take about 4 hours of each day for both males and females [44]. We observed foraging behavior throughout the day. An earlier study [44] found that adult males feed juveniles for 5–9 months, and our subjective observations are consistent with this estimate. Another report [43] notes that juveniles leave the shore and swim out to parents from which they food beg, which also comports with our observations.

Tide height influences the behavior of numerous marine organisms [17, 21], including some cormorants [4, 35, 50, 53]. Our data, however, suggest that tide height plays only a weak role in the timing of behavior by flightless cormorants (Tables A5.1 and A5.2). Tide height was highly correlated with time of day and to a lesser degree with weather variables, but each of these variables was held constant in our regression analysis on the effects of tide. Thus, we are confident that tide height imposes little influence on the timing of flightless cormorant behavior. Flightless cormorants can dive to depths of at least 73 m for relatively long periods of time [51]. Moreover, the tidal range at Cabo Douglas during our time of study was only 1.6 m (Table 5.1). It appears, then, that the impact of tide height on the behavior of these birds in these usually food-rich waters [7, 44] was minimal. In contrast to intertidal and surface-feeding marine birds which are tide-dependent [17, 21], flightless cormorants appear to arrange their daily activities around time of day and weather variables.

5.4.1 Conclusion

Strong geographical isolation and unique oceanographic and climatic conditions have shaped the evolutionary history of flightless cormorants. Flightlessness, a unique trait among the more than three dozen cormorant species, constrains the behavior patterns of these birds. *Colony occupancy and courtship behavior were influenced mostly by time of day, whereas preen, wing spread, rest posture, and back sleep were primarily influenced by weather variables. The incidence of gular flutter, which accompanies heat stress, increased with solar flux, time of day, and barometric pressure but decreased with wind speed. Tide exerted little influence on the behavior of flightless cormorants, in contrast to its influence on the behavior of many other marine birds.*

This study was carried out during a 7-day interval close to the beginning of the most active breeding period; it also coincided with La Niña conditions. In view of the threats posed by global warming, the timing of flightless cormorant behavior should be studied during other seasons and during non-La Niña years. Time budget analyses on members of this species should be made to assess how thermoregulation in response to solar flux impacts reproductive activity and success at other times of the year and under a variety of other environmental circumstances.

5.5 Appendix: Flightless Cormorant Responses to Environmental Factors

Table A5.1: Responses of colony occupancy, preen, wings-spread, and courtship behavior by flightless cormorants to seven environmental factors, with factor importance and model-averaged coefficient estimates β. Positive (or negative) coefficients correspond to an increase (or decrease) in the response variable given an increase in the factor. Confidence intervals that are completely positive or negative indicate significance at the $\alpha = 0.05$ level. For a given c-unit increase in a significant factor, $e^{\beta c}$ is the count ratio for occupancy and the odds ratio (OR) for behaviors. The percent change in occupancy or odds of a behavior is the difference between $e^{\beta c}$ and 1. Asterisks (*) identify significant factors (confidence intervals that are completely positive or negative)

Factor	Importance	Coefficient β	95% confidence interval	c	$e^{\beta c}$	%$\triangle$
Colony occupancy						
(Intercept)	1	1.8103	(−9.2443, 12.87)			
TIME*	1.0	0.03557	(0.02099, 0.05015)	3 hrs	1.11	11%↑
TEMP	0.71	−0.02945	(−0.08511, 0.02621)			
HUM	0.45	−0.002447	(−0.01112, 0.006221)			
SOLEL	0.42	0.0005458	(−0.001316, 0.002407)			
TIDE	0.38	0.01759	(−0.04400, 0.07919)			
WIND	0.29	−0.0003617	(−0.003160, 0.002436)			
BAR	0.28	0.002413	(−0.01181, 0.01663)			
Preen						
(Intercept)	1	4.318	(−22.37, 31.00)			
WIND*	1.0	−0.02638	(−0.04132, −0.01144)	$10 \frac{m}{sec}$	0.77	23%↓
SOLEL	0.74	0.002989	(−0.001775, 0.007754)			
TIME	0.68	0.01840	(−0.01568, 0.05248)			
TEMP	0.38	0.01208	(−0.03777, 0.06194)			
BAR	0.32	−0.007415	(−0.04232, 0.02749)			
HUM	0.31	−0.001025	(−0.007404, 0.005354)			
TIDE	0.28	0.0002704	(−0.05543, 0.05597)			
Wing-spread						
(Intercept)	1	−53.51	(−155.7, 48.67)			
BAR	0.64	0.06529	(−0.06820, 0.1988)			
SOLEL	0.59	0.003400	(−0.004165, 0.01097)			
TEMP	0.59	0.05962	(−0.07539, 0.1946)			
WIND	0.59	−0.01189	(−0.03854, 0.01476)			
TIME	0.42	0.01327	(−0.02979, 0.056320)			
HUM	0.33	−0.0001343	(−0.01076, 0.01049)			
TIDE	0.29	0.01947	(−0.1009, 0.1399)			
Courtship						
(Intercept)	1	22.58	(−61.48, 106.6)			
TIME*	1.0	−0.2031	(−0.2849, −0.1213)	3 hrs	0.54	46%↓
TIDE	0.73	−0.3411	(−0.9043, 0.2221)			
TEMP	0.59	−0.08050	(−0.2638, 0.1028)			
HUM	0.37	0.002039	(−0.01966, 0.02374)			
BAR	0.34	−0.02812	(−0.1378, 0.08153)			
SOLEL	0.34	0.001207	(−0.005346, 0.007760)			
WIND	0.31	0.003631	(−0.01423, 0.02150)			

Table A5.2: Responses of rest posture, back sleep, and gular flutter by flightless cormorants to eight environmental factors, with factor importance and model-averaged coefficient estimates β. Positive (or negative) coefficients correspond to an increase (or decrease) in the response variable given an increase in the factor. Confidence intervals that are completely positive or negative indicate significance at the $\alpha = 0.05$ level. For a given c-unit increase in a significant factor, $e^{\beta c}$ is the odds ratio (OR) of the behavior. The percent change in odds of the behavior is the difference between $e^{\beta c}$ and 1. Asterisks (*) identify significant factors (confidence intervals that are completely positive or negative)

Factor	Importance	Coefficient β	95% confidence interval	c	$e^{\beta c}$	%△
Rest posture						
(Intercept)	1	−28.94	(−100.5, 42.65)			
SOLF*	1.0	0.0005524	(0.0001928, 0.0009121	200 Wm^{-2}	1.12	12%↑
TIME	0.82	0.03382	(−0.01234, 0.07998)			
BAR	0.51	0.03597	(−0.05798, 0.1299)			
TIDE	0.48	−0.07094	(−0.2654, 0.1235)			
SOLEL	0.39	−0.001286	(−0.005771, 0.003199)			
TEMP	0.34	0.01121	(−0.03353, 0.05595)			
WIND	0.32	0.001837	(−0.006108, 0.009783)			
HUM	0.27	−0.0002678	(−0.004982, 0.004446)			
Back sleep						
(Intercept)	1	−1.194	(−33.54, 31.15)			
SOLF*	1.0	−0.001668	(−0.002256, −0.001081)	200 Wm^{-2}	0.72	28%↓
WIND*	0.99	0.03930	(0.01275, 0.06584)	10 $msec^{-1}$	1.48	48%↑
TIDE	0.75	0.2438	(−0.1379, 0.6254)			
HUM	0.66	−0.02114	(−0.06355, 0.02127)			
TEMP	0.63	−0.1071	(−0.3359, 0.1217)			
SOLEL	0.33	0.001224	(−0.003971, 0.006420)			
TIME	0.33	−0.006751	(−0.03389, 0.02039)			
BAR	0.28	0.004938	(−0.03534, 0.04522)			
Gular flutter						
(Intercept)	1	−189.2	(−367.8, −10.57)			
WIND*	1.0	−0.06590	(−0.09721, −0.03459)	10 $msec^{-1}$	0.52	48%↓
SOLF*	1.0	0.002068	(0.001109, 0.003026)	200 Wm^{-2}	1.51	51%↑
TIME*	1.0	0.2391	(0.1113, 0.3669)	3 hrs	2.05	105%↑
BAR*	0.91	0.2405	(0.006625, 0.4744)	3 mmHg	2.06	106%↑
SOLEL	0.55	0.006887	(−0.009748, 0.02352)			
TEMP	0.40	0.03927	(−0.09488, 0.1734)			
HUM	0.35	−0.004529	(−0.02364, 0.01459)			
TIDE	0.29	0.03643	(−0.1624, 0.2353)			

References

1. Alava JJ, Haase B (2011) Waterbird biodiversity and conservation threats in coastal Ecuador and theGalápagos Islands. In: Grillo O (ed) Ecosystems biodiversity. InTech, New York, pp 271–314. www.cdn.intechweb.org/pdfs/25332.pdf

2. Amlaner CJ, Ball NJ (1983) A synthesis of sleep in wild birds. Behaviour 87:85–119.

3. Amlaner CJ, McFarland DJ (1981) Sleep in the herring gull (*Larus argentatus*). Anim Behav 29:551–556
4. Anderson CD, Roby DD, Collis K (2004) Foraging patterns of male and female double-crested cormorants nesting in the Columbia River estuary. Can J Zool 82:541–554
5. Axum 6.0 (1999) MathSoft Inc., Cambridge, Massachusetts, USA
6. Bartholomew GA, Lasiewski RC, Crawford EC (1968) Patterns of panting and gular flutter in cormorants, pelicans, owls, and doves. Condor 70:31–34
7. Boersma PD (1978) Breeding patterns of Galápagos Penguins as an indicator of oceanic conditions. Science 200:1481–1483
8. Burnham KP, Anderson DR (2002) Model selection and multi-model inference: a practical information-theoretic approach, 3rd (ed). Springer, New York
9. Calcagno V (2012) glmulti: model selection and multimodel inference made easy. R package version 1.0.6. CRAN.R-project.org/package=glmulti
10. Cornwall C, Horiuchi A, Lehman C (2012) Solar position calculator. U.S. Department of Commerce, National Oceanic and Atmospheric Administration, Earth System Research Lab, Boulder, Colorado, USA. www.srrb.noaa.gov/highlights/sunrise/azel.html
11. Davis M (1938) Partial nidification of the flightless cormorant. Auk 55:596–597
12. Duffie CV, Glenn TC, Vargas FH, Parker PG (2009) Genetic structure within and between island populations of the flightless cormorant (*Phalacrocorax harrisi*). Mol Ecol 18:2103–2111
13. Edgar GJ, Banks SA, Brandt M, Bustamante RH, Chiriboga A, Earle SA, Garske LE, Glynn PW, Grove JS, Henderson S, Hickman CP, Miller KA, Rivera F, Wellington GM (2010) El Niño, grazers and fisheries interact to greatly elevate extinction risk for Galápagos marine species. Glob Chang Biol 16:2876–2890
14. Grémillet D, Boulinier T (2009) Spatial ecology and conservation of seabirds facing global climate change: a review. Mar Ecol Prog Ser 391:121–137
15. Harris MP (1974) A complete census of the flightless cormorant (*Nannopterum harrisi*). Biol Conserv 6:188–191
16. Harris MP (1979) Population dynamics of the flightless cormorant *Nannopterum harrisi*. Ibis 121:135–146
17. Hayward JL, Henson SM, Tkachuck RD, Tkachuck CM, Payne BG, Boothby CK (2009) Predicting gull/human conflicts with mathematical models: a tool for management. Nat Resour Model 22:544–563

18. Hayward JL, Megna LC, Payne BG, Velastegui Chávez SR, Henson SM (2013) Temporal and environmental effects on the behavior of flightless cormorants. Wilson J Ornith 125:790–799

19. Hennemann WW (1984) Spread-winged behaviour of double-crested and flightless cormorants *Phalacrocorax auritus* and *P. harrisi*: wing drying or thermoregulation? Ibis 126:230–239

20. Henson SM, Galusha JG, Hayward JL, Cushing JM (2007) Modeling territory attendance and preening behavior in a seabird colony as functions of environmental conditions. J Biol Dyn 1:95–107

21. Henson SM, Hayward JL, Burden CM, Logan CJ, Galusha JG (2004) Predicting dynamics of aggregate loafing behavior in glaucous-winged gulls (*Larus glaucescens*) at a Washington colony. Auk 121:380–390

22. Henson SM, Weldon LM, Hayward JL, Greene DJ, Megna LC, Serem MC (2012) Coping behaviour as an adaptation to stress: post-disturbance preening in colonial seabirds. J Biol Dyn 6:17–37

23. Hosmer DW, Lemeshow S (2000) Applied logistic regression, 2nd edn. Wiley, New York

24. International Union for the Conservation of Nature (IUCN) (2013) The IUCN red list of threatened species. Version 2013.1. www.iucnredlist.org

25. Jiménez-Uzcátegui G, Valle CA, Vargas FH (2012) Longevity records of flightless cormorants *Phalacrocorax harrisi*. Mar Ornithol 40:127–128

26. Jiménez-Uzcátegui G, Vargas FH (2007) Censo del pingüino de Galápagos y cormorán no volador en 2007. Informe Técnico para la Fundación Charles Darwin y Parque Nacional Galápagos, Puerto Ayora

27. Johnsgard PA (1993) Cormorants, Darters, and Pelicans of the World. Smithsonian Institution Press, Washington, DC

28. Kennedy M, Valle CA, Spencer HG (2009) The phylogenetic position of the Galápagos cormorant. Mol Phylogenet Evol 53:94–98

29. Livezey BC (1992) Flightlessness in the Galápagos cormorant (*Compsohalieus* [*Nannopterum*] *harrisi*)—heterochrony, giantism and specialization. Zool J Linnean Soc 105:155–224

30. MATLAB® (2012) MATLAB® version 7.14.0. The MathWorks, Inc., Natick, Massachusetts, USA

31. McCullagh P, Nelder JA (1989). Generalized linear models, 2nd edn. Chapman and Hall/CRC, Boca Raton, Florida

32. Porter WP, Kearney M (2009) Size, shape, and the thermal niche of endotherms. Proc Natl Acad Sci 106(Supplement 2):19666–19672

33. R Development Core Team (2012) R: a language and environment for statistical computing. R Foundation for Statistical Computing, Vienna, Austria. www.R-project.org/

34. Reebs SG (1986) Sleeping behavior of black-billed magpies under a wide range of temperatures. Condor 88:524–526

35. Richner H (1995) Wintering cormorants *Phalacrocorax carbo carbo* in the Ythan estuary, Scotland: numerical and behavioural responses to fluctuating prey availability. Ardea 83:193–197

36. Rosenberg DK, Valle CA, Coulter MC, Harcourt SA (1990) Monitoring Galápagos penguins and flightless cormorants in the Galápagos Islands. Wilson Bulletin 102:525–532

37. Rothschild W (1898) Phalacrocorax harrisi. Bull Br Ornithol Club 7:52

38. Rothschild W, Hartert E (1899) A review of the ornithology of the Galápagos Islands. With notes on the Webster-Harris expedition. Novitates Zoologicae 6:5–205

39. Rothschild W, Hartert E (1902) Further notes on the fauna of the Galápagos Islands. Novitates Zoologicae 9:373–418

40. Sellers RM (1995) Wing-spreading behavior of the cormorant *Phalacrocorax carbo*. Ardea 83:27–36

41. Shaffery JP, Ball NJ, Amlaner CJ (1985) Manipulating daytime sleep in herring gulls (*Larus argentatus*). Anim Behav 33:566–572

42. Snodgrass R, Heller E (1904) Papers of the Hopkins-Stanford Galápagos Expedition, 1899, XVI: Birds. Proceedings of the Washington Academy of Sciences 5:231–372

43. Snow BK (1966) Observations on the behavior and ecology of the flightless cormorant *Nannapterum harrisi*. Ibis 108:265–280

44. Tindle RW (1984) The evolution of breeding strategies in the flightless cormorant (*Nannopterum harrisi*) of the Galápagos. Biol J Linn Soc 21:157–164

45. Townsend CH (1929) The flightless cormorant in captivity. Auk 46:211–213

46. Travis EK, Vargas FH, Merkel J, Gottdenker N, Miller RE (2006) Hematology, plasma chemistry, and serology of the flightless cormorant (*Phalacrocorax harrisi*) in the Galápagos Islands, Ecuador. J Wildl Dis 42:133–141

47. Valle CA (1994) The ecology and evolution of sequential polyandry in Galápagos cormorants (*Compsohalieus* (*Nannopterum*) *harrisi*). PhD dissertation, Princeton University, Princeton, NJ

48. Valle CA (1995) Effective population-size and demography of the rare flightless Galápagos cormorant. Ecol Appl 5:601–617

49. Valle CA, Coulter MC (1987) Present status of the flightless cormorant, Galápagos penguin and greater flamingo populations in the Galápagos Islands, Ecuador, after the 1982–83 El Niño. Condor 89:276–281

50. van Tets GF (1959) A comparative study of some social communication patterns in the Pelecaniformes. Ornithological Monograph No. 2, American Ornithologists' Union, Washington, DC

51. Wilson RP, Vargas FH, Steinfurth A, Riordan P, Ropert-Coudert Y, MacDonald DW (2008) What grounds some birds for life? Movement and diving in the sexually dimorphic Galápagos cormorant. Ecol Monogr 78:633–652

52. Winkler H (1983) Das Flugelspreitverhalten der Mohrenscharbe *Phalacrocorax niger*. J Ornithol 124:177–186

53. Zeenath C, Zacharias VJ (2010) Foraging behavior and diving pattern of little cormorant *Phalacrocorax niger* (Vieillot)(Pelecaniformes: Phalacrocoracidae) at Kallampara backwaters, Kerala, India. J Threatened Taxa 2:1382–1386

Chapter 6

Coping Behavior as an Adaptation to Stress in Gulls

In this chapter we employ logistic regression and Darwinian dynamics to explore how a behavior with a crucial physiological function might be co-opted to function as a coping behavior in response to stress generated by the presence of predators. This study was initially reported in the Journal of Biological Dynamics *in 2012 by senior researchers Shandelle Henson, Lynelle Weldon, and James Hayward and students Daniel Greene, Libby Megna, and Maureen Serem [23].*

6.1 Introduction

Here we pose a mathematical framework for modeling microevolutionary change in animal behavior. Specifically, we evaluate how a physiologically advantageous behavior could evolve into a coping behavior that could help calm an animal in times of stress. The animals we studied were gulls, classic model organisms for the study of behavior. We consider whether the common behavior of feather preening (Fig. 6.1) by glaucous-winged gulls (*Larus glaucescens*) could be adapted as a coping behavior ("comfort preening") in response to increasing levels of disturbance by bald eagles (*Haliaeetus leucocephalus*).

6.1.1 Coping Behavior as an Evolutionary Adaptation to Stress

Coping behavior in humans is defined as "a characteristic and often automatic action or set of actions taken in dealing with stressful or threatening situations" [37] as a part of stress reduction and to minimize personal or interpersonal conflict [35, 43, 45]. For example, "comfort sucking" can sooth an infant [1], and "comfort eating" helps adults cope with stress [13]. Preening by birds (Fig. 6.1), important for flight feather maintenance, parasite

J. M. Cushing et al., *Modeling Behavior and Population Dynamics*, Interdisciplinary Applied Mathematics 57, https://doi.org/10.1007/978-3-031-34283-7_6

removal, and thermoregulation [10, 34], may function as a coping behavior to calm birds following a stressful event [14, 28, 39]. We refer to this hypothetical form of preening as "comfort preening."

It is impossible to provide conclusive evidence that nonhuman animals engage in coping behaviors because we are unable to measure the emotional state of animals directly. Physiological responses to stressors, however, can be measured, and it is possible in the laboratory to measure how such responses are modified by behaviors which are analogous to human coping responses. A less conclusive approach in a natural setting is to compare the incidence of a hypothetical coping behavior before a point disturbance with the incidence of the same behavior after the disturbance, which is the method used in this study.

Figure 6.1: Preening functions to keep a bird's feathers clean, oiled, and neatly arranged. The incidence of preening in glaucous-winged gulls increases following bald eagle disturbances on the Protection Island gull colony, suggesting that this behavior functions as a stress releaser (Photo by J. L. Hayward)

Natural selection has shaped many behaviors into heritable phenotypic traits [18]. The mechanism by which an extant trait is converted by natural selection into one with a new purpose is called co-option [3]. An extant behavior with a physiological function such as sucking in infants or preening in birds might become a coping behavior through co-option. The primary advantages of behaviors such as sucking by infants and preening by birds are apparent, but the benefits of using these behaviors as a means of coping with stressful events are less obvious. An example of such a benefit might include comfort preening in colonial seabirds in response to an eagle disturbance which leads to a reduction in the number of territorial disputes following the disturbance; conversely, the same behavior could be harmful if it reduced an alertness to predators. In this study we build a mathematical model that describes the evolutionary results of trade-offs such as these.

The postulates that form the basis for the theory of natural selection are

(1) Populations tend to grow exponentially with resultant competition.
(2) Phenotypic traits are at least partially heritable.
(3) These traits impact the outcome of competitions.

These postulates have been extensively tested in both the laboratory and the field [17].

It is often difficult, however, to establish a genetic basis for a particular trait, particularly for a behavioral trait. Thus, in this study we utilize the methodology of Darwinian dynamics [42], which tracks mean phenotypic traits in a dynamic population rather than tracking the gene frequencies of a trait in a population.

6.1.2 Preening Behavior as a Function of Environmental Correlates

Gulls have served as classical and convenient models of animal behavior—their breeding colonies contain large numbers of individuals, they exhibit relatively complex behavioral repertoires, and they are easily observed [36].

Preening (Fig. 6.1) occurs when a bird draws feathers through its bill or slides its head over a portion of the body [29]. Preening occupies a significant proportion of the time a bird is not in flight—for example, first-year mourning doves engage in this behavior 23% of their time [27], and glaucous-winged gulls spend 27% of their waking hours engaged in this activity [28]. Preening occurs within a variety of circumstances with several suggested functions. Gulls preen after flying [15]. In herring gulls (*L. argentatus*), sandwich terns (*Thalasseus sandvicensis*), and common terns (*Sterna hirundo*), the incidence of preening increases if the feathers have become damp [38–40]. In glaucous-winged gulls, increases in relative humidity increase the occurrence of preening [22]. Increases in light intensity and sound intensity and decreases in temperature are correlated with preening in day-old domestic poultry [4]. Increases in preening by incubating ringed turtle doves (*Streptopelia risoria*) are correlated with increases in egg temperature [16].

Our study on glaucous-winged gulls uses logistic regression to quantify the incidence of functional preening in relation to humidity, temperature, solar elevation, wind speed, tide height, barometric pressure, and breeding season stage.

6.1.3 Preening as a Hypothetical Coping Behavior Following Eagle Disturbances

Bald eagles flying over the gull colony and preying on adults, chicks, and eggs constitute the disturbances to which our gull subjects respond. These disturbances produce stress in the gulls [21, 31, 44]. When bald eagle populations plummeted due to DDT and other threats, North American seabird colonies on islands experienced little predation, isolated as they often were from ground predators. But over the past several decades, given the outlawing of DDT and the reduction of other threats, bald eagle populations have increased rapidly, as have their disturbances to seabird colonies [19, 21, 25, 31, 44] (Fig. 6.2).

Figure 6.2: Glaucous-winged gulls on Protection Island National Wildlife refuge become stressed in response to large disturbances created by bald eagle flyovers and predation events. During the event depicted in this photo, ten eagles are stirring up the colony (Photo by J. L. Hayward)

In previous studies preening has been associated with both environmental and psychological factors. Early ethologists suggested that preening is a form of "displacement behavior" in response to an internal conflict but irrelevant to the circumstances bringing about the conflict [24]. Chaffinches (*Fringilla coelebs*), for example, preen when they are thought to experience an equal tendency to approach and flee [41], and European starlings (*Sternus vulgaris*) sometimes preen vigorously during fights with conspecifics [24]. And in concert with our hypothesis, preening by gulls and terns has been interpreted as a soothing activity following lengthy periods of disturbance [14, 28, 39].

Although preening occurs commonly in birds, this behavior incurs costs. The visual field of dunlin (*Calidris alpina*), for example, is obscured during preening which raises their vulnerability to predation [32]. Moreover, preening utilizes about twice as much energy as utilized by basal metabolism [12, 20]. Significant benefits, however, accrue with the cleaning and repairing of feathers, including the enhancement of thermoregulation and flight [34], the reduction in numbers of ectoparasites [8, 9, 11] which enhances mating success [7], and improvements in social communication [5, 30].

We hypothesize that preening behavior, which has a direct physiological function, could be co-opted as a coping behavior in response to bald eagle disturbances. If this is true, any declines in fitness due to the occurrence of preening must be offset by significant benefits [28].

This study has four objectives:

(1) Development of a general model framework for the microevolution of a coping behavior that combines logistic regression models and Darwinian dynamic ordinary differential equation models
(2) Determination of a logistic regression model for preening behavior from empirical data collected at a large glaucous-winged gull colony
(3) Construction of a specific Darwinian dynamic model for the evolution of comfort preening by gulls in response to bald eagle disturbances
(4) Conduction of an independent empirical test of the occurrence of comfort preening in the gull colony after eagle disturbances by comparing the occurrence of preening before and after the occurrence of bald eagle fly overs

6.2 General Modeling Framework

The modeling framework is constructed in three stages. We:

(1) Pose a logistic regression model for behavior B under the assumption that the coefficients are subject to natural selection
(2) Create a differential equation population model with parameters that depend on the regression coefficients
(3) Formulate a Darwinian dynamics model that unites the population model with dynamic equations for the evolving regression coefficients.

We begin with a logistic regression model for a behavior B performed in response to both biotic and abiotic cues. Logistic regression [26] is frequently used to measure the effects of factors on a binary variable—in this case whether the behavior occurs or does not occur. The log-odds of the incidence of behavior B is regressed on a vector $x = (x_1, x_2.....x_n)$ of factors:

$$\ln\left(\frac{p}{1-p}\right) = \beta_0 + \sum_{j=1}^{n} \beta_j x_j, \tag{6.1}$$

where p is the probability of behavior B. The intercept β_0 quantifies the baseline occurrence of behavior B when all factors are zero, and the regression slope coefficients $\beta = (\beta_1, ..., \beta_n)$ quantify the response of behavior B to changes in the factors. "Interaction terms" of the form $\beta_{ij} x_i x_j$ can be included in the right side of Eq. (6.1). These are necessary when the effect of factor x_i depends on the value of factor x_j.

The slope coefficients β possess a convenient interpretation. For non-interaction terms, when x_i increases by c units while all other factors remain constant, the log-odds changes by $\beta_i c$ units:

$$\ln\left(\frac{p_2}{1-p_2}\right) - \ln\left(\frac{p_1}{1-p_1}\right) = \left(\beta_0 + \beta_i(x_i + c) + \sum_{\substack{i=1 \\ j\neq i}}^{n} \beta_j x_j\right) - \left(\beta_0 + \sum_{j=1}^{n} \beta_j x_j\right) = \beta_i c, \tag{6.2}$$

where p_1 and p_2 are the probabilities of behavior B before and after the change, respectively. Therefore, the "odds ratio" (OR) is

$$OR = \frac{p_2/(1-p_2)}{p_1/(1-p_1)} = e^{\beta_i c}. \tag{6.3}$$

Thus, given an increase in factor x_i by c units with all other factors held constant, the odds of behavior B are $e^{\beta_i c}$ times the previous value. (The computation for interaction terms is similar but somewhat more complicated.) Consequently, if $\beta_i > 0$ ($\beta_i < 0$), the odds ratio is greater than (less than) one and the odds of behavior B have increased (decreased). For our model, we assume that the regression coefficients β_i, which quantify the response of behavior B to changes in the factors, are mean heritable phenotypic traits subject to natural selection.

Next, we pose a population model with parameters that depend on the trait vector β. We assume for the sake of simplicity that population growth is logistic and modeled as

$$\frac{dN}{dt} = b(\beta)N - d(\beta)N^2, \tag{6.4}$$

where $b(\beta)$ is the net per capita rate of change at small population sizes, which is the difference between birth and death rates for members of the population having trait vector β in the absence of crowding, and $d(\beta)N$ is the density-dependent per capita death rate due to the effects of crowding on members of the population with trait vector β.

Finally, we pose a Darwinian dynamics model that tracks microevolutionary alterations in mean heritable traits β_i on the time scale of the population dynamics [42]. The Darwinian dynamics model is the $n+1$-dimensional system

$$\begin{aligned} \frac{dN}{dt} &= b(\beta)N - d(\beta)N^2 \\ \frac{d\beta_i}{dt} &= s_i^2 \frac{\partial G}{\partial \beta_i} + \sum_{j\neq i} \delta_{ij}^2 \frac{\partial G}{\partial \beta_j}, \quad i = 1, 2, ..., n, \end{aligned} \tag{6.5}$$

where

$$G = \frac{1}{N}\frac{dN}{dt}$$

is the "fitness generating function" (or simply "fitness"), i.e., the per capita growth rate [42]. Each s_i is a positive constant and s_i^2 (the variance of trait β_i in the population) is termed the "speed of evolution." Coefficient δ_{ij}^2 is the covariance of traits β_i and β_j [42].

We denote a solution by the vector

$$(N(t), \beta(t)) = (N(t), \beta_1(t), \beta_2(t), ..., \beta_{n-1}(t), \beta_n(t)).$$

Under standard smoothness assumptions on G, there exists a unique solution (for at least an interval of t surrounding 0) for each initial condition $(N(0), \beta(0))$. Since, by the fundamental theorem of calculus, the differential equation for $N(t)$ can be equivalently written as the integral equation

$$N(t) = N(0) \exp\left(\int_0^t (b(\beta(s)) - d(\beta(s)) N(s))\, ds\right),$$

we see that $N(0) > 0$ implies $N(t) > 0$ for as long as the solution exists.

Application of model (6.5) to a particular system requires the specification of the:

(1) Logistic regression model (6.1) for the behavior
(2) Population rates $b(\beta)$ and $d(\beta)$ as functions of the behavioral trait vector β

In the next two sections, we address these tasks in the specific context of the evolution of comfort preening in gulls.

6.3 Specification of the Logistic Regression Model for Preening in Gulls

Here we identify environmental correlates for preening in gulls, and we measure the response of preening to changes in these correlates. Our objective is to specify the logistic regression model (6.1) and estimate its parameters.

6.3.1 Data

Data on preening behavior by glaucous-winged gulls were collected at Protection Island National Wildlife Refuge ($48°08'N, 122°55'W$), Jefferson County, Washington, USA, situated in the southeast corner of the Strait of Juan de Fuca. Violet Point is a gravel spit that extends to the southeast (Fig. 6.3). During this study, the gull colony on Violet Point contained more than 2,400 pairs. Five sample plots in the colony collectively measured 4,205 m^2 and contained 259 and 238 nests with eggs in 2006 and 2007, respectively, the years of our study.

At the top of each hour from 05:00 to 20:00 hours (LST here and below) and using a 20–60x spotting scope, scan counts of the behaviors of resident birds in five sample plots

(Fig. 6.3, plots A–E) were recorded by voice from the top of a 33-m-high bluff bordering the west end of Violet Point. The voice recordings were subsequently transcribed. If a sample plot contained more than 50 birds, scanning typically stopped at 50 birds. Scans were carried out during three stages of the breeding season: May 19–29, 2006 (nest-building and egg-laying), June 13–21 and 23, 2006 (egg-laying and incubation), and July 10–13 and 15–19, 2007 (incubation and chick-rearing).

Tide heights and solar elevations were obtained from online data posted by the National Oceanographic and Atmospheric Administration (NOAA) from the Port Townsend buoy (Station PTWW1). Hourly values of relative humidity, temperature, wind speed, and barometric pressure were recorded by a weather station mounted 2 m above site elevation in the northwest part of Violet Point. No rainfall was recorded during the study.

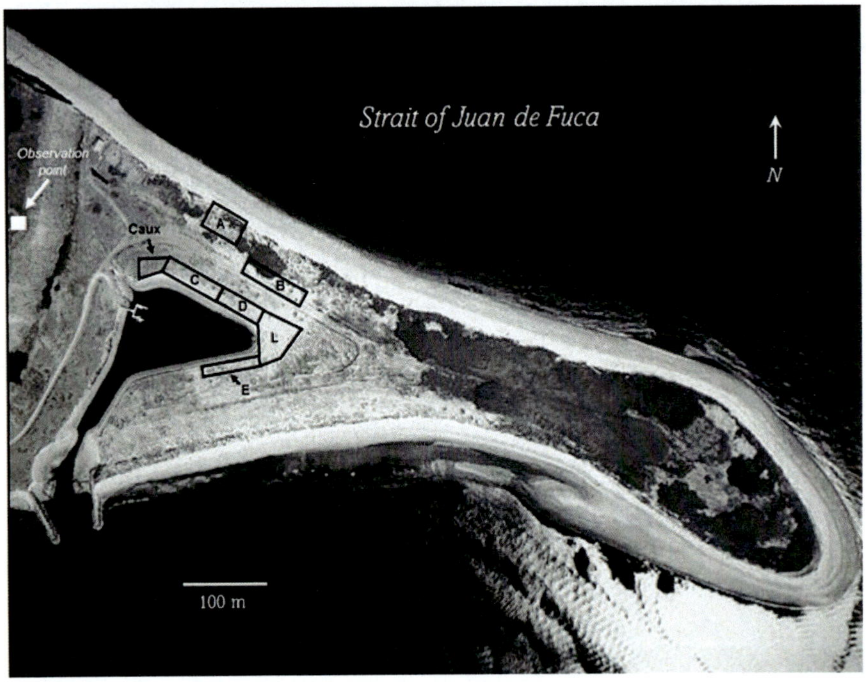

Figure 6.3: Aerial photograph of Violet Point, Protection Island, Washington. The seven areas or plots referred to in this chapter and in Chaps. 7, 8, and 9 are indicated by the polygons. The observation point atop the 33-m-high bluff from which we sometimes observed, counted, and recorded behaviors is shown at the left edge of the photo. The vast majority of nesting gulls occupied the light-shaded areas north and east of the triangle-shaped marina. The dark-shaded areas dominating the eastern half of the point were covered with dense American dune grass (*Leymus mollis*) and avoided by the gulls

6.3.2 Regression Analysis

We included seven environmental variables in our global logistic regression model: percent relative humidity (x_1 = HUM), ambient temperature in degrees Celsius (x_2 = TEMP), solar elevation in degrees above the horizon (x_3 = SUN), wind speed in m/sec (x_4 = WIND), tide height in meters (x_5 = TIDE), barometric pressure in mmHg (x_6 = BAR), and stage of breeding season using two design variables (D_{71} and D_{72} coded at three levels = STAGE; see Table 6.1). Twelve interactions between environmental variables and design variables were included in the global model:

$$\ln\left(\frac{p}{1-p}\right) = \beta_0 + \sum_{j=1}^{6}\beta_j x_j + \sum_{k=1}^{2}\beta_{7k}D_{7k} + \sum_{k=1}^{2}\sum_{j=1}^{6}\beta_{jk}x_j D_{7k}, \tag{6.6}$$

where p equals the probability of preening.

Table 6.1: Coding of design variables for STAGE, coded at three levels. Nest-building/egg-laying period is the reference stage

Data collection interval	Stage	D_{71}	D_{72}
May 19–29, 2006	Nest-building/egg-laying	0	0
June 13–21, 23, 2006	Egg-laying/incubation	1	0
July 10–13, 15–19, 2007	Incubation/chick-rearing	0	1

The global model (6.6) and all its submodels (in which various coefficients are set to zero) can be viewed as a set of alternative hypotheses. Model selection and parameter estimation were accomplished using an information-theoretic approach [2]. To rank the nested models, we used the Akaike information criterion (AIC), adjusted for overdispersion. Overdispersion happens when the variance of the samples is higher than the theoretical variance, which can occur when individual responses lack independence, a frequent problem with behavior scans which may include the same animals in repeated scans. Using Akaike weights, we ranked the factors HUM, TEMP, SUN, WIND, TIDE, and BAR in order of importance [2]. See Chap. 7 for details on AIC, factor importance, and Akaike weights.

The global model (6.6) was best (ΔAIC = 0) for the whole data set. Four additional models had $\triangle$AIC < 10 (Table 6.2). The overdispersion parameter estimate ($\hat{\tau}$ =1.95; df = 2371) for the global model was within the range (1–4) that indicates a small measure of extrabinomial variation as a result of demographic stochasticity and lack of independence rather than inadequate model structure [2]. Model-averaged coefficients (see Chap. 7), unconditional standard errors, and model-averaged correlation coefficients are listed in Table 6.3, and odds ratios with 95% confidence intervals are shown in Table 6.4. The ranking of factor importance is shown in Table 6.5; the most important factors are the same as those that have the most significant effect on the odds ratio (Table 6.4).

Table 6.2: Summary of AIC statistics for the five models with $\Delta < 10$, with second-order Akaike information criterion for overdispersed data ($QAIC_c$), $QAIC_c$ differences (Δ_i), and Akaike weights (w_i)

Model	$QAIC_c$	Δ_i	w_i
Global	41710.723	0	0.767
Without BAR	41714.988	4.264	0.091
Without TEMP	41715.086	4.363	0.087
Without WIND	41716.776	6.052	0.037
Without TIDE	41719.027	8.304	0.012

The logistic regression analysis indicates that alterations in the abiotic environment correlate significantly with the incidence of preening by glaucous-winged gulls. When considered over the entire breeding season, solar radiation, humidity, tide height, wind speed, temperature, and barometric pressure all were important in predicting the likelihood of preening. The most important factor, however, was breeding season stage. Depending on the stage, two to four other factors were relatively important.

The nonsignificance of a factor in Table 6.4 does not imply that a corresponding factor in Table 6.3 can be eliminated from the model. Consider, for example, humidity. This factor is not significant during the egg-laying/incubation stage, but this does not allow us to remove the term $\beta_{11}\text{HUM} \times D_{71}$ from the regression model. To understand why this is true, note the three humidity terms with coefficient estimates in Table 6.3:

$$0.0283 \times \text{HUM} - 0.0226 \times \text{HUM} \times D_{71} - 0.0131 \times \text{HUM} \times D_{72}.$$

When $D_{71} = 1$ and $D_{72} = 0$ during the egg-laying/incubation stage, the combined coefficient on HUM is near zero. This is consistent with the fact that humidity is not a significant factor during that stage of the breeding season, and the middle term is required to cancel out the first term. Therefore, none of the terms may be deleted from the global regression model (6.6).

Table 6.3: Model-averaged estimates for the global model. Coefficient estimates are averages of the maximum likelihood (ML) coefficients over all models; unconditional standard error (SE) estimates include information from all models containing the parameter and incorporate model uncertainty; correlation coefficient estimates of (r) are model-averaged over models containing the interaction terms, $N = 79{,}286$

Term	Coefficient	SE	$\overline{r}$
β_0	0.8497	5.641	
β_1HUM	0.0283	0.0093	
β_2TEMP	−0.1353	0.0454	
β_3SUN	0.0008	0.0020	
β_4WIND	−0.0336	0.0283	
β_5TIDE	0.0701	0.0462	
β_6BAR	−0.0051	0.0075	
$\beta_{71}D_{71}$	1.338	7.851	
$\beta_{72}D_{72}$	20.59	12.05	
β_{11}HUM $\times$ D_{71}	−0.0226	0.0103	−0.9061
β_{21}TEMP $\times$ D_{71}	0.1477	0.0499	−0.9095
β_{31}SUN $\times$ D_{71}	−0.0056	0.0025	−0.8096
β_{41}WIND $\times$ D_{71}	0.0545	0.0361	−0.7744
β_{51}TIDE $\times$ D_{71}	−0.0127	0.0535	−0.8648
β_{61}BAR $\times$ D_{71}	−0.0004	0.0107	−0.7053
β_{12}HUM $\times$ D_{72}	−0.0131	0.0110	−0.8890
β_{22}TEMP $\times$ D_{72}	0.1594	0.0501	−0.9154
β_{32}SUN $\times$ D_{72}	−0.0013	0.0024	−0.8318
β_{42}WIND $\times$ D_{72}	−0.0670	0.0428	−0.6510
β_{52}TIDE $\times$ D_{72}	0.0008	0.0532	−0.8713
β_{62}BAR $\times$ D_{72}	−0.0303	0.0136	−0.5528

Table 6.4: Odds ratios (ORs) and 95% confidence intervals (CIs). Note that c is the unit of increase in the variable that generates the OR. For the nest-building/egg-laying stage, N = 13,384; for the egg-laying/incubation stage, N = 35,505; for the incubation/chick-rearing stage, N = 30,397. An asterisk (*) indicates that the variable is significant

Factor	c	OR	95% CI
Nest-building / egg-laying stage ($D_{71} = D_{72} = 0$)			
HUM*	20	1.8	(1.223, 2.533)
TEMP*	5	0.51	(0.3258, 0.7936)
SUN	60	0.95	(0.7551, 1.205)
WIND	5	0.85	(0.6405, 1.115)
TIDE	3	1.2	(0.9407, 1.619)
BAR	10	0.95	(0.8203, 1.102)
Egg-laying / incubation stage ($D_{71} = 1$ and $D_{72} = 0$)			
HUM	20	1.1	(0.9446, 1.328)
TEMP	5	1.1	(0.8683, 1.304)
SUN*	60	0.68	(0.5753, 0.8075)
WIND	5	1.1	(0.8872, 1.389)
TIDE*	3	1.2	(1.014, 1.391)
BAR	10	0.95	(0.8165, 1.098)
Incubation / chick-rearing stage ($D_{71} = 0$ and $D_{72} = 1$)			
HUM*	20	1.4	(1.111, 1.648)
TEMP	5	1.1	(0.9260, 1.375)
SUN	60	0.88	(0.7535, 1.031)
WIND*	5	0.60	(0.4397, 0.8312)
TIDE*	3	1.2	(1.061, 1.442)
BAR*	10	0.70	(0.5618, 0.8774)

Table 6.5: Relative variable importance. Evidence for the importance of each variable is based on sums of Akaike weights (w_+) over models in which the variable occurs. Factors are listed in decreasing order of importance for the nest-building/egg-laying stage (NE), egg-laying/incubation stage (EI), incubation/chick-rearing stage (IC), and the entire breeding season (All). For each stage the calculations were based on an overdispersion parameter estimate from the data for that stage (NE $\hat{\tau} = 1.86$; EI $\hat{\tau} = 1.91$; IC $\hat{\tau} = 2.08$). Significant variables (*) are arbitrarily designated as those with $w_+ > 0.7$

NE	w_+	EI	w_+	IC	w_+	All	w_+
TEMP*	0.993	SUN*	1.000	BAR*	0.989	STAGE*	1.00000
HUM*	0.986	TIDE*	0.875	HUM*	0.987	SUN*	0.99988
TIDE	0.668	HUM	0.660	WIND*	0.980	HUM*	0.99979
WIND	0.437	BAR	0.528	TIDE*	0.972	TIDE*	0.98587
SUN	0.318	TEMP	0.339	SUN	0.498	WIND*	0.95850
BAR	0.312	WIND	0.329	TEMP	0.358	TEMP*	0.91156
						BAR*	0.90538

The results of logistic regression by season are summarized as follows: In the nest-building/egg-laying stage, the odds of preening rose 80% with each 20% rise in relative humidity, but the odds decreased 49% with every 5°C rise in temperature. In the egg-laying/incubation stage, the odds of preening rose 20% at high tide relative to low tide, but the odds decreased 32% at midday (SUN = 60°) from sunrise and sunset. In the incubation/chick-rearing stage, the odds of preening rose 40% with each 20% increase in humidity, as well as 20% at high tide as compared with low tide; the odds of preening decreased 40% with each 5 m/s increase in wind speed and 30% with each 10 mmHg rise in barometric pressure.

6.4 Specification of Population Rates $b(\beta)$ and $d(\beta)$

Here we construct $b(\beta)$ and $d(\beta)$ as functions of the behavioral trait vector β in order to pose a specific form of model (6.5) for our application to comfort preening. We do so with four general assumptions:

> (1) Comfort preening is co-opted from preening that has a physiological function.
> (2) Comfort preening can be harmful to physiological health because it redirects energy from physiological functions.
> (3) A reduction in alertness while comfort preening could lead to death during eagle flyovers and other stressful events.
> (4) Density-dependent mortality can be reduced by comfort preening, for instance, by soothing gulls' conflict with colony neighbors during or following stressful events.

For notational simplicity, we rename the 20 slope coefficients in the regression model (6.6) to be $\beta_1, \beta_2, \cdots, \beta_{20}$, in the order listed in Table 6.3. For clarity, however, when we discuss a specific slope coefficient such as β_6 (the coefficient on BAR), we write it as β_{bar}. These 20 slope coefficients measure the response of preening to the environmental variables; in other words, they quantify the physiological function of preening. In addition, *we introduce a new coefficient,* β_{21}, *that quantifies the response of preening to eagle disturbance.* Thus, in the Darwinian model (6.5), $n = 21$ and the behavioral trait vector is

$$\beta = (\beta_1, \beta_2, ..., \beta_{20}, \beta_{21}).$$

6.4.1 Specification of $b(\beta)$

We assume that $b(\beta)$ can be expressed as

$$b(\beta) = a(\beta) - c(\beta) \tag{6.7}$$

where $a(\beta)$ equals the net per capita rate of change (the balance of birth and death rates) in small populations without predation, and $c(\beta)$ represents a death rate in small populations due to predation. In our system, $a(\beta)$ represents a measure of the physiological health of a gull with trait vector β, and $c(\beta)$ constitutes a measure of the risk of predation to such an individual.

Specification of General Physiological Health $a(\beta)$ We assume that deviation of the coefficients $\beta_1, \beta_2, ..., \beta_{20}$ from their optimal values $\alpha_1, \alpha_2, ..., \alpha_{20}$ results in a decline in general physiological health in the absence of stress. We also assume that any change in the incidence of preening behavior in response to stress causes a decline in general physiological health, due to diverting energy away from physiological functions. Specifically, we assume

$$a(\beta) = a_0 \exp\left(-\frac{1}{2}\sum_{j=1}^{20}\left(\frac{\beta_j - \alpha_j}{\sigma_j}\right)^2\right)\exp\left(-\frac{1}{2}\left(\frac{\beta_{21}}{\sigma_{21}}\right)^2\right) \tag{6.8}$$

for $a_0 > 0$, where $\beta_{21} > 0$ ($\beta_{21} < 0$) indicates an increase (decrease) in preening when there is stress. Note that general physiological health $a(\beta)$ is optimal (with value a_0) if and only if $\beta_i = \alpha_i$ for $i = 1, 2, ..., 20$ and $\beta_{21} = 0$.

Specification of Death Rate $c(\beta)$ **Due to Predation** We assume that use of preening as a coping behavior ($\beta_{21} > 0$) directly after an eagle disturbance might lead to a decrease in alertness and, thus, a higher death rate due to predation. This is especially true because departing eagles often change course and suddenly stoop into the colony. Additionally, we assume that a reduction in preening in response to disturbance ($\beta_{21} < 0$) may lead to more alertness and a lower death rate. Specifically, we assume $c(\beta)$ to be a monotonically increasing function of β_{21} with values on the interval $(0, c_0)$ for $c_0 \geq 0$:

$$\begin{aligned} c(\beta) &= c_0 h(\beta_{21}) \\ h &\in C^1(\mathbf{R}, (0,1)) \\ h'(x) &> 0 \ \forall x \in \mathbf{R} \\ \lim_{x\to-\infty} h(x) &= 0 \quad \text{and} \quad \lim_{x\to\infty} h(x) = 1. \end{aligned} \tag{6.9}$$

The function

$$h(x) = \frac{1}{2} + \frac{1}{\pi}\arctan(\alpha_h x), \quad \alpha_h > 0$$

is an example of a function h that satisfies these properties, one that we will use in Sect. 6.6.

6.4.2 Specification of Density-Dependent Death Rate $d(\beta)$

We assume that the density-dependent death rate $d(\beta)$ in Eq. (6.4) can be divided into three parts:

(1) Environmental effects, due to environmental resource limitation caused by crowding. We assume the environmental effect component is constant $d_0 > 0$ and is not influenced by preening.
(2) Psychological effects, due to interindividual aggression that is ameliorated by coping behavior. We assume the psychological effect component is a monotonically decreasing function of β_{21} with values on the interval $(0, d_1)$, for $d_1 > 0$. Specifically, we assume psychological effects have the form $d_1 f(\beta_{21})$, where f satisfies

$$\begin{aligned} f &\in C^3(\mathbf{R}, (0,1)) \\ f'(x) &< 0 \quad \forall x \in \mathbf{R}, \end{aligned} \tag{6.10}$$

$$\lim_{x\to-\infty} f(x) = 1, \quad \lim_{x\to\infty} f(x) = 0, \quad \lim_{x\to\infty} f'(x) = 0.$$

In our system, psychological effects would include territory disputes which at times lead to fatality and the killing of chicks that enter territories of neighbors when disturbed. A bird that comfort preens following a disturbance may exhibit a lower probability of mortality due to these causes.
(3) Physiological effects, due to compromised physical health in the presence of crowding. Preening at rates higher or lower than what is optimal for general physiological health might place an individual at risk in territorial disputes. We assume this component of the density-dependent death rate has the form

$$d_2 \left(1 - \frac{a(\beta)}{a_0}\right).$$

with $d_2 \geq 0$ and $a_0 > 0$.

The function

$$f(x) = \frac{1}{2} - \frac{1}{\pi} \arctan(\alpha_f x), \quad \alpha_f > 0 \tag{6.11}$$

is an example of a function f that satisfies these properties, one that we will use in Sect. 6.6.

From these assumptions, we obtain

$$d(\beta) = d_0 + d_1 f(\beta_{21}) + d_2 \left(1 - \frac{a(\beta)}{a_0}\right), \tag{6.12}$$

where $d_0, d_1, d_2 \geq 0$ and f satisfies (6.10).

The immediate effects of stress and coping behavior are incorporated within the assumptions given above. The interaction between stress and the endocrine system is complex, and it is possible that the cumulative effect of stress on birth and death rates remain unaccounted for in these assumptions. Our general approach might be modified to handle other effects, but in this study we make the assumption that all other effects are higher order and can be ignored.

6.5 Some Preliminary Dynamics of the Darwinian Model

The fitness-generating function [42] for the Darwinian model (6.5) in this case is

$$G(\beta, N) = a(\beta) - c(\beta) - d(\beta)N. \tag{6.13}$$

In this section we provide some baseline dynamics for an analytically tractable case of the model.

First, we assume there are no covariances among the traits so that $\delta_{ij} = 0$. Under this assumption, Eqs. (6.5) and (6.13) imply that the components $\beta_i(t)$ for $i = 1, ..., 20$ of a solution $(N(t), \beta(t)) = (N(t), \beta_1(t), \beta_2(t), ..., \beta_{20}(t), \beta_{21}(t))$ satisfy the differential equations

$$\frac{d\beta_i}{dt} = -s_i^2 a(\beta)\left(1 + \frac{d_2}{a_0}N\right)\left(\frac{\beta_i - \alpha_i}{\sigma_i^2}\right), \ i = 1, 2, ..., 20.$$

Thus, regardless of the dynamics of $N(t)$ and $\beta_{21}(t)$, each component $\beta_i(t)$ for $i = 1, ..., 20$ in a solution is increasing when $\beta_i(t) < \alpha_i$ and decreasing when $\beta_i(t) > \alpha_i$. This implies $\lim_{t\to\infty} \beta_i(t) = \alpha_i$ and, moreover, that α_i is an equilibrium for every $i = 1, 2, ..., 20$. Thus, the hyper-(N, β_{21}) plane $(N, \alpha_1, \alpha_2, ...\alpha_{20}, \beta_{21})$ in 22-dimensional Euclidean space is an invariant attractor of the Darwinian model (6.5) and the dynamics of the model are determined by the dynamics on that two-dimensional hyperplane, which are described by the equations in (6.5) for N and β_{21} with $\beta_i = \alpha_i$ for $i = 1, 2, ..., 20$. This two-dimensional system is studied in [23] under two additional special assumptions, namely, when eagle predation and the physiological effects described in (3) are both absent, i.e., when the coefficients $c_0 = d_2 = 0$ and we have

$$\begin{aligned}
\frac{dN}{dt} &= a_0 \exp\left(-\frac{u^2}{2\sigma^2}\right)N - (d_0 + d_1 f(u))N^2 \\
\frac{du}{dt} &= -s^2\left(a_0\frac{u}{\sigma^2}\exp\left(-\frac{u^2}{2\sigma^2}\right) + d_1 f'(u)N\right),
\end{aligned}$$

where, in order to simplify the notation, we have set $u = \beta_{21}$, $\sigma = \sigma_{21}$, and $s = s_{21}$. If we utilize (6.11) this plane autonomous system

$$\begin{aligned}
\frac{dN}{dt} &= a_0 \exp\left(-\frac{u^2}{2\sigma^2}\right)N - \left(d_0 + d_1\left(\frac{1}{2} - \frac{1}{\pi}\arctan(\alpha_f u)\right)\right)N^2 \\
\frac{du}{dt} &= -s^2\left(\frac{a_0}{\sigma^2}u\exp\left(-\frac{u^2}{2\sigma^2}\right) - \frac{d_1\alpha_f}{\pi}\frac{1}{1+\alpha_f^2 u^2}N\right)
\end{aligned}$$

has the equilibrium $(N, u) = (0, 0)$ and it is straightforward to see from the equilibrium equations that this is the only "extinction" equilibrium (i.e., equilibrium with $N = 0$).

This equilibrium is unstable, as is seen by the linearization principle from the Jacobian evaluated at the equilibrium:

$$\begin{pmatrix} a_0 & 0 \\ s^2 \alpha_f \frac{d_1}{\pi} & -s^2 \frac{a_0}{\sigma^2} \end{pmatrix}$$

which has the positive eigenvalue $a_0 > 0$. With regard to "survival" equilibria (N, u), $N > 0$, part (b) in the following theorem is proved in [23].

Theorem 1 *Assume* $\delta_{ij} = c_0 = d_2 = 0$ *and* $f(x)$ *is given by (6.11).*

(a) The only extinction equilibrium is $(N, \beta) = (0, \alpha_1, \alpha_2, ...\alpha_{20}, 0)$ *and it is unstable.*

(b) There exists a survival equilibrium $(N, \beta) = (N_e, \alpha_1, \alpha_2, ...\alpha_{20}, u_e)$, $N_e > 0$ *and* $u_e > 0$, *and it is unique and (locally asymptotically) stable for all sufficiently large* $\sigma^2 > 0$.

This theorem addresses the special case when predation risk is not increased by comfort preening ($c_0 = 0$) and density-dependent mortality is not influenced by compromised health due to comfort preening ($d_2 = 0$). It guarantees that, in this special case, there is a positive stable state to which coping behavior can evolve and gives us a baseline from which more general results can be obtained. As we see in the next section, simulations suggest that this situation is fairly robust in the general case as well.

6.6 Simulations of the Darwinian Model

Here we discuss a typical numerical simulation of model (6.5) to suggest if and how comfort preening might evolutionarily invade the behavior system of gulls.

In Fig. 6.4 we allow preening to evolve without stress and then we apply stress to see how this affects the final states. In Eq. (6.8) we set the parameters α_i, which denote the optimal values of the preening coefficients, equal to the values in Table 6.3 empirically estimated from Protection Island data. The caption beneath Fig. 6.4 provides the other parameters in the model. Stress level is set at zero ($c_0 = d_1 = 0$) during time $t = 0$ to $t = 3000$. During this time the population size approaches a carrying capacity and the non-comfort preening coefficients $\beta_1, \beta_2, \cdots, \beta_{20}$ approach their optimal values. Only β_{hum} and β_{temp} are shown in Fig. 6.4; they approach their optimal values α_{hum} and α_{temp} shown in Table 6.3 (0.0283 and -0.1353, respectively). The coefficient β_{stress} fluctuates because of the covariances $\delta_{ij}^2 > 0$ in model (6.8) and then approaches its non-stressed optimal value of zero. In other words, *when there is no stress, comfort preening does not invade the system.*

At time $t = 3000$, stress is initiated by setting $c_0, d_1 > 0$ (Fig. 6.4). At this point, when eagle disturbances begin, the comfort preening coefficient β_{stress} begins to increase, and the non-comfort preening coefficients fluctuate slightly around optimal values. The gull population size decreases and then increases to a new, but now lower, carrying capacity. Non-comfort preening coefficients return to their optimal values, and the coefficient β_{stress}

approaches a positive value ($\approx$0.05). Thus, *comfort preening is predicted to invade the system in the presence of stress.*

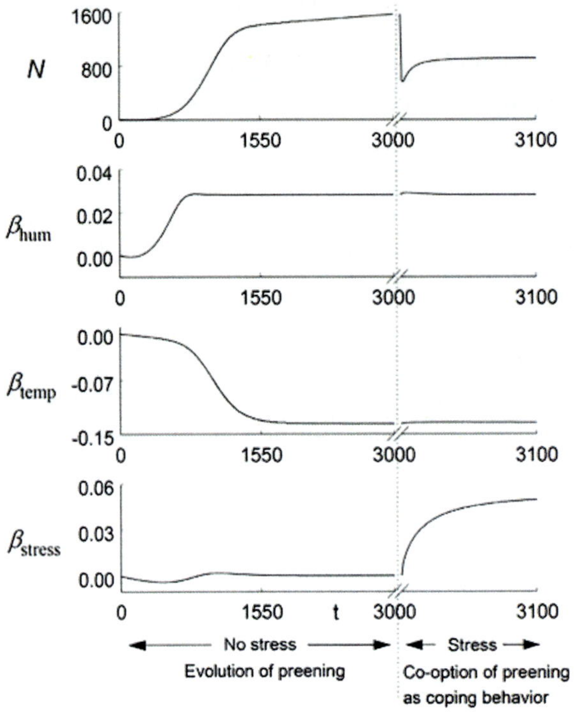

Figure 6.4: Simulation of model (6.5)–(6.13) with the behavior coefficients from model (12). The initial population size is $N(0) = 5$ and all preening coefficients are initially zero. For $t = 0 - 3000$ stress is set to zero ($c_0 = d_1 = 0$). The population size N approaches the carrying capacity, the 20 non-comfort preening coefficients (of which only β_{hum} and β_{temp} are shown) approach their optimal values given in Table 6.3, and the comfort preening coefficient β_{stress} approaches zero. For $t - 3000 - 3100$ the population is stressed ($c_0 = 0.0005$, $d_1 = 0.005$). The population size drops and then rises to a lower carrying capacity than before, the non-comfort preening coefficients return to their optimal values, and the comfort preening coefficient approaches a positive limit. The other parameters are $a_0 = 3$, $d_0 = d_2 = c_0 = 0.0005$, $s_i = 0.01$, $\delta_{ij} = 0.005$. We used $h(x) = 1/2 + (1/\pi)\arctan(70x)$ and $f(x) = 1/2 - (1/\pi)\arctan(80x)$ in Eqs. (6.9) and (6.12). The optimal values α_i for the non-comfort preening coefficients in Eq. (6.8) are set to the estimated coefficients in Table 6.3, and the σ_i for the non-comfort preening coefficients were set to $\sigma_i = |\alpha_i|$. The σ_{stress} for comfort preening (σ_n in Eq. (6.8)) is $\sigma_{stress} = 0.1$. (Originally published in Henson et al., "Coping behaviour as an adaptation to stress: post-disturbance preening in colonial seabirds," *Journal of Biological Dynamics* 6:17–37. Copyright © 2012, Taylor & Francis. Used with permission.)

Thus, given increased levels of eagle activity, models (6.8)–(6.13) predict that preening will be co-opted as a coping behavior. If we interpret the limiting value $\beta_{stress} \approx 0.05$ to be a regression coefficient for a binary factor STRESS (set at 1 if there is an eagle disturbance; otherwise, set to 0), the odds ratio is $e^{0.05} \approx 1.05$; thus, the odds of preening are predicted to be 5% greater following an eagle disturbance than before. The actual value of the odds of preening depends, of course, on the values of the parameters, particularly of σ_{stress} in Eq. (6.8). We do not have estimates for the σ_i, so the 5% prediction is qualitative, not quantitative.

6.7 Empirical Test for Comfort Preening

Here we pose an independent test to determine if gulls residing in the Protection Island gull colony engage in comfort preening following eagle disturbances.

Bald eagles frequently perch along the 33-m bluff that borders the west end of Violet Point. They also sit on refuge signs along the beach and logs at the east end of Violet Point in the vicinity of a harbor seal (*Phoca vitulina*) haulout area. Flights between these two locations create disturbances of the gull colony. The disturbances have increased in frequency as a result of rebounding eagle populations. We noted very few eagle disturbances in the 1980s, but during a single day in 1995, we counted 213 such disturbances [21].

We could observe the approach of an eagle flying from east to west and monitor the wave of disturbance as the eagle passed over the colony. At the beginning of a disturbance, individual gulls assume an "upright alert" posture, followed by an increased volume of sound by a cacophony of "long calls." Gulls situated beneath the direct path of the eagle flight may take flight, whereas those along the periphery may remain on territory. If an eagle suddenly swoops down toward the colony surface, the colony erupts with large numbers of resident gulls taking flight in a massive cyclonic flight pattern [21].

In a blind at the top of the bluff, we mounted five digital video cameras together linked to operate synchronously. The linked cameras simultaneously recorded the behaviors of gulls in the five sample plots during disturbances. When the camera operator saw an eagle beginning to fly over the colony, the cameras were activated to record the behaviors of the gulls before, during, and after the disturbance. We defined the onset of disturbance for a sample plot as the time when gulls closest to the approaching eagle assumed an upright alert posture.

Because it was difficult for the camera operator to anticipate the beginning of a disturbance, most recordings had to be eliminated. Two recordings, however, were usable for analysis. These were made at 11:00 and 14:30 hours on June 17, 2006. For recording, we counted the total number of gulls in the sample plots, as well as the number of gulls preening in that plot. Counts were made at the beginning of each 15-second interval. To test the null hypothesis that preening frequency was the same before and after a disturbance, one-tailed paired t-tests at the 0.05 significance level were used to compare the average before and after frequencies.

The 11:00-hour disturbance enveloped nearly the entire colony; 171 gulls of 182 total gulls (94%) counted in the five plots fled (Fig. 6.5a). The average proportion of gulls preening at 15-second intervals during the 2-minute period just prior to the disturbance was considered the "before" preen frequency for each sample plot. Three minutes past the start of the disturbance, almost all the gulls had returned to their territories. The average proportion of gulls preening at 15-second intervals during the next 2 minutes was considered the "after" preen frequency for each sample plot. Mean (±SD) preen frequency averaged over all five sample plots was significantly higher for after ($\overline{X} = 0.41 \pm 0.07$) than before ($\overline{X} = 0.31 \pm 0.06$) (paired $t = 2.80, 4\ \text{df}, P = 0.02$).

The disturbance at 14:30 hours PST was localized over Plot A, where 32 of 33 (97%) gulls fled the plot (Fig. 6.5a). But in plots B–E, only 5 of 164 (3%) gulls fled (Fig. 6.5b). Given the large difference in response between the two plots, data from plot A were omitted from the t-test comparison. Doing this enabled us to determine if "before" and "after" preen frequency was different for gulls that remain on their territories during a disturbance. Also, given that most of the gulls in plots B–E did not fly, they could preen earlier than the gulls that flew. Only 1 minute 45 seconds was captured on video before the start of the disturbance; thus, the average "before" frequency was based on a shorter sample period. One minute after the start of the disturbance, the average proportion of gulls preening at 15-second intervals during the next 2 minutes was assumed to be the "after" preen frequency. Once again, the mean (±SD) preen frequency averaged over plots B–E was significantly higher for "after" ($\overline{X} = 0.37 \pm 0.02$) than for "before" ($\overline{X} = 0.31 \pm 0.03$) (paired $t = 2.98, 3\ \text{df}, P = 0.03$). The odds of preening were 31% higher "after" the disturbance ($\overline{X}/(1 - \overline{X}) \approx 0.59$) than "before" ($\overline{X}/(1 - \overline{X}) \approx 0.45$). Plot A was located approximately 17 m from the nearest edge of the closest sample plot B, and no territories linked plot A with any of the other four plots (Fig. 6.3). Consequently, we are confident that social facilitation by gulls in plot A was not the cause of the increase in incidence of preen after the disturbance in the other four plots.

Preening by gulls occurs frequently after flight, probably to rearrange flight feathers [15]. Thus, a rise in the frequency of preening after birds have flown is not surprising. The results from the 14:30-hour disturbance, however, indicate that rearrangement of the flight feathers was not the cause of the increase in preening following a disturbance. *This result comports with our hypothesis that preening can function as a coping behavior.*

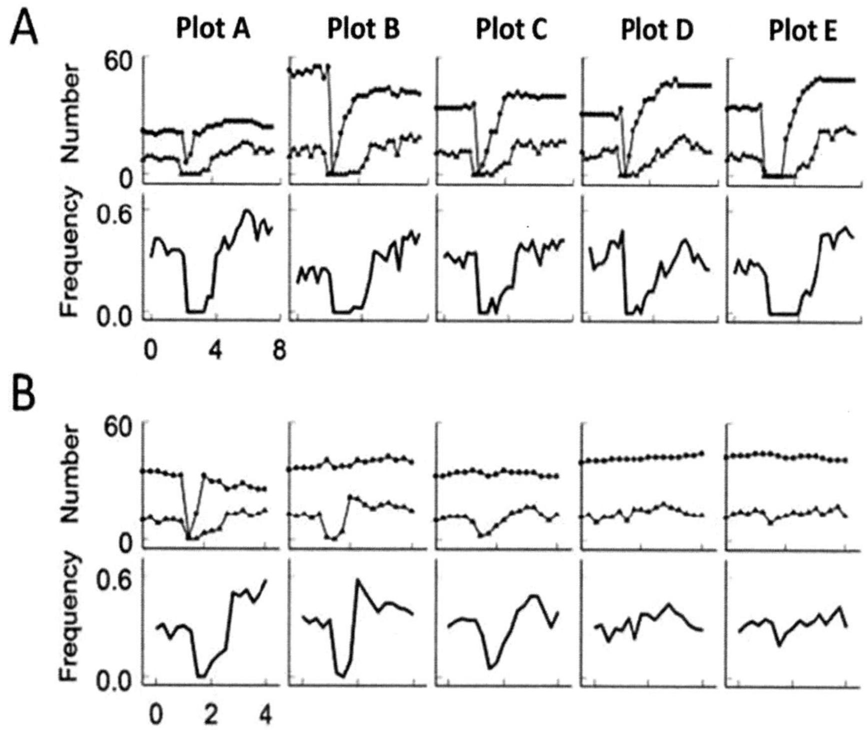

Figure 6.5: Observed occupancies (dark circles), numbers of gulls preening (triangles), and preening frequency (line) before, during, and after eagle disturbances in the five sample areas. Units on the horizontal axes are minutes. (**a**) Disturbance recorded at 11:00 a.m. PST on June 17, 2006. (**b**) Disturbance recorded at 2:30 p.m. PST on June 17, 2006. (Originally published in Henson et al., "Coping behaviour as an adaptation to stress: post-disturbance preening in colonial seabirds," *Journal of Biological Dynamics* 6:17–37. Copyright © 2012, Taylor & Francis. Used with permission.)

6.8 Discussion

The results of our study provide four contributions:

(1) A technique for modeling the microevolution of animal behavior combining the methods of logistic regression and Darwinian dynamics and allowing for a theoretical exploration of the adaptiveness of a stress-induced coping behavior
(2) An examination of environmental correlates of preening behavior in gulls using logistic regression
(3) A theoretical model investigation into if and how preening behavior might be co-opted as a coping behavior

(4) Conduction of an empirical test of the hypothesis that preening can function as a coping behavior in gulls

However, we point out several caveats.

Inferential Caveats This is the first study to demonstrate and measure an increase in the incidence in preening in birds that were aroused by the presence of a predator but did not fly. While this result is consistent with the hypothesis that gulls use preening to cope with stress, a causal link between disturbance and preening is not established. Moreover, the psychological state of the birds remains unknown.

Even if a causal link could be established, preening might simply be the result of a learned response rather than an evolutionary adaptation. Other work has demonstrated a strong genetic basis for the frequency of grooming behavior in mice [6], however, as is likely true for birds as well.

Finally, it is important to realize that the environmental correlates identified by the logistic regression analysis may not actually drive the occurrence of preening behavior.

Mathematical Caveats The logistic regression model (6.1) may be a fairly robust method for the quantification of behavior, but a logistic differential equation population model (6.5) provides a crude understanding of population dynamics. A more accurate population model would, for example, account for the fact that glaucous-winged gulls reach sexual maturity only after 4 years. Also, stress and coping responses most likely have cumulative effects on the population; our population model includes only proximal effects under the assumption that higher-order effects could be ignored. Moreover, our model assumes a "closed" population with no immigration. Banding studies have shown that dispersion of gulls from the Protection Island colony is primarily to other locations in the Salish Sea and that most immigration to the island is from other colonies in the same region. Consequently, our assumption of a closed population is most likely accurate for the larger regional population [33].

We assumed an unrealistically abrupt onset of eagle disturbances in our simulations (Fig. 6.4) by setting $c_0 = d_1 = 0$ for times $t = 0 - 3000$ and then setting $c_0, d_1 > 0$ at time $t = 3000$. Future studies of eagles and gulls should consider a predator-prey model in which the coefficients c_0 and d_1 are functions of numbers of eagles in the system.

Finally, Darwinian dynamic modeling assumes that the trait variances s_i^2 and covariances δ_{ij}^2 are constant over time, an assumption that may not always be true, for instance, following a population bottleneck. The trait variances and covariances perhaps could be estimated directly from the population as a means of testing this assumption and to approximate the rate of evolution in the system.

Biological Caveats During our data collection period, rain was infrequent, episodic, and unmeasurable by the weather station. Consequently, we did not include rain as an environmental variable in the logistic regression analysis. Light rain did occur during the time period of our study, however. We evaluated its effect by comparing the mean frequency of preening during rain, during the hour before rain, and during hours without rain. We compared mean frequencies with a one-way ANOVA carried out at the 0.05 significance level. The mean (±SD) proportions of bird preening during rain hours ($\overline{X} = 0.21 \pm 0.16, N = 41$) with hours before rain ($\overline{X} = 0.12 \pm 0.08, N = 24$), hours directly after rain ($\overline{X} = 0.15 \pm 0.11, N = 30$), and hours on days with no rain ($\overline{X} = 0.17 \pm 0.01, N = 503$) were significantly different ($F = 3.127, 3$ df, $597, P = 0.0254$) on the basis of square-root-transformed data. A Bonferroni multiple comparison post hoc test showed that the proportion of gulls preening during times of rain was higher than for hours before rain.

The data for this study were collected during two different years, 2006 and 2007, the latter of which was an El Niño year. But because the 2007 data set and the incubation/chick-rearing data set were the same, we were unable to consider the effect of year in the logistic regression analysis. Thus, some results linked to the incubation/chick-rearing stage may actually have been due to the effects of El Niño. Our experience with these birds on this colony for several decades, including multiple El Niño years, suggest, however, that maintenance behaviors such as preening are not greatly affected by these events.

In addition to its other functions, preening is critical for the removal of ectoparasites [10], but our study did not address this function.

Finally, our independent, empirical test for comfort preening yielded only two recordings of disturbances for analysis, despite efforts over multiple days to make such recordings. The number of gulls involved in each of these events was large, however, and our counts yielded important information.

6.9 Conclusion

The theoretical and empirical approaches in this chapter suggest that preening can serve as a coping behavior in gulls.

Theory and data are integrated, although not rigorously so, in this study. Thus, we were unable to provide rigorous estimates of all the model parameters; functional effects of behavior on fitness and population dynamics were not quantified but were hypothetical; and the population model was highly simplified biologically. Nonetheless, the partial integration of data and theory provided suggests an approach to examine the effects of stress on the microevolution of coping behavior, one that in the future could be carried out with more rigor.

References

1. Barrows EM (2001) Animal behavior desk reference: a dictionary of animal behavior, ecology, and evolution, 2nd edn. CRC Press LLC, Boca Raton, FL

2. Burnham KP, Anderson DR (2002) Model selection and multi-model inference: a practical information-theoretic approach, 2nd edn. Springer, New York

3. Buss DM, Haselton MG, Shackelford TK, Bleske AL, Wakefield JC (1998) Adaptations, exaptations, and spandrels. Am Psychol 53:533–548

4. Caldwell DY, Moore RW, Caldwell DJ, Hargis BM (2001) Effect of photointensity, sound intensity, and ambient temperature on preening behavior and ingestion of spray-applied biologics. J Appl Poult Res 10:99–106

5. Caryl PG (1976) Sexual behavior in the zebra finch *Taeniopygia guttata*: response to familiar and unfamiliar partners. Anim Behav 24:93–107

6. Chen SC, Tvrdik P, Peden E, Cho S, Wu S, Spangrude G, Capecchi MR (2010) Hematopoietic origin of pathological grooming in Hoxb8 mutant mice. Cell 141:775–785

7. Clayton DH (1990) Mate choice in experimentally parasitized rock doves: lousy males lose. Am Zool 30:251–262

8. Clayton DH (1991) Coevolution of avian grooming and ectoparasite avoidance. In: Loye JE, M. Zuk M (eds) Bird-Parasite interactions: ecology, evolution and behaviour. Oxford University, Oxford, pp 258–289

9. Clayton DH, Lee PLM, Tompkins DM, Brodie ED (1999) Reciprocal natural selection on host-parasite phenotypes. Am Nat 154:261–270

10. Clayton DH, Walther BA (2001) Influence of host ecology and morphology on the diversity of Neotropical bird lice. Oikos 94:455–467

11. Cotgreave P, Clayton DH (1994) Comparative analysis of time spent grooming by birds in relation to parasite load. Behaviour 131:171–187

12. Croll DA, McLaren E (1993) Diving metabolism and thermoregulation in common and thick-billed murres. J Comp Physiol B 163:160–166

13. Dallman MF, Pecoraro N, Akana SF, la Fleur SE, Gomez F, Houshyar H, Bell ME, Bhatnagar S, Laugero KD, Manalo S (2003) Chronic stress and obesity: a new view of "comfort food." Proc Natl Acad Sci USA 100:11696–11701

14. Delius JD (1988) Preening and associated comfort behavior in birds. Ann N Y Acad Sci 525:40–55

15. Forsyth DA (1994) A comparison of the behavior of resident and intruder glaucous-winged gulls (*Larus glaucescens*) in a colony during territory formation. M.S. thesis, Walla Walla College, College Place, WA

16. Franks EC (1967) The responses of incubating ringed turtle doves (*Streptopelia risoria*) to manipulated egg temperatures. Condor 69:268–276

17. Freeman S, Herron JC (2007) Evolutionary analysis, 4th edn. Pearson/Prentice Hall, Upper Saddle River, NJ

18. Fuller JL, Thompson WR (1960) Behavior genetics, Wiley, New York

19. Galusha JG, Hayward JL (2002) Bald eagle activity at a gull colony and seal rookery on Protection Island, Washington. Northwest Nat 83:23–25

20. Goldstein DL (1988) Estimates of daily energy expenditure in birds: the time-energy budget as an integrator of laboratory and field studies. Am Zool 28:829–844

21. Hayward JL, Galusha JG, Henson SM (2010) Foraging-related activity of bald eagles at a Washington seabird colony and seal rookery. J Raptor Res 44:19–29

22. Henson SM, Galusha JG, Hayward JL, Cushing JM (2007) Modeling territory attendance and preening behavior in a seabird colony as functions of environmental conditions. J Biol Dyn 1:95–107

23. Henson SM, Weldon LM, Hayward JL, Greene DJ, Megna LC, Serem MC (2012) Coping behaviour as an adaptation to stress: post-disturbance preening in colonial seabirds. J Biol Dyn 6:17–37

24. Hinde RA (1970) Animal behaviour: a synthesis of ethology and comparative psychology, 2nd edn. McGraw-Hill, New York

25. Hipfner JM, Blight LK, Lowe, RW, Wilhelm SI, Robertson GJ, Barrett RT, Anker-Nilssen T, Good TP (2012) Unintended consequences: how the recovery of sea eagle *Haliaeetus* spp. populations in the northern hemisphere is affecting seabirds. Marine Ornithology 40:39–52

26. Hosmer DW, Lemeshow S (2000) Applied logistic regression, 2nd edn. Wiley, New York

27. Losito MP, Mirarchi RE, Baldassarre GA (1990) Summertime activity budgets of hatching-year mourning doves. Auk 107:18–24

28. Northam M (2001) Analysis of preening behavior in glaucous-winged gulls (*Larus glaucescens*). M.S. thesis, Walla Walla College, College Place, WA

29. Olango GJ (1998) Time budgets and behavior of glaucous-winged gull (*Larus glaucescens*) chicks in a breeding colony. M.S. thesis, Walla Walla College, College Place, WA

30. Palestis GP, Burger J (1998) Evidence for social facilitation of preening in the common tern. Anim Behav 56:1107–1111

31. Parrish JK, Marvier M, Paine RT (2001) Direct and indirect effects: interactions between bald eagles and common murres. Ecol Appl 11:1858–1869

32. Redpath S (1988) Vigilance levels in preening dunlin *Calidris alpine*. Ibis 130:555–557

33. Reid WV (1988) Population dynamics of the glaucous-winged gull. J Wildl Manag 52:763–770

34. Rowell CHF (1961) Displacement grooming in the chaffinch. Anim Behav 9:38–63

35. Simmons KEL (1985) Comfort behavior. In: Campbell B, Lack E (eds) A dictionary of birds, pp 101–104. T. and Poyser AD, Calton, Staffordshire, Published for the British Ornithologists' Union by Buteo Books, Vermillion, SD

36. Snyder CR (ed) (1999) Coping: the psychology of what works. Oxford University Press, New York

37. Tinbergen N (1989) The study of instinct. Clarendon Press, Oxford.

38. VandenBos GR (2007) APA dictionary of psychology, 1st edn. American Psychological Association, Washington, DC

39. van Iersel JJA, Bol ACA (1958) Preening of two tern species: a study on displacement activities. Behaviour 13:1–88

40. van Rhijn JG (1977a) The patterning of preening and other comfort movements in a herring gull. Behaviour 63:71–109

41. van Rhijn JG (1977b) Processes in feathers caused by bathing in water. Ardea 65:126–147

42. Vincent TL, Brown JS (2005) Evolutionary game theory, natural selection, and darwinian dynamics. Cambridge University Press, Cambridge

43. Weiten W, Lloyd MA (2006) Psychology applied to modern life. Thomson Wadsworth, Belmont, CA

44. White AF, Heath JP, Gisborne B (2006) Seasonal timing of bald eagle attendance and influence on activity budgets of glaucous-winged gulls in Barkley Sound, British Columbia. Waterbirds 29:497–500

45. Zeidner M, Endler NS (eds) (1996) Handbook of coping: theory, research, applications. Wiley, New York

Part II

Reproductive Behavior and Population Dynamics

Nothing is more fundamental to the life of an animal than reproduction. In fact, behavioral ecologists view everything an animal does in terms of its contribution to fitness, the relative capacity to pass one's genes to future generations. Thus, biologists are interested in all aspects of reproductive behavior.

Reproductive behavior occurs within an interplay of environmental variables. Some of these variables foster reproductive output, whereas others are detrimental to it. The relative capacity of an animal to capitalize on positive environmental inputs and to circumvent the negative inputs determines its fitness.

Currently, rapid climate change challenges the reproductive lives of many organisms. Finding adequate habitat in which to bear young, for example, and then finding sufficient food to feed these young strain the behavioral capacities of many species. These difficulties, in turn, impact population trends.

In Chaps. 7–9 of this section, we use mathematics and statistics to explore how climate change may be increasing the incidence of egg cannibalism in a gull breeding colony at Protection Island, Washington, and the unexpected way in which female gulls seem to be counteracting the effects of cannibalism. These chapters provide empirical background for the theoretical models developed in Chaps. 11–13.

In Chap. 10 we look at the impact of bald eagle predation on the population of breeding gulls on Protection Island and the interrelated population dynamics of these two species in the context of a modified Lotka-Volterra predator-prey model.

Chapter 7

Egg Cannibalism in Glaucous-Winged Gulls

Climate change results in a broad range of changes in animals including physiological responses, predator-prey relations, and feeding tactics. In this chapter we explore the impact of changes in sea surface temperature on egg cannibalism, a feeding tactic used by gulls. We use logistic regression and information-theoretic methods of model selection in the assessment of this tactic. This chapter provides background for empirical and theoretical analyses highlighted in much of the remainder of the book. The original publication of this research appeared in The Condor: Ornithological Applications *in 2014, by senior authors James Hayward, Lynelle Weldon, and Shandelle Henson and student authors Libby Megna, Brianna Payne, and Andre Moncrieff [37].*

7.1 Introduction

Cannibalism, or intraspecific predation, involves the killing and eating another member of the same species at any stage of life. Cannibalism is widespread among organisms, having been documented in some 1,300 species, from protozoans to humans [24, 26, 61].

Several ecological and evolutionary consequences result from cannibalism. These include changes in population size, alterations in community energy flow, lowered reproductive success, adjustments in social behavior, kin selection, and nonlinear population dynamics including chaos [5, 8, 9, 16, 20, 23, 31, 41, 46, 72, 79].

Ecological and behavioral factors that contribute to the occurrence of cannibalism include crowding, unusual behavior patterns by vulnerable individuals, stress, victim availability, age and size differences, and genetics [4, 26, 67]. Food shortages and poor food quality, however, are the most reported factors [21]. For example, the cannibalism of eggs and larvae by Peruvian anchovy (*Engraulis ringens*) [2] and Peruvian hake (*Merluccius*

J. M. Cushing et al., *Modeling Behavior and Population Dynamics*, Interdisciplinary Applied Mathematics 57, https://doi.org/10.1007/978-3-031-34283-7_7

gayi peruanus) [27] is enhanced by El Niño-Southern Oscillation (ENSO) events which lead to prolonged increases in sea surface temperature (SST).

Every few years ENSO events create food shortages for marine birds. These events lead to increased SSTs, lowered thermoclines, and weakened upwellings, all of which lower surface water productivity and elicit the movement of forage fish from surface waters to lower, cooler, and more productive levels. Gulls favor fish as a food source but cannot dive, so they are unable to prey on fish at these deeper levels (Fig. 7.1). Consequently, during these times they take longer to forage, obtain lower-quality food, and feed lower-quality food to chicks. All this leads to decreases in reproductive success and population declines [3, 49, 65, 68]. Climate change may be leading to these same ecological conditions, except over a longer time scale [30].

Many studies have associated ENSO events with reduced reproductive success in marine birds [1, 70, 75], but our study is the first to link ENSO events with cannibalism in birds. In this study we test the null hypothesis that ENSO-related rises in SST were unrelated to egg cannibalism and reproductive success in glaucous-winged gulls (*Larus glaucescens*) nesting on Protection Island National Wildlife Refuge, Washington, from 2006 to 2011.

Figure 7.1: A primary food source for gulls is small fish. When sea surface temperatures are high, as during El Niño-Southern Oscillation (ENSO) events, plankton and fish move lower in the water column. Gulls are too buoyant to dive so they have difficulty finding enough fish. Egg cannibalism on the Protection Island gull colony increases during these times (Photo by J. L. Hayward)

7.2 Methods

7.2.1 Study Site and Data Collection

Data were collected from 2006 to 2011 on Protection Island National Wildlife Refuge ($48°07'40''N, 122°55'3''W$), Jefferson County, Washington, located at the southeast corner of the Strait of Juan de Fuca. Violet Point, a gravel spit that extends from the main island to the southeast, supported a breeding colony of several thousand glaucous-winged gulls during our study.

We selected five rectangular sample plots (Chap. 6, Fig. 6.3, plots A–E), with dimensions that varied to include most of the nests aggregated within specific areas of the colony. The five plots collectively covered 4,205 m^2 and contained from 199 to 267 nests during the six years of study.

Each afternoon, from late May to mid-July, we monitored all nests in each sample plot. At the time of clutch initiation, we marked each nest with a numbered, wooden stake placed within 1 m of the nest but positioned to minimize interference with the movements of territory occupants. Staked nests were checked daily for new eggs. Each new egg was marked with permanent ink at its apex to indicate its position in the laying sequence, and each egg was monitored daily until it could be recorded as cannibalized, eagle depredated, addled, dead during pipping, hatched, punctured, flooded, or rolled from the nest.

We measured the distance of each nest center to the nest center of its "nearest neighbor" [57] using a tape measure or laser rule (2006 to 2010), or a Trimble GPS and ArcGIS Desktop 10 with the "pointdistances" tool in Geospatial Modeling Environment software (2011). Nest habitat was recorded (A) as short, sparse, or lacking vegetation, (B) by a shrub (usually gumweed, *Grindelia integrifolia*) or log not on the beach, (C) on the beach, or (D) beside or within American dune grass (*Leymus mollis*). The mass of each egg was determined in 2007 to 2011 on the day it was laid using a 400-g capacity Ohaus Scout Pro SP401 portable electronic balance.

The mean SST for September to May before each reproductive season was determined from National Oceanographic and Atmospheric Administration (NOAA) data recorded by its Port Townsend, Washington, buoy (PTWW1), 12 km east of the gull colony. We used the September to May average for the following reasons: (A) breeding success in gulls is dependent in part on resource availability prior to egg-production [6, 7, 42, 48], and (B) a time lag exists between changes in the physical environment and responses by seabirds to those changes, in part due to bottom-up control in plankton-based food webs [30, 78]. The duration of the time lag could be days or decades, depending on the system [7]. Gulls exit Protection Island in late September and October, return in February and March, and begin egg production in late May [36, 40]. We had no specific data concerning the time lag for this colony, but we considered the average SST from September to May a reasonable estimate of the pre-breeding oceanic conditions impacting the local marine food web [62].

The 2007 and 2010 breeding seasons (Table 7.1) were preceded by warm ENSO conditions. Local SSTs were also higher prior to the 2006 breeding season, even though this period was not officially considered an El Niño event by NOAA's Oceanic Niño Index [52].

Table 7.1: Mean laying dates, sea surface temperatures (SSTs), and fates of gull eggs at the Protection Island breeding colony, 2006–2011. SSTs are averages from September to May prior to egg laying. Percentages of year totals are shown in parentheses

Year	Mean laying day	SST (C°)	Cannib	Depred	Addled	Died pipping	Other	Hatched	N
2006	June 16	9.12	96 (20.6)	31 (6.7)	24 (5.2)	0 (0)	18 (3.9)	297 (63.7)	466
2007	June 12	9.18	200 (41.8)	24 (5.0)	24 (5.0)	3 (0.6)	16 (3.3)	211 (44.1)	478
2008	June 17	8.61	94 (16.8)	11 (2.0)	48 (8.6)	2 (0.4)	27 (4.8)	378 (67.5)	560
2009	June 9	8.68	93 (14.4)	65 (10.0)	48 (7.4)	6 (0.9)	1 (0.0)	436 (67.3)	649
2010	June 13	9.33	144 (22.5)	100 (15.6)	35 (5.5)	6 (0.9)	19 (3.0)	335 (52.4)	639
2011	June 9	8.78	100 (16.5)	11 (1.8)	40 (6.6)	9 (1.5)	17 (2.8)	429 (70.8)	606

7.2.2 Identification of Egg Predators

Bald eagles (*Haliaeetus leucocephalus*), cannibalistic gulls, and northwestern crows (*Corvus brachyrhynchos caurinus*) are the only Protection Island predators known to feed on gull eggs. When bald eagles prey on a nest, they usually eat all the eggs; sometimes they destroy the eggs in all the nests within a large area of the colony [35, 71]. Typically, areas around eagle-predated nests contain the fragments of eggshells left behind and sometimes even the remains of adult gulls. Gull egg predators, in contrast, often fly to nests not their own, grab a single egg, and fly the egg back to their own territory where it is devoured (Fig. 7.2). Thus, gull-depredated nests typically lose only a single egg during a predation event; moreover, eggshell fragments are not scattered around the depredated nest but are concentrated in the territory of the egg cannibal [38] (Fig. 7.3). Northwestern crows also remove single eggs from gull nests and fly them to another spot before feeding on them [75, 76]. During thousands of hours of observation of the Violet Point colony, however, we never observed a crow preying on an egg, leading us to believe that crow predation of gull eggs is a relatively rare event on this colony. This conclusion comports with the results of an experiment with herring gulls (*L. argentatus*) in which 44 or 45 eggs placed in artificial nests were taken by gulls, and only one egg was taken by a crow [11]. It also is consistent with observations by Good [29] on a glaucous-winged x western gull (*L. occidentalis*) colony who reported extensive depredation of gull eggs by conspecifics. All this evidence led us to consider that losses of single gull eggs on the Violet Point colony were due to egg cannibals, even though the possibility exists that eagles and crows took small numbers of these eggs.

Figure 7.2: Two eggs nearly fulfill the daily energy requirement of one adult glaucous-winged gull. During the egg-laying and incubation period on the Protection Island gull colony, some egg cannibals take more than two eggs per day. (**a**) This egg cannibal has just stolen an egg. (**b**) After stealing an egg, the cannibal flies it back to his territory where it is broken open and eaten (Photos by J. L. Hayward)

7.2.3 Logistic Regression Analysis

We used logistic regression [43] to analyze the binary outcomes of cannibalism (0 not cannibalized, 1 cannibalized) and hatching success (0 not hatched, 1 hatched) of 2,932 eggs monitored from 2007 to 2011. Cannibalism and hatching success were considered functions of SST, egg mass (MASS), nearest-neighbor distance (NN), number of days before or after the mean laying date (Table 7.1) for the season (DAYS), sample area (PLOT), habitat type (HAB), egg order (ORDER), and total number of eggs laid in the nest (CSIZE). The log-odds of cannibalism and hatching were regressed on these eight variables and on 17 interaction terms between PLOT and HAB (three of the 20 possible combinations did not occur, Table 7.2) with an intercept term. The 2006 data were not included in this part of the analysis because we did not measure egg masses that year.

We obtained a suite of alternative models for cannibalism and separately for hatching success by taking all possible linear combinations of the eight factors, with intercepts. For those models with both PLOT and HAB variables, interaction terms were included. We used an information-theoretic approach for parameter estimation and model selection [15]. Specifically, we determined parameter estimates and computed the Akaike information criterion (AIC) for the global model and all submodels. Using Akaike weights [15], we ranked SST, MASS, NN, DAYS, ORDER, CSIZE, PLOT, and HAB in order of importance. We also used Akaike weights to compute a weighted mean for each parameter estimate over all the models containing that parameter. Model-averaged parameter estimates are more stable than estimates associated with any one model, for they account for the uncertainty in model selection [15]. We used model-averaged parameter estimates to compute odds ratios. See Chap. 1 for details on Akaike weights, factor importance, and model-averaged parameter estimates.

Figure 7.3: The territories of glaucous-winged gull egg cannibals are littered with large numbers of eggshell fragments, as on this territory on Protection Island National Wildlife Refuge, Washington. Only male gulls are known to steal eggs (Photo by J. L. Hayward)

Table 7.2: Number of eggs by plot and habitat type, 2007–2011

Plot	Short/sparse vegetation	Beside shrub/log	Beach	Beside/in tall grass	N
A	14	107	14	277	412
B	153	150	0	179	482
C	340	216	437	14	1,007
D	202	239	91	0	532
E	239	214	46	0	499
Total	948	926	588	470	2,932

7.2.4 Model Validation

We utilized the AIC to choose the best cannibalism and hatching success models from the respective sets of parameterized alternatives. In this way we were able to test the capability of model prediction. We followed the procedure recommended by [28] for logistic regression models to measure goodness-of-fit and to validate the selected models. Given that MASS appeared in none of the best cannibalism or hatching success models (see Results), we incorporated the 2006 egg data, for which MASS was not available, to obtain a larger data set ($N = 3,398$ for 2006–2011) for validation. This procedure increased the power of hypothesis tests.

The 2006–2011 egg data were split into two random samples: 75% for parameter estimation ($N = 2,548$) and 25% for model validation ($N = 850$). The selected models were fitted to the larger sample followed by measurement of model performance on both samples. Utilizing a larger sample for fitting helps to avoid difficulties with quasi-complete variable separation. The data-splitting process was repeated 100 times. Details on measurement of model performance are in [28, 37].

7.3 Results

7.3.1 Egg Fates

The most frequent cause of egg loss during all years was cannibalism, followed by eagle predation, addling, death during pipping, and other causes. The highest rate of cannibalism occurred in 2007, which also was the only year in which hatching success was less than 50%. The year 2010 also experienced a high rate of cannibalism and a low rate of hatching success, and these two breeding seasons followed nine months of higher-than-average ENSO-related SSTs. The relatively high rate of cannibalism also occurred in 2006, which also followed nine months of higher-than-average SSTs (Table 7.1).

7.3.2 AIC, Factor Importance, and Odds Ratios

The factors used in the best hatching and cannibalism models, as well as the factors involved in the factor importance analysis, were almost identical (PLOT, HAB, CSIZE, DAYS, and SST) but with opposite effects (Tables 7.3, 7.4, 7.5, and 7.6). The cannibalism model with $\Delta\text{AIC} = 0$ for 2007–2011 contained all the factors except ORDER, NN, and MASS. The hatching success model with $\Delta\text{AIC} = 0$ for those years contained all the factors except ORDER. The model that omitted ORDER and MASS, however, had one less parameter and yielded similar results ($\Delta\text{AIC} = 0.87$). For comparative purposes, we list all models with any support ($\Delta\text{AIC} < 10$) in Table 7.3.

Overdispersion parameter estimates from the global models (1.048 for cannibalism and 1.015 for hatching success) were well within the range of values (1 to 4) that indicate a small amount of extrabinomial variation due to lack of independence and demographic

Table 7.3: Logistic regression models of cannibalism and hatching success that have any support from the data ($\Delta QAIC_c<10$) with Akaike weights (w_i), the number of parameters including σ^2 (κ), and the log likelihood (L). The $\Delta QAIC_c$ is the second order quasi-likelihood AIC. The global model includes the variables SST, MASS, NN, DAYS, ORDER, CSIZE, PLOT, and HAB, as well as the interaction terms between PLOT and HAB. (Note: w/o = without)

Model	$\Delta QAIC_c$	w_i	κ	L
Cannibalism				
w/o ORDER, NN, MASS	0.00	0.22	22	−1248
w/o NN	0.12	0.20	25	−1245
w/o NN, MASS	0.26	0.19	24	−1246
w/o ORDER, NN	1.10	0.12	23	−1248
w/o ORDER, MASS	2.01	0.08	23	−1248
Global model	2.12	0.07	26	−1245
w/o MASS	2.27	0.07	25	−1246
w/o ORDER	3.10	0.05	24	−1248
Hatching Success				
w/o ORDER	0.00	0.41	24	−1736
w/o ORDER, MASS	0.87	0.26	23	−1738
Global model	2.56	0.11	26	−1736
w/o ORDER, NN	3.24	0.08	23	−1739
w/o MASS	3.97	0.05	25	−1737
w/o ORDER, NN, MASS	4.20	0.05	22	−1741
w/o NN	5.78	0.02	25	−1738
w/o NN, MASS	7.30	0.01	24	−1740

stochasticity rather than poor model structure [15]. Tables 7.4 and 7.5 display odds ratios and 95% confidence intervals.

For cannibalism, relative factor importance based on Akaike weights showed that factors PLOT, HAB, CSIZE, DAYS, and SST were important, ORDER and MASS were somewhat unimportant, and NN was unimportant. For hatching success, PLOT, HAB, CSIZE, SST, and DAYS were important, NN was somewhat important, MASS was somewhat unimportant, and ORDER was unimportant (Table 7.6).

Odds ratios demonstrated the effects of the various factors on cannibalism (Table 7.4) and hatching success (Table 7.5). *The odds of cannibalism increased in response to increases in average SST prior to egg laying, whereas the odds of hatching success decreased under the same conditions.* When keeping all other factors constant, the odds of cannibalism

Table 7.4: Cannibalism, odds ratios (OR) with 95% confidence intervals (CI) associated with a c unit increase in the factor or, in the case of design variables, relative to the given reference variable. Habitats are short or sparse vegetations (SV), beside shrub or log (SL), beach (BC), and beside or in tall grass (TG). An asterisk (*) indicates the effect of the variables is significant

Factor	c/reference	OR	95% CI
MASS	5 grams	0.96	(0.89, 1.03)
SST*	0.1 deg	1.10	(1.06, 1.13)
NN	100 cm	0.99	(0.92, 1.07)
DAYS*	1 day	1.09	(1.06, 1.11)
CSIZE	1 egg		
2 eggs*		0.13	(0.07, 0.21)
>2 eggs*		0.09	(0.05, 0.15)
ORDER	1st egg		
2nd egg		0.81	(0.64, 1.03)
3rd egg		0.77	(0.57, 1.03)
PLOT A	HAB SV		
HAB SL		0.90	(0.14, 5.92)
HAB BC		2.05	(0.20, 20.90)
HAB TG		1.20	(0.20, 7.04)
PLOT B	HAB SV		
HAB SL*		0.28	(0.23, 0.33)
HAB TG*		0.06	(0.04, 0.09)
PLOT C	HAB SV		
HAB SL*		0.54	(0.45, 0.63)
HAB BC*		0.68	(0.62, 0.75)
HAB TG		0.79	(0.30, 2.11)
PLOT D	HAB SV		
HAB SL*		0.28	(0.26, 0.32)
HAB BC*		0.40	(0.34, 0.48)
PLOT E	HAB SV		
HAB SL*		0.39	(0.35, 0.42)
HAB BC*		0.68	(0.54, 0.87)

increased by 10% for each increase of 0.1°C. Higher odds of hatching success and lower odds of cannibalism were associated with larger clutch sizes: an egg from a three-egg clutch was 91% less likely to be cannibalized and more than seven times more likely to hatch than an egg from a one-egg clutch; an egg from a two-egg clutch was 87% less likely to be cannibalized and more than six times more likely to hatch than an egg from a one-egg

Table 7.5: Hatching success odds ratios (ORs) with 95% confidence intervals (CIs) associated with a c unit increase in the factor or, in the case of design variables, relative to the given reference variable. Habitats are short or sparse vegetations (SV), beside shrub or log (SL), beach (BC), and beside or in tall grass (TG). An asterisk (*) indicates the effect of the variables is significant

Factor	c/reference	OR	95% CI
MASS	5 grams	1.05	(0.99, 1.11)
SST*	0.1 deg	0.90	(0.88, 0.93)
NN*	100 cm	1.07	(1.01, 1.13)
DAYS*	1 day	0.93	(0.90, 0.95)
CSIZE	1 egg		
2 eggs*		6.11	(3.51, 10.63)
>2 eggs*		7.38	(4.31, 12.65)
ORDER	1st egg		
2nd egg		1.11	(0.92, 1.35)
3rd egg		1.09	(0.86, 1.38)
PLOT A	HAB SV		
HAB SL		0.27	(0.05, 1.50)
HAB BC		0.15	(0.02, 1.07)
HAB TG*		0.13	(0.02, 0.69)
PLOT B	HAB SV		
HAB SL*		2.82	(2.46, 3.24)
HAB TG*		2.17	(1.93, 2.44)
PLOT C	HAB SV		
HAB SL*		1.99	(1.80, 2.20)
HAB BC*		0.77	(0.73, 0.81)
HAB TG		0.95	(0.49, 1.86)
PLOT D	HAB SV		
HAB SL*		3.30	(3.02, 3.60)
HAB BC		0.96	(0.84, 1.11)
PLOT E	HAB SV		
HAB SL*		2.19	(2.02, 2.38)
HAB BC*		0.62	(0.48, 0.80)

clutch. An egg laid near to the average laying date was less likely to be cannibalized and more likely to hatch, and every day further removed from this date increased the odds of cannibalism by 9% and decreased the odds of hatching by 7%. The odds of hatching

Table 7.6: Relative factor importance for cannibalism and hatching success

Factor	Importance
Cannibalism	
PLOT	1.00
HAB	1.00
CSIZE	1.00
DAYS	1.00
SST	1.00
ORDER	0.54
MASS	0.45
NN	0.27
Hatching success	
PLOT	1.00
HAB	1.00
CSIZE	1.00
SST	1.00
DAYS	1.00
NN	0.84
MASS	0.62
ORDER	0.20

increased for an egg by approximately 7% with each increase of 1 m away from its nest from the nearest neighbor, although the odds of cannibalism were unaffected.

Odds ratios also demonstrated the effect of habitat on cannibalism (Table 7.4) and hatching success (Table 7.5) in the five plots. Eggs laid in non-beach areas adjacent to a shrub or log exhibited the lowest odds of being cannibalized except in plots A and B. In plot A, the decreased odds of cannibalism were insignificant, whereas in plot B eggs in tall grass were even less likely to be cannibalized. An egg in a nest adjacent to a shrub or log exhibited the highest odds of hatching, except in plot A, where there was a nonsignificant decrease in the odds of hatching. Beach habitat was the next best habitat for avoiding cannibalism, except in plot B where there was no beach, and in plot A where no habitat was significantly superior or inferior. Beach habitat was always superior to sparse vegetation habitat. Next to shrub or log habitat, sparse vegetation exhibited the best hatching success, except in plot B where tall grass habitat was better. In plots C and E, beach habitat was significantly inferior for hatching success, and in plot A tall grass was disadvantageous.

7.3.3 Model Validation

We carried out validation procedures on one cannibalism model and on one hatching success model. For cannibalism, the best model did not contain ORDER, NN, and MASS (ΔAIC= 0; Table 7.3). For hatching success, the best model did not contain ORDER and MASS (ΔAIC= 0.87 and one less parameter than the model with $\Delta\text{AIC} = 0$; Table 7.3). None of the models included MASS; thus, we combined the 2006 data (which did not include measurements of mass) with the 2007–2011 data for model validation. The two models performed well with good parameter stability.

Note that the odds ratios and factor importance analysis were based on model averaging over all 2007–2011 data to reflect the still-present level of model uncertainty. By contrast, the model performance analysis for validation and prediction was based on the selected, individually parameterized models.

7.4 Discussion

Our results from a colony of glaucous-winged gulls demonstrate a positive relationship between local SST and egg cannibalism and a negative relationship between local SST and hatching success. Nest location, habitat type, clutch size, and number of days to or from the average hatching date are also influential.

The fitness of individuals and the dynamics of populations are directly influenced by cannibalism [21]. Field data and modeling studies demonstrate that cannibalism can affect population dynamics in ways that include cycles and chaotic fluctuations [18, 33, 50, 54]. Moreover, cannibalism of juveniles by adults can allow populations to survive times of resource scarcity by redirecting reproductive effort to better times [24]. Mathematical models demonstrate that cannibalism can function as a “lifeboat” mechanism when resources for adults are scarce and juveniles consume resources inaccessible to adults [17, 73]. Modeling also demonstrates that cannibalism can allow a population to survive times of low resource availability even if adults and juveniles utilize the same resource [39].

Adult gulls commonly engage in cannibalism by stealing eggs and chicks from other colony residents [11, 20, 22, 55–58, 72]. According to Stanback and Koenig [67], nutritional benefit is a motivating factor for egg and chick cannibalism, although they did not link this behavior directly to low food supply. Fordham and Cormack [25] found that when food supplies were low, adult kelp gulls (*L. dominicanus*) killed more chicks. Hunt and McLoon [44] suggested that increased infanticide was due to the tendency of chicks to wander into neighboring territories when they were unable to obtain food from their parents. To our knowledge, our study is the first to link avian egg cannibalism to low food supply.

The 2007 and 2010 nesting seasons were preceded by higher-than-average SSTs linked with ENSO episodes; these two years also coincided with the highest percentages of egg cannibalism. Although 2006 was not designated as an official ENSO episode, SSTs were higher than average and the 2006 breeding season experienced a moderately high level of egg cannibalism. Eggs provide locally available, low-cost energy sources for food-stressed gulls. The average mass of a glaucous-winged gull egg on this colony was 93.6 g ($N = 2,932$ eggs weighed 2007–2011). Assuming an energy content of 6.7 kJ/g for whole gull eggs [64], each cannibalized egg contained an estimated 627 kJ of food energy. Based on estimates for similarly sized herring gulls [53], this amount of energy is more than half the daily energy requirement of an adult gull.

Logistic regression analysis of our data shows that a 0.1°C *rise in SST raised the odds of cannibalism by* 10%. From 1950 to 1998, sea surface temperatures in the Strait of Juan de Fuca increased about 1°C [69]. This region has warmed at a higher rate than the global warming trend; predictions suggest this trend will continue [66]. A 1°C rise in average SST is beyond the range of variability in our data set, but extrapolation from the parameterized model yields an odds ratio of 2.50 for this increase. This suggests that this level of increase would increase the odds that a given egg would be cannibalized by 150%.

Few data have been available in the literature on the rate of egg cannibalism among gulls. An exception are data from Mandarte Island, British Columbia. Vermeer [77] noted that 5.0% (83 of 1,652) sample glaucous-winged gull eggs were lost or eaten in 1961 and 1962, and Verbeek [74] at the same colony noted that 5.6% (26 of 466) sample eggs were lost or eaten in 1979 and 1980. By contrast, from 2006 to 2011, 28.5% (967 of 3,397) sample eggs were lost or eaten during our Protection Island study, including 240 lost eggs attributed to eagle predation (Table 7.1). It is not certain, however, that the greatly increased loss on Protection Island was due to rising SST levels.

Eggs laid early or late in relation to the average laying date were less likely to hatch and more likely to be cannibalized, a finding consistent with Darling's hypothesis that synchronized egg laying satiates predators [19]. Moreover, younger parents on this colony tend to lay later in the season with eggs that experience lower hatching success [63].

Our study found that eggs in larger clutches exhibited lower odds of cannibalism and higher hatching success, a finding consistent with an earlier study of this species [51]. Reid [63] showed that older, more experienced parents produced slightly larger clutch sizes on average than younger birds, so possibly these effects are due to better parenting. Alternately, the advantage associated with larger clutch sizes may be due to differential predation rates on eggs in nests with different numbers of eggs. If a cannibal, for example, randomly chooses a nest to predate, regardless of clutch size, an egg in a nest with a larger clutch size has less chance of being taken. Assuming this second possibility, two facts must be noted. First, the clutch size factor was measured as the number of eggs laid in the nest, not how many eggs were in the nest when it was raided. Second, the egg, not the nest, was the experimental unit; if all factors were held constant, each egg had an equal chance of being cannibalized. We suspect the benefit of larger clutch size is due to better parenting, because cannibals tend to take eggs opportunistically from poorly guarded nests, especially during disturbances. Regardless, all confounding effects leading to the nonindependence of eggs in the same clutch were addressed by the overdispersion analysis. A relatively small effect from a lack of independence (see Results) was indicated by overdispersion parameter estimates.

Nest habitat strongly affected the rate of cannibalism and hatching success. Logs, shrubs, and other landscape features offered shading and wind protection to eggs, which enhanced hatching success [59, 60]. Eggs laid in nests adjacent to these features were also less exposed visually than those laid in more open habitats [10, 29], and these features inhibit rapid movement by cannibals trying to avoid being attacked by territory owners. Plot B contained more tall grass habitat than any of the plots. Eggs in nests situated at the ends of grass tunnels or located at the edge of this habitat were less likely to be cannibalized than eggs in other habitats (see also [12–14, 29, 34, 38]), but these eggs were more vulnerable to bald eagles which were attracted to nests in this habitat presumably because the tall grass protected their heads from diving gulls [35].

Even when we controlled for habitat types, we found that location of a plot within the colony was an important factor. Complex interactions between number of eagle disturbances and proportions of habitat types may have played a role. Moreover, access to the beach and water may have played a part; closely related western gulls (*L. occidentalis*) that nest close to water were able to quickly drink and bathe without leaving eggs and chicks exposed to harm for long periods [32].

The highest fraction of egg deaths during each of the 6 years of this study was from cannibalism (Table 7.1). This was true even during years of minimal food stress, but food stress increased the frequency of this behavior. Gulls nesting in the colony exhibit individualistic feeding preferences: some birds specialized on cockles, others on blue mussels, and yet others on crustaceans. A study of regurgitated gull boli on the colony indicated that landfill products occupied only a small part of the diets of these birds [47]. Fewer than 1% of colony residents specialized on cannibalizing eggs from other residents, and the territories of these cannibals are easily identified because they are littered with fragments of eggshell [38]. The quantity of eggshell on individual cannibal territories was not higher during years of high levels of cannibalism, so it appears that more gulls engaged in cannibalism during these stressful years instead of individual cannibals taking more eggs during those years. Indeed, a well-substantiated prediction of foraging theory states that consumers broaden their dietary intake when food is scarce [21]. Gulls cannibalize eggs opportunistically [11, 29], so increases in egg cannibalism may be due in part to a reduction in territory defense by gulls on longer foraging trips off the colony when stressed by food.

Gulls that nest in the marine environment function as important indicator species [4, 45]. We do not know if higher SSTs are associated with higher levels of cannibalism in other marine birds. Also, we do not know how higher cannibalism rates in gulls will influence their overall population dynamics. Gulls are multi-year breeders and often have more than one opportunity to breed. The possibility exists, however, that persistently high levels of egg cannibalism, without range shifts or other compensatory responses, could lead to population declines.

References

1. Ainley DG, Carter HR, Anderson DW, Briggs KT, Coulter MC, Cruz F, Cruz JB, Valle CA, Fefer SI, Hatch SA, Schreiber EA, Schreiber RW, Smith NG (1986) Effects of the 1982–83 El Niño-Southern Oscillation on Pacific Ocean bird populations. Acta XIX Congressus Internationalis Ornithologici 2:1747–1758

2. Alheit J, Niquen M (2004) Regime shifts in the Humboldt Current ecosystem. Prog. Oceanogr 60:201–222

3. Barber RT, Chavez FP (1983) Biological consequences of El Niño. Science 222:1203–1210

4. Baur B, Baur A (1986) Proximate factors influencing egg cannibalism in the land snail *Arianta arbustorum* (Pulmonata, Helicidae). Oecologia 70:283–287

5. Beaver R (1974) Intraspecific competition among bark beetle larvae (Coleoptera: Scolytidae). J Anim Ecol 43:455–467

6. Blight LK (2011) Egg production in a coastal seabird, the glaucous-winged gull (*Larus glaucescens*), declines during the last century. PLoS One 6(7):22027

7. Boersma PD (1998) Population trends of the Galápagos penguin: impacts of El Niño and La Niña. Condor 100:245–253

8. Brouwer A, Spaans AL (1994) Egg predation in the herring gull *Larus argentatus*: why does it vary so much between nests? Ardea 82:223–231

9. Brower LP (1961) Experimental analysis of egg cannibalism in the monarch and queen butterflies, *Danaus plexippus* and *D. gilippus berenice*. Physiol Zool 34:287–296

10. Burger J (1972) The use of a fish-eye lens to study nest placement in Franklin's gulls. Ecology 53:262–264

11. Burger J (1980) Territory size differences in relation to reproductive stage and type of intruder in herring gulls (*Larus argentatus*). Auk 97:733–741

12. Burger J, Gochfeld M (1981) Nest site selection by kelp gulls in southern Africa. Condor 83:243–251

13. Burger J, Gochfeld M (1985) Nest site selection by laughing gulls (*Larus atricilla*): comparison of tropical colonies (Culebra, Puerto Rico) with temperate New Jersey colonies. Condor 87:364–373

14. Burger J, Shisler J (1978) Nest-site selection and competitive interactions of herring and laughing gulls in New Jersey. Auk 95:252–266

15. Burnham KP, Anderson DR (2002) Model selection and multi-model inference: a practical information-theoretic approach, 2nd edn. Springer, New York

16. Bygott JD (1972) Cannibalism among wild chimpanzees. Nature 238:410–411

17. Cushing JM (1991) A simple model of cannibalism. Math Biosci 107:47–71

18. Cushing JM, Costantino RF, Dennis B, Desharnais RA, Henson SM (2002) Chaos in ecology: experimental nonlinear dynamics. Academic Press, New York

19. Darling F (1938) Bird flocks and the breeding cycle. Cambridge University Press, Cambridge

20. Davis JWF, Dunn EK (1976) Intraspecific predation and colonial breeding in lesser black-backed gulls *Larus fuscus*. Ibis 118:65–77

21. Dong Q, Polis GA (1992) The dynamics of cannibalistic populations: a foraging perspective. In: Elgar MA, Crespi BJ (eds) Cannibalism: ecology and evolution among diverse taxa. Oxford University Press, Oxford, pp 13–37

22. Drent RH (1970) Functional aspects of incubation in the herring gull. In: Baerends GP, Drent RH (eds) The herring gull and its egg. E. J. Brill, Leiden, The Netherlands, pp 1–125

23. Eickwort K (1973) Cannibalism and kin selection in *Labidomera clivicollis* (Coleoptera: Chrysomelidae). Am Nat 107:452–453

24. Elgar MA, Crespi BJ (1992) Ecology and evolution of cannibalism. In: Elgar MA, Crespi BJ (eds) Cannibalism: ecology and evolution among diverse taxa. Oxford University Press, Oxford, pp 1–12

25. Fordham RA, Cormack RM (1970) Mortality and population change of Dominican gulls in Wellington, New Zealand: with a statistical appendix. J Anim Ecol 39:13–27

26. Fox LR (1975) Cannibalism in natural populations. Annual Review of Ecology and Systematics 6:87–106

27. Geuvara-Carrasco R, Lleonart J (2008) Dynamics and fishery of the Peruvian hake: between nature and man. J Mar Syst 71:249–259

28. Giancristofaro RA, Salmaso L (2003) Model performance analysis and model validation in logistic regression. Statistica 63:375–396

29. Good TP (2002) Breeding success in the western gull x glaucous-winged gull complex: the influence of habitat and nest-site characteristics. Condor 104:353–365

30. Grémillet D, Boulineir T (2009) Spatial ecology and conservation of seabirds facing global climate change: a review. Mar Ecol Prog Ser 391:121–137

31. Hamai M, Nishida T, Takasaki H, Turner LA (1992) New records of within-group infanticide and cannibalism in wild chimpanzees. Primates 33:151–162

32. Hand JL, Hunt GL, Warner M (1981) Thermal stress and predation: influences on the structure of a gull colony and possibly on breeding distribution. Condor 83:193–203

33. Hastings A, Costantino RF (1987) Cannibalistic egg-larva interactions in *Tribolium*: an explanation for the oscillations in population numbers. Am Nat 130:36–52

34. Hayward JL (1993) Nest-site selection and reproductive success of ring-billed gulls at Sprague Lake, Washington. Northwest Nat 74:67–76

35. Hayward JL, Galusha JG, Henson SM (2010) Foraging-related activity of bald eagles at a Washington seabird colony and seal rookery. J Raptor Res 44:19–29

36. Hayward JL, Henson SM, Bové J, Bové C, Gregory CJ (2017) Daily and annual habitat use and habitat-to-habitat movement by glaucous-winged gulls at Protection Island, Washington. Northwest Nat 98:180–189

37. Hayward JL, Weldon LM, Henson SM, Megna LC, Payne BG, Moncrieff AE (2014) Egg cannibalism in a gull colony increases with sea surface temperature. The Condor Ornithol Appl 116:62–73

38. Hayward JL, Zelenitsky DK, Smith DL, Zaft DM, Clayburn JK (2000) Eggshell taphonomy at modern gull colonies and a dinosaur clutch site. Palaios 15:343–355

39. Henson SM (1997) Cannibalism can be beneficial even when its mean yield is less than one. Theor Popul Biol 51:109–117

40. Henson SM, Hayward JL, Cushing JM, Galusha JG (2010) Socially induced synchronization of every-other-day egg laying in a seabird colony. Auk 127:571–580

41. Hiraiwa-Hasegawa M (1992) Cannibalism among non-human primates. In: Elgar MA, Crespi BJ (eds) Cannibalism: ecology and evolution among diverse taxa. Oxford University Press, Oxford, pp 323–338

42. Hodder J, Graybill MR (1985) Reproduction and survival of seabirds in Oregon during the 1982–1983 El Niño. Condor 87:535–541

43. Hosmer DW, Lemeshow S (2000) Applied logistic regression. 3rd edn. Wiley, New York

44. Hunt GL, McLoon SC (1975) Activity patterns of gull chicks in relation to feeding by parents: their potential significance for density-dependent mortality. Auk 92:523–527

45. Kershner J, Samhouri JF, James CA, Levin PS (2011) Selecting indicator portfolios for marine species and food webs: a Puget Sound case study. PLoS One 6(10):e25248

46. Kirkpatrick TW (1957) Insect life in the tropics. Longmans, London

47. Lindborg VA, Ledbetter JF, Walat JM, Moffett C (2012) Plastic consumption and diet of glaucous-winged gulls (*Larus glaucescens*). Mar Pollut Bull 64:2351–2356

48. Marra PP, Hobson KA, Holmes RT (1998) Linking winter and summer events in a migratory birds by using stable carbon isotopes. Science 282:1884–1886

49. McGowan JA, Cayan DR, Dorman LM (1998) Climate-ocean variability and ecosystem response in the northeast Pacific. Science 281:210–217

50. Mertz DB, Davies RB (1968) Cannibalism of the pupal stage by adult flour beetles: an experiment and a stochastic model. Biometrics 24:247–275

51. Murphy EC, Hoover-Miller AA, Day RH, Oakley KL (1992) Intracolony variability during periods of poor reproductive performance at a glaucous-winged gull colony. Condor 94:598–607

52. National Oceanographic and Atmospheric Administration (2012) Cold and warm episodes by season: changes to the Oceanic Niño Index (ONI). www.cpc.ncep.noaa.gov/products/analysis_monitoring/ensostuff/ensoyears.shtml

53. Norstrom RJ, Clark TP, Kearney JP, Gilman AP (1986) Herring gull energy requirements and body constituents in the Great Lakes. Ardea 74:1–23

54. Orr BK, Murdoch WW, Bence JR (1990) Population regulation, convergence, and cannibalism in *Notonecta* (Hemiptera). Ecology 71:68–82

55. Parsons J (1971) Cannibalism in herring gulls. British Birds 64:228–237

56. Parsons J (1975) Seasonal variation in the breeding success of the herring gull: an experimental approach to pre-fledging success. J Anim Ecol 44:553–573

57. Patterson IJ (1965) Timing and spacing of broods in the black-headed gull *Larus ridibundus*. Ibis 107:433–459

58. Paynter RA, Jr (1949) Clutch size and the egg and chick mortality of Kent Island herring gulls. Ecology 30:146–166

59. Pierotti R (1982) Habitat selection and its effect on reproductive output in the herring gull in Newfoundland. Ecology 63:854–868

60. Pierotti RJ, Good TP (1994) Herring gull (*Larus argentatus*). In: Poole A (ed) The birds of North America online. Cornell Lab of Orthithology, Ithaca. http://bna.birds.cornell.edu.bnaproxy.birds.cornell.edu/bna/species/124

61. Polis GA (1981) The evolution and dynamics of intraspecific predation. Annual Review of Ecology and Systematics 12:225–251

62. Ramos JA, Maul AM, Ayrton V, Bullock I, Hunter J, Bowler J, Castle G, Mileto R, Pacheco C (2002) Influence of local and large-scale weather events and timing of breeding on tropical roseate tern reproductive parameters. Mar Ecol Prog Ser 243:271–279

63. Reid WV (1988) Age-specific patterns of reproduction in the glaucous-winged gull: increased effort with age? Ecology 69:1454–1465

64. Ricklefs RE (1977) Composition of eggs of several bird species. Auk 94:350–356

65. Schreiber RW, Schreiber EA (1984) Central Pacific seabirds and the El Niño Southern Oscillation: 1982 to 1983 perspectives. Science 225:713–716

66. Snover AK, Mote PW, Whitelly Binder L, Hamelet AF, Mantua NJ (2005) Uncertain future: climate change and its effects on puget sound. In: A Report for the Puget Sound Action Team by the Climate Impacts Group, Center for Science in the Earth System, Joint Institute for the Study of the Atmosphere and Oceans. University of Washington, Seattle

67. Stanback MT, Koenig WD (1992) Cannibalism in birds. In: Elgar MA, Crespi BJ (eds) Cannibalism: ecology and evolution among diverse taxa. Oxford University, Oxford, pp 277–298

68. Stenseth NC, Mysterud A, Ottersen G, Hurrell JW, Chan K-S, Lima M (2002) Ecological effects of climate fluctuations. Science 297:1292–1296

69. Strom A, Francis RC, Mantua NJ, Miles EL, Peterson DL (2004) North Pacific climate recorded in growth rings of geoduck clams: a new tool for paleoenvironmental reconstruction. Geophys Res Lett 31:L06206:1–4

70. Surman CA, Nicholson LW (2009) The good, the bad and the ugly: ENSO-driven oceanographic variability and its influence on seabird diet and reproductive performance at the Houtman Abrolhos, Eastern Indian Ocean. Marine Ornithology 37:129–138

71. Thompson SP (1989) Observations of bald eagles eating glaucous-winged gull eggs in western Washington. Northwest Nat 70:13–14

72. Tinbergen N (1961) The Herring Gull's World: a study of social behavior of birds. Basic Books, New York

73. Van Den Bosch F, De Roos AM, Gabriel W (1988) Cannibalism as a lifeboat mechanism. J Math Biol 26:619–633

74. Verbeek NAM (1986) Aspects of the breeding biology of an expanded population of glaucous-winged gulls in British Columbia. J Field Ornith 57:22–33

75. Verbeek NAM (1988) Differential predation of eggs in clutches of glaucous-winged gulls *Larus glaucescens*. Ibis 130:512–518

76. Verbeek NA, Butler RW (1999) Northwestern Crow (*Corvus caurinus*). In: Poole A (ed) The birds of North America online. Cornell Lab of Ornithology, Ithaca, New York. http://bna.birds.cornell.edu.bnaproxy.birds.cornell.edu/bna/species/407. https://doi.org/10.2173/bna.407

77. Vermeer K (1963) The breeding ecology of the glaucous-winged gull (*Larus glaucescens*) on Mandarte Island, BC. Occasional Papers of the British Columbia Provincial Museum, Number 13

78. Walther G-R, Post E, Convey P, Menzel A, Parmesan C, Beebee TJC, Fromentin J-M, Hoegh-Guldberg O, Bairlein F (2002) Ecological responses to recent climate change. Nature 416:389–395

79. Watts DP, Mitani JC (2000) Infanticide and cannibalism by male chimpanzees at Ngogo, Kibale National Park, Uganda. Primates 41:357–365

Chapter 8

Egg-Laying Synchrony in Gulls

A wide variety of natural systems exhibit spontaneous oscillator synchrony. In this chapter we use discrete-time mathematical modeling and Monte Carlo methods to provide the first demonstration of synchronous, every-other-day egg laying in seabirds. Along with Chap. 7, this chapter provides background for empirical and theoretical analyses that appear in several of the remaining chapters. The results of this work were initially published in The Auk *in 2010 by Shandelle Henson, James Hayward, J. M. Cushing, and Joseph Galusha [11] and in* The Wilson Journal of Ornithology *in 2017 and 2021 by Gordon Atkins, Ashley Reichert, Hayward, and Henson [1, 2].*

8.1 Introduction

Oscillator synchronization can result from environmental forcing or can spontaneously occur as a phenomenon of self-organization. In the case of environmental forcing, synchrony is mediated by an external periodic cue such as the regulation of feeding times by low tide in shorebirds. In the case of spontaneous self-organization, interacting oscillators spontaneously adjust the period or phase and begin to cycle together. Spontaneously synchronized oscillators include pendulum clocks, Malaysian fireflies, cardiac pacemaker cells, laser photo emitters, and the estrous cycles of rats [14–16, 20, 24].

Birds and mammals exhibit hormone oscillations that regulate ovulation. Mammalian ovulation cycles are interrupted by fertilization, followed by gestation and birthing of an entire brood. Many birds, by contrast, ovulate and lay eggs at regular intervals until clutch completion [27]. Ovulation and egg laying in birds are associated with periodic surges of luteinizing hormone (LH). Domestic turkeys and chickens, for example, experience LH surges approximately every 24 hours. The surge is followed about 30 minutes later by release of the ovum from a mature follicle into the oviduct [12], and the egg from the previous ovulation is laid. Glaucous-winged gulls lay eggs approximately every 2 days [26],

J. M. Cushing et al., *Modeling Behavior and Population Dynamics*, Interdisciplinary Applied Mathematics 57, https://doi.org/10.1007/978-3-031-34283-7_8

presumably caused by every-other-day ovulation cycles. We detected what seemed to be an every-other-day pulse in the number of new eggs laid in the glaucous-winged gull colony at Protection Island, suggestive that ovulation cycles were synchronized. We tested this observed egg-laying pattern against patterns generated by chance alone.

8.2 Data

We monitored egg laying at the Protection Island glaucous-winged gull colony daily, in the afternoon, from late May through June 2006–2008. Data were collected (Fig. 8.1) from every nest within five sample plots (A–E; Chap. 6, Fig. 6.3) near the west end of the colony, a spatially heterogeneous area consisting of densely populated, socially coherent neighborhoods separated by areas with few or no territories. We used the mean nearest-neighbor distance (NN, [19]) as a measure of the nest density in each sample plot.

Figure 8.1: Glaucous-winged gull nests in sample areas of the Protection Island breeding colony were marked with numbered stakes on the day the first day an egg appeared in each nest. Thereafter, each staked nest was monitored until the last egg in the nest was hatched or lost to predation (Photo by J. L. Hayward)

Nest totals across all plots ranged from 235 to 251 over the 3 years, with mean clutch size ranging from 2.41 to 2.46 (±SD 0.67–0.83) eggs and intra-clutch mean laying intervals ranging from 2.18 to 2.46 (±SD 0.73–0.96) days. The mean nearest-neighbor distances ranged from 1.94 to 3.54 (±SD 0.73–2.16) meters over the 3 years. Previous studies of this species report similar ranges [26, 32].

The colony experienced La Niña conditions in 2008. The mean March–June air temperature was 2.3°C and 0.9°C cooler than in 2006 and 2007, respectively (National Oceanographic and Atmospheric Administration's National Data Buoy Center, Station 46088, New Dungeness), and the mean sea-surface temperature was 0.7°C degrees cooler than in both 2006 and 2007 (National Oceanographic and Atmospheric Administration's National Data Buoy Center, Station PTWW1-9444900, Port Townsend). Average clutch initiation occurred 4.4 days later than in 2006, and 6.4 days later than in 2007.

8.3 Measuring Synchrony

Glaucous-winged gulls typically lay clutches of three eggs produced at approximately 2-day intervals [26]. If a social group of these birds were perfectly synchronized, numbers of eggs laid per day would exhibit an oscillation period of 2 days; every-other-day "highs" would be of irregular height, and "lows" would equal zero (Fig. 8.2a). The mean egg-laying interval is slightly greater than 2 days [26], however, and there could be temporal gaps during which time no eggs were laid; thus, a perfectly synchronous time series could sometimes experience "skips" consisting of two or more consecutive zeros (Fig. 8.2a). Perfect synchrony would be simple to identify, but in reality ecological stochasticity forces us to quantify the level of synchrony in noisy time series data (Fig. 8.2b–e).

We defined the degree of synchrony z for the time series $E_1, E_2, E_3, \ldots, E_m$ in which E_i is the number of eggs laid on day i, and days 1 and m refer to the first and last days, respectively, on which eggs were laid:

$$z = \frac{\sum_{i=0}^{m} |E_{i+1} - E_i|}{2\sum_{i=1}^{m} E_i}, \tag{8.1}$$

where E_0 and E_{m+1} are defined to be zero. The "up-and-down" variation of the time series graph is measured by the numerator, and the denominator measures the up-and-down variation for any time series exhibiting perfect synchrony with the same total number of eggs

$$N = \sum_{i=1}^{m} E_i.$$

The value $z = 1$ indicates perfect synchrony, and $z < 1$ indicates a departure from perfect synchrony.

We computed the observed z-value for the number of clutch initiations per day (number of first eggs laid in a nest per day), expecting that if the laying of first eggs were synchronized, then the laying of subsequent eggs every other day would also be synchronized through entrainment. We verified this by also computing z for the total number of eggs laid per day. Here we consider only clutch initiation synchrony; the details for overall egg-laying synchrony are in [11].

8.4 Null Model

To test whether the observed time series exhibited a level of synchrony significantly different from expectation due to chance, we constructed a null Monte Carlo model based on the following null hypothesis:

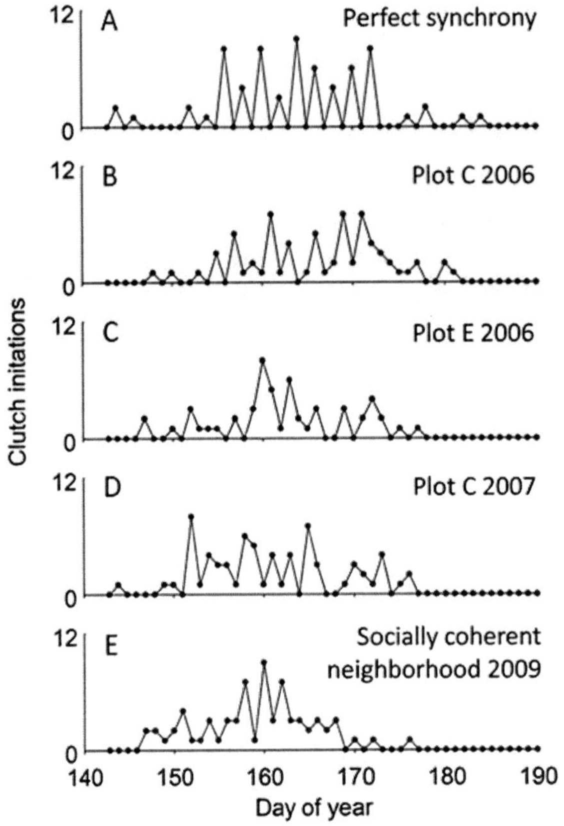

Figure 8.2: Number of clutch initiations per day among glaucous-winged gulls at Violet Point, Protection Island National Wildlife Refuge, Washington. (**a**) Perfect synchrony in simulated time series. (**b**) Observed data, plot C 2006. Significant level of process synchrony ($P = 0.0006$ for H_N). (**c**) Observed data, plot E 2006. Significant level of process synchrony ($P = 0.001$ for H_N). (**d**) Observed data, plot C 2007. Significant level of process synchrony ($P = 0.001$ for H_N). (**e**) Observed data from a socially coherent neighborhood from 2009. Some evidence of process synchrony ($P = 0.077$ for H_N). (Originally published in Henson et al., "Socially induced synchronization of every-other-day egg laying in a seabird colony," *The Auk* 127:571–580. Copyright © Oxford University Press, 2010. Used with permission.)

H_N: *Clutch initiations were random-normally distributed in time.*

For each observed time series of clutch initiations, the total number of clutches initiated (N), the number of days occupied by the clutch initiation period (m), and the sample mean clutch initiation date and standard deviation (d and s, respectively) were determined. For each Monte Carlo simulation based on hypothesis H_N, we randomly selected N real numbers from a truncated normal distribution $Norm(d, s^2)$ lying within the time interval $[0, m]$, and then we binned the N numbers by 1-day intervals to produce a simulated clutch initiation time series.

We compared z_{obs} for the observed time series to the distribution of z-values estimated from 10^6 simulated time series under the null hypothesis. P-values were determined as the proportion of simulations satisfying $z \geq z_{obs}$. Each P-value was computed five times to check convergence of the result. We also determined the critical value z_{crit} corresponding to significance level $\alpha = 0.01$.

The null distribution of z-values generated by the null model depends on model parameters (N, m, d, and s in model H_N). Given that the observed parameters were different for each of the 15 data sets (five sample plots over 3 years), we calculated the null distribution separately for each set.

For each time series, null model rejection indicates a level of synchrony unlikely to occur by chance. Such a time series exhibits what we term "process synchrony."

We also tested the following version of the null hypothesis:

(H_U) *Clutch initiations were random-uniformly distributed in time.*

Simulations of H_U were determined as they were for H_N, except the truncated normal distribution was replaced by a uniform distribution on the same time interval $[0, m]$.

8.5 Process Model

The beginning and end times of the reproductive window for most temperate-breeding avian species are determined by photoperiod. When daylight increases in the spring, light stimulation of the avian hypothalamus stimulates dramatic growth of the gonads. Prolonged exposure to long days in many species leads to a loss of reproductive responsiveness to light, or "photorefractoriness." Photoperiod, most likely independent of temperature, sets the peak and decline timings of baseline LH levels. The onset of ovulation cycling within the reproductive window, however, initiated in each female by an LH surge, may depend on other cues including temperature [3, 6–9, 18, 28, 29]. Low ambient temperatures, for example, can delay the onset of ovulation in gulls [22, 25].

The 2-day ovulation cycle in gulls can be divided into two segments. The first day consists of the LH surge and ovulation. The second day involves movement of the egg down the oviduct during which the albumen and shell membrane are deposited [6]. On the third day (which is day one of the next cycle) and after the next LH surge, the shell is formed and the egg is laid. Ovulation of the next follicle quickly follows oviposition.

We constructed a mathematical model to probe the possible mechanisms underlying every-other-day clutch initiation synchrony. The model is based on the following three assumptions:

(M_1) Baseline LH levels reach a high point determined by photoperiod and independent of temperature and then decline.
(M_2) The chance that a given female will experience the initial LH surge is proportional to its baseline LH level and the average ambient temperature, given that cool weather can delay the initial LH surge.
(M_3) Ovulating females exhibit every-other-day LH surges. LH surges in neighbors synchronize as a result of social stimulation, the strength of which is proportional to the baseline LH level and inversely proportional to the nearest-neighbor distance.

We qualitatively characterized the dynamics of the baseline LH level with a quadratic curve starting from a peak value of one that declines to zero:

$$R_t = \max\left[1 - \left(\frac{t}{r}\right)^2, 0\right], \tag{8.2}$$

where $r > 0$ is the number of days from peak LH level to complete refraction, and $\max[\cdot, 0]$ constrains the value of R_t to be nonnegative (Fig. 8.6a; assumption M_1). We characterized the qualitative behavior of spring mean ambient temperature by the increasing function

$$T_t = b\left(\max\left[(t-a), 0\right]\right)^2, \tag{8.3}$$

where $a \geq 0$ is the number of days warm weather is delayed and $b > 0$ (Fig. 8.3a).

We assumed the probability that a female is prepared to experience the initial LH surge is proportional to both temperature and the baseline LH level:

$$f_t = \min\left[R_t T_t, 1\right], \tag{8.4}$$

where min $[\cdot, 1]$ constrains the value of f to be ≤ 1 (Fig. 8.3b; assumption M_2). Finally, we modeled the strength of social stimulation c as proportional to baseline LH and inversely proportional to mean nearest-neighbor distance:

$$c_t = \frac{\widetilde{c} R_t}{NN}, \tag{8.5}$$

where $\widetilde{c} > 0$ is the constant of proportionality (assumption M_3) and NN is nearest-neighbor distance.

The discrete-time process synchrony model, depicted by the diagram in Fig. 8.4, is

$$\begin{aligned} w_{t+1} &= (1 - f_t)w_t + f_t(1 - e^{-c_t x_t})w_t \\ x_{t+1} &= f_t w_t e^{-c_t x_t} + q y_t e^{-c_t x_t} \\ y_{t+1} &= x_t + q(1 - e^{-c_t x_t})y_t, \end{aligned} \tag{8.6}$$

where the discrete-time step is 1 day and $t = 0$ corresponds to the day of peak baseline LH levels. Here, w_t is the number of birds on day t that have not begun to cycle, x_t is the number of birds in the first phase of the ovulation cycle on day t, and y_t is the number of birds in the second phase of the ovulation cycle on day t. The fraction f_t is the probability that a noncycling individual is ready to experience the initial LH surge and begin cycling, and $(1 - e^{-c_t x_t})$ is the probability that a cycling individual or one ready to cycle will "skip" a day in order to synchronize, where c_t is the strength of social stimulation. The fraction $1 - q$ is the probability that a bird in the second phase of the cycle will stop cycling and begin to incubate. The number w_{t+1} of noncycling individuals at the next time step is the number $(1 - f_t)w_t$ that are not ready to cycle plus the number $f_t(1 - e^{-c_t x_t})w_t$ that are ready to cycle but delay so as to synchronize. The number x_{t+1} of individuals in the first cycle phase at the next time step is the number $f_t w_t e^{-c x_t}$ that are ready to start cycling and do not delay, plus the number $q y_t e^{-c_t x_t}$ that return to the first phase from the second phase without delay. We assume that an individual in the first phase always enters the second phase at the next time step; hence, the number y_{t+1} of individuals in the second phase at the next time step is the number x_t leaving the first phase plus the number $q(1 - e^{-c_t x_t})y_t$ in the second phase that delay returning to the first phase.

According to the third model assumption M_3, an egg is laid 2 days following its ovulation. Thus, the number E_{t+1} of eggs laid on day $t+1$ is the total number x_{t-1} of ovulations on day $t-1$, and the number I_{t+1} of clutches initiated on day $t+1$ is the number of first-time recruits to the ovulating class x_{t+1} on day $t - 1$:

$$\begin{aligned} E_{t+1} &= x_{t-1} \\ I_{t+1} &= f_{t-2} w_{t-2} e^{-c_{t-2} x_{t-2}}. \end{aligned} \tag{8.7}$$

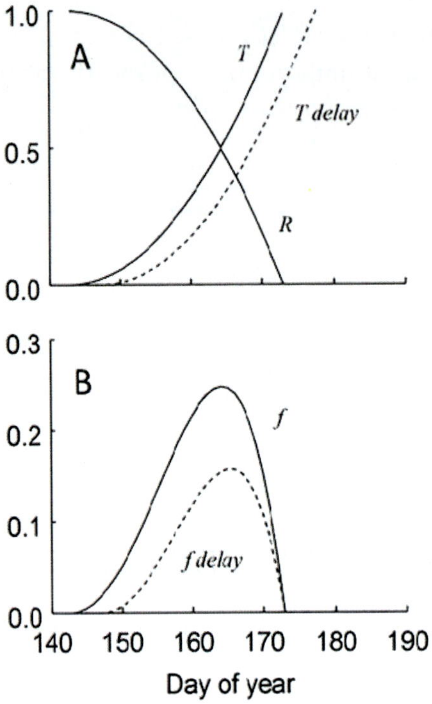

Figure 8.3: Submodels for baseline luteinizing hormone (LH) and temperature. (**a**) Baseline LH function R, temperature function T, and temperature function with delay T_{delay}, as in Models 2 and 3. (**b**) Probability of initial LH surge, with and without temperature delay (f_{delay} and f, respectively), as in Model 4. (Originally published in Henson et al., "Socially induced synchronization of every-other-day egg laying in a seabird colony," *The Auk* 127:571–580. Copyright © Oxford University Press, 2010. Used with permission.)

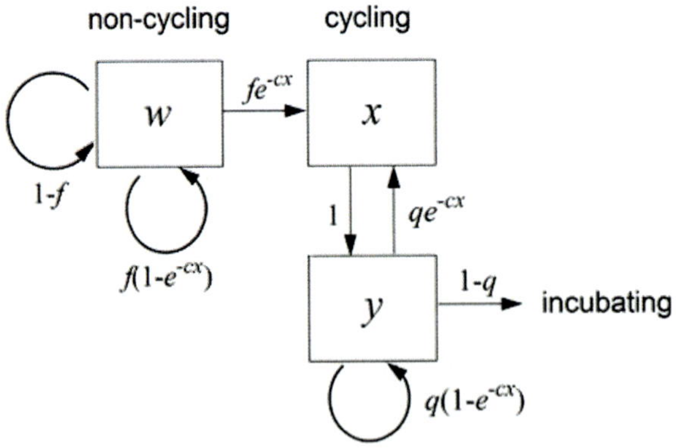

Figure 8.4: Conceptual diagram for model (8.6). (Originally published in Henson et al., "Socially induced synchronization of every-other-day egg laying in a seabird colony," *The Auk* 127:571–580. Copyright © Oxford University Press, 2010. Used with permission.)

We accounted for demographic stochasticity by adding noise on the square root scale [5] and accounted for "lattice effects" (dynamic outcomes due to the discrete nature of animal numbers) by rounding the right-hand sides of the model equations at each iteration [10]. The stochastic model equations, algebraically simplified, are

$$\begin{aligned}
w_{t+1} &= round\left[\left(\sqrt{(1-f_t e^{-c_t x_t})\,w_t} + \sigma_w \varepsilon_{t1}\right)^2\right] \\
x_{t+1} &= round\left[\left(\sqrt{(f_t w_t + q y_t)\,e^{-c_t x_t}} + \sigma_x \varepsilon_{t2}\right)^2\right] \\
y_{t+1} &= round\left[\left(\sqrt{x_t + q y_t (1-e^{-c_t x_t})} + \sigma_y \varepsilon_{t3}\right)^2\right] \\
I_{t+1} &= round\left[\left(\sqrt{f_{t-2} w_{t-2} e^{-c_{t-2} x_{t-2}}} + \sigma_I \varepsilon_{t4}\right)^2\right] \\
E_{t+1} &= round\left[\left(\sqrt{x_{t-1}} + \sigma_E \varepsilon_{t5}\right)^2\right],
\end{aligned} \tag{8.8}$$

where the sigmas are nonnegative constants and the epsilons are standard normal random variables (mean = 0, SD = 1) uncorrelated in time. The total number of eggs laid on day $t+1$ is the number of ovulations 2 days earlier, with little variation; hence, we assumed $\sigma_E = 0$. The quantities inside the squares are taken to be zero if negative.

Lacking data for state variables w and y, we were unable to conduct rigorous maximum likelihood parameter estimation (as in [4]). Instead, we estimated the four parameters $\widetilde{c}, r, b, q$ using data from plot C 2006 by setting the observed values of parameters $a = 0$, NN $= 2.07$, $w_0 = 68$, $x_0 = 0$, and $y_0 = 0$ (Table 8.1) and minimizing the sums of squared departures (on the square root scale [4]) of predicted I- and E-values from observed clutch initiation values. The predictions at each time step were conditioned as far as possible on the observed values at the previous step. This resulted in the parameters given in the caption of Fig. 8.6. The generalized R^2 values [4] for plot C 2006 data were $R_I^2 = 0.89$ and $R_E^2 = 0.94$. We used model (8.8), parameterized for plot C 2006, to make predictions for the other data sets without reparameterizing. Plot C 2006 was chosen as the model-fitting data set because it had the lowest P-value in the test of the null model and therefore showed the strongest evidence of process synchrony (Table 8.1).

8.6 Model Predictions

8.6.1 Null Model

Null model predictions consist of distributions of z-values generated by the Monte Carlo simulations (Fig. 8.5a). The null model makes the following two predictions:

(N1) The expected z-value and the critical value z_{crit} are decreasing functions of the number of nests N (Figs. 8.6a,c).

Table 8.1: Test of null hypotheses H_N

Year	Variable	Plot A (877 m^2)	Plot B (1,394 m^2)	Plot C (836 m^2)	Plot D (705 m^2)	Plot E (393 m^2)
2006	Number of clutches	42	31	68	34	53
	Number of eggs	104	74	164	81	119
	NN distance[a]	2.26	3.13	2.07	2.35	2.19
	z (P)	0.50 (0.22)	0.52 (0.62)	0.59 (0.0006[b])	0.44 (0.65)	0.60 (0.004[b])
2007	Number of clutches	29	38	68	41	53
	Number of eggs	65	89	166	97	124
	NN distance[a]	2.94	3.44	2.03	2.57	2.51
	z (P)	0.48 (0.68)	0.45 (0.53)	0.57 (0.001[b])	0.54 (0.23)	0.38 (0.75)
2008	Number of clutches	39	46	62	41	45
	Number of eggs	95	110	158	108	101
	NN distance[a]	2.79	3.54	1.94	2.16	2.55
	z (P)	0.49 (0.37)	0.37 (0.72)	0.31 (0.68)	0.46 (0.36)	0.38 (0.69)

[a]Mean nearest-neighbor (NN) distances are in meters.
[b]Significant at the $\alpha = 0.01$ level; P values reported to the accuracy obtained by five sets of 10^6 Monte Carlo simulations.

This prediction occurs because, in the null model, random occurrences of synchrony are likely if the number of clutches N is small, but if N is large, they are rare. Because the observed values of m, d,and s varied a small amount among the sample plots, the expected and critical z-values z_{crit} for the 15 data sets do not lie exactly on the associated curves in Fig. 8.6a).

(N2) The expected z-value is an increasing function of nearest-neighbor distance (Fig. 8.6d).

This prediction occurs because, although the null model does not depend explicitly on the number of nests N, in our study plots N turned out to be a decreasing function of average nearest-neighbor distance (Fig. 8.6b; $n = 15$; $P_{trend} = 0.017$). This finding, in concert with prediction N1, implies prediction N2.

8.6.2 Process Model

The process model (8.8) makes the following three predictions, the first two of which are in direct opposition to null model predictions N1 and N2:

(P1) If the colony density (nearest-neighbor distance) is held constant, the expected z-value is an increasing function of the number of nests (w_0 in model (8.6)).

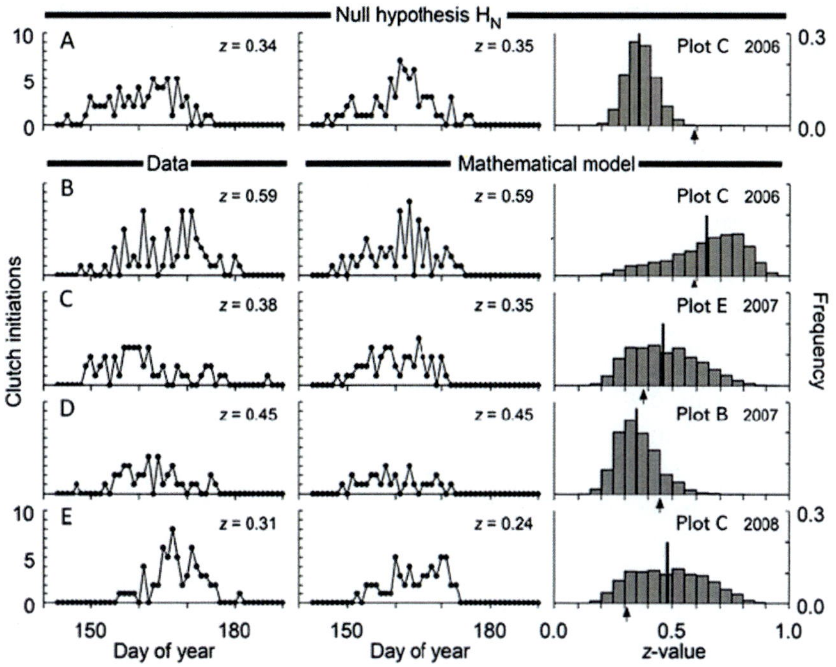

Figure 8.5: Clutch initiation data for glaucous-winged gulls at Violet Point, Protection Island National Wildlife Refuge, Washington (2006–2008), simulations and distributions of z-values generated from simulations. Vertical bar indicates distribution mean; arrow indicates observed z-value (Table 8.1). (**a**) Two Monte Carlo simulations and distribution of z-values for 200,000 Monte Carlo simulations of null hypothesis (H_N) for plot C 2006. Mean = 0.36. Null hypothesis rejected ($P = 0.0006$). (**b**)–(**e**) Simulations are from Model 8. Parameters estimated for plot C 2006 are $w_0 = 68$, $x_0 = 0$, $y_0 = 0$, $a = 0$, NN = 2.07, $\bar{c} = 0.78$, $r = 29$, $b = 0.0011$, and $q = 0.62$. The estimation procedure resulted in crude estimates of variances <1. We used $\sigma_w = 0.1$, $\sigma_x = 0.3$, $\sigma_y = 0.1$, $\sigma_I = 0.3$, and $\sigma_t = 0.0$. Distributions are for 10,000 simulations. (**b**) Plot C 2006 ($w_0 = 68$,NN = 2.07). Mean = 0.64. (**c**) Plot E 2007; parameters adjusted for this plot ($w_0 = 53$, NN= 2.51). Mean = 0.46. (**d**) Plot B 2007; parameters adjusted for this plot ($w_0 = 38$, NN = 3.44). Mean = 0.35. (**e**) Plot C 2008; parameters adjusted for this plot (NN $= 1.94, a = 4.4$). Mean = 0.48. (Originally published in Henson et al., "Socially induced synchronization of every-other-day egg laying in a seabird colony," *The Auk* 127:571–580. Copyright © Oxford University Press, 2010. Used with permission.)

This prediction occurs as a result of the fact that, under the assumption of demographic stochasticity, deterministic signals are clarified at larger population sizes [5]. Under this circumstance the deterministic model signal is synchronous, so the expected model synchrony level increases with the number of nests.

(P2) The expected z-value is a decreasing function of nearest-neighbor distance, given that the number of nests (w_0) is held constant.

This prediction occurs because social stimulation in the mathematical model is assumed to be inversely proportional to the average nearest-neighbor distance, as expressed in Eq. (8.5).

(P3) A cool-weather delay in the onset of egg laying decreases the expected z-value.

This prediction follows from the fact that in the process model, sufficient temperature, the "final cue" for reproduction, occurs later in the season even though the endpoints of the reproductive window remain fixed. The process model predicts a decreased level of ovulation synchrony on a daily time scale, because the smaller time window provides insufficient time for the group of two-periodic oscillators to synchronize. One might intuit, however, that a cool-weather delay in the onset of egg laying would *increase* synchrony on a yearly time scale, that is, reduce the length of the reproductive pulse. In fact, as we show in Chap. 9, the length of the reproductive pulse is short when sea surface temperature is low.

8.7 Results

8.7.1 Null Model

The observed z-values (open circles) with the critical values z_{crit} (solid circles) for null model H_N are compared in Fig. 8.6a. Observed z-values and P-values for null model H_N are listed in Table 8.1; results for H_U were equivalent in terms of significance and are not listed.

For 2006 and 2007 data, the null models H_N and H_U were rejected at the $\alpha = 0.01$ significance level ($P < 0.005$ for H_N) whenever mean NN distance was <2.20 m (plots C and E in 2006 and plot C in 2007; Fig. 8.2b–d). Thus in 2006 and 2007, clutch initiation showed strong evidence of process synchrony in dense areas of the colony, but not elsewhere. Plot C, which contained the most nests, exhibited the strongest evidence of process synchrony in clutch initiation ($P = 0.0006$ in 2006 and $P = 0.001$ in 2007 for H_N).

In 2008 when laying onset was delayed, neither H_N nor H_U could be rejected for any plot; none of the plots showed process synchrony in clutch initiation.

8.7.2 Process Model

The data are consistent with predictions of the process model. The predictions of the null model H_N and the process model (8.8) fitted to data from plot C 2006 are compared in Fig. 8.5a and b; arrows indicate the observed z-value and the black vertical bars indicate the means of the distributions of z-values generated by the null model (Fig. 8.5a) and

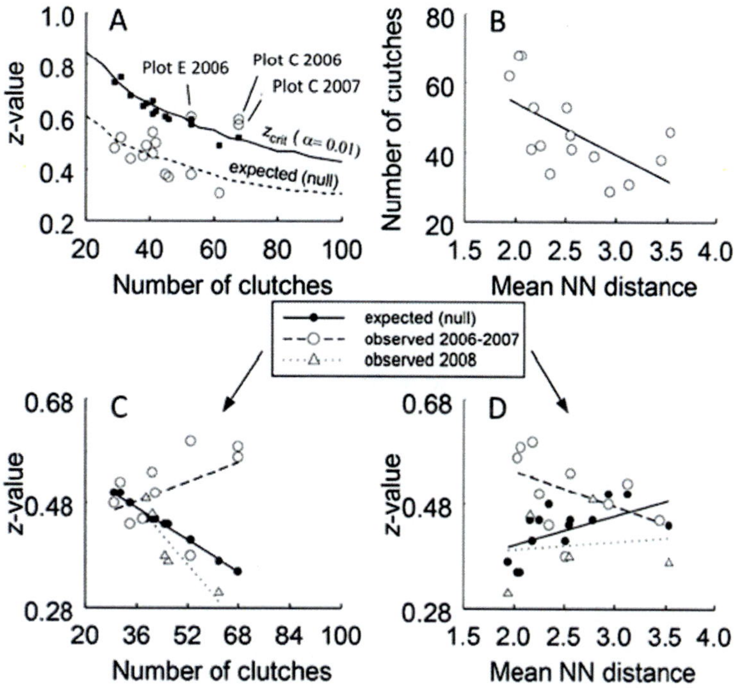

Figure 8.6: Data z-values, numbers of glaucous-winged gull clutches, and mean NN distances. (**a**) Critical z- values for the $\alpha = 0.01$ significance level (H_N, solid squares) and actual z-values (open circles) for all 15 data sets. For Monte Carlo parameters based on plot C 2006–2007 (H_N, $m = 34$, $\bar{d} = 17$, $s = 7.5$), z_{crit} (solid line) and mean Monte Carlo z- value (dashed line) are shown as a function of the number of clutches N. (**b**) Number of clutches and mean NN distance ($P_{trend} = 0.017$; $n = 15$). (**c**) Mean Monte Carlo z-values (H_N, solid circles) decrease with N ($P_{trend} = 0.00$; $n = 15$). Observed clutch initiation z-values (open circles) showed the opposite trend in 2006–2007 (dashed line, $P_{trend} = 0.16$; $N = 10$). In 2008, observed values decreased with N ($P = 0.041$; $n = 5$). There was no trend in 2006–2008 (open circles and open triangles, $P_{trend} = 0.69$; $n = 15$). (**d**) Mean Monte Carlo z-values (H_N, solid circles) increase with mean NN distance ($n = 15$, $P_{trend} = 0.019$). Observed clutch initiation z-values (open circles) showed the opposite trend in 2006–2007 (dashed line, $P_{trend} = 0.18$; $n = 10$) and essentially no trend in 2008 (open triangles, $P_{trend} = 0.84$; $n = 5$). (Originally published in Henson et al., "Socially induced synchronization of every-other-day egg laying in a seabird colony," *The Auk* 127:571–580. Copyright © Oxford University Press, 2010. Used with permission.)

simulations of the stochastic process model (Fig. 8.5b). The distribution arising from the null model and the process model overlap (Figs. 8.5a and b). The overlap indicates that rejecting the null model when it is true (a Type I error, or false positive) is unlikely given the nature of the Monte Carlo analysis, and failing to reject the null model when it is false (a Type II error, or false negative) is quite likely because of demographic noise in the system. In other words, a value of z observed in a process-synchronous system might occur in the left-hand tail of the histogram in Fig. 8.5b and in the middle of the null histogram in Fig. 8.5a. If this occurred, the null model would not be rejected.

In Fig. 8.5c–e, the data and predictions of the mathematical models for three other plots and years are compared to those of plot C (Fig. 8.5b). The expected model z-value decreased notably (from 0.64 to 0.46) for plot E in 2007, which had fewer clutches and larger mean NN distance (Fig. 8.5c; $w_0 = N = 53$ clutches; NN = 2.51) than plot C 2006 (Fig. 8.5; $NN = 2.07$). For plot B in 2007, which had even fewer clutches and larger mean nearest-neighbor distance (Fig. 8.5d; $w_0 = N = 38$; $NN = 3.44$), the expected model z-value decreased even more (to 0.35) than for plot E. For plot C in 2008 (Fig. 8.5e; delay $a = 4.4$ days), the model predicts a decreased expected z-value (0.48) as compared to plot C in 2006 (Fig. 8.5b; delay $a = 0$ days). These specific outcomes are consistent with the three general predictions P1–P3) of the process model (8.8).

8.7.3 Test of Predictions N1 and P1

The 2006 and 2007 data were consistent with prediction P1, and the 2008 data were consistent with prediction N1. Linear regression of observed clutch initiation z-values on N showed a positive trend in 2006 and 2007 (Fig. 8.6c; $n = 10$; $P_{trend} = 0.16$) and a negative trend in 2008 (Fig. 8.6c; $n = 5$; $P_{trend} = 0.04$).

8.7.4 Test of Predictions N2 and P2

The 2006 and 2007 data were consistent with prediction P2. Linear regression of observed clutch initiation z-values on the average NN distance showed a negative trend in 2006 and 2007 (Fig. 8.6d; $n = 10$; $P_{trend} = 0.18$) and essentially no trend in 2008 (Fig. 8.6d; $n = 5$; $P_{trend} = 0.84$).

8.8 Concluding Remarks

8.8.1 General Tenets

The following general tenets arise from our modeling approach:

(1) In socially coherent neighborhoods with few nests, one likely cannot distinguish process synchrony from randomness.
(2) In socially coherent neighborhoods exhibiting process synchrony, an increase in nest density increases the expected level of synchrony if the number of nests is held constant.

(3) In socially coherent neighborhoods exhibiting process synchrony, an increase in nest number increases the expected level of synchrony if nest density is held constant.
(4) In socially coherent neighborhoods exhibiting process synchrony, a delay in the onset of egg laying may decrease the expected level of synchrony. (See the sixth entry under "Caveats" below.)
(5) Because a Type II error (false negative) is likely, an observer of a system with a synchronizing mechanism should not expect to identify process synchrony in every time series. Because a Type I error (false positive) is unlikely, the observer of a system without a synchronizing mechanism is unlikely to infer process synchrony in any time series.

8.8.2 Looking for Synchrony

Our study suggests that egg-laying synchrony may be difficult to detect in other colonies or species. How might a researcher collect data in a way that will maximize the chances of detecting synchrony if it exists?

(1) Nest checks must occur at an interval suitable for the egg-laying interval. If the egg-laying interval is greater or equal to 2 days, nest checks can occur daily at the same time of day. Alternately, if the egg-laying interval is 1 day, nest checks must occur more frequently.
(2) Data must be collected from socially coherent neighborhoods. The geographic areas containing socially coherent neighborhoods change from year to year; thus, one might need to alter the geographic area monitored from 1 year to the next.
(3) There must be a sufficient number of nests in the monitored neighborhoods. As Table 8.1 suggests, a sufficient number of nests in a glaucous-winged gull neighborhood might be approximately 50. Comparing results from nearest-neighbor pairs to a random selection of pairs using pairwise analysis cannot be expected to exhibit significance.
(4) The nests within the socially coherent neighborhoods must be sufficiently close to one another. Table 8.1 suggests that for glaucous-winged gulls, a nearest-neighbor distance approximately 2.2 m or less would be necessary.
(5) The neighborhoods should experience average weather conditions for the region and minimal disturbance, given that stress can modulate hormone levels.
(6) Significant process synchrony will not always be observed even if the system contains appropriate mechanisms for process synchrony.

We conducted a test of the above guidelines in 2009. We identified a large, dense socially coherent neighborhood that experienced minimal disturbance. It included an area west of plot C_{aux} plus about half of the west end of plot C (Fig. 6.3, Chap. 6). The neighborhood contained 69 nests and a mean nearest-neighbor distance of 1.8 meters (SD ± 0.73). Clutch initiation data from this neighborhood showed some evidence of synchrony ($P = 0.077$ for H_N; Fig. 8.2e). Including the rest of plot C in the data set yielded $P = 0.045$, with 109 clutches and mean NN distance 1.86 m (SD ± 0.72).

Using the above guidelines, we also documented egg-laying synchrony in a colony of ring-billed gulls (*Larus delawarensis*) in Michigan, thus suggesting reproductive synchrony may be distributed more widely among larid species [21].

Our hypotheses underlying assumptions M_1– M_4 on which the mathematical model is based involve complicated systems of endocrine activity. These assumptions were motivated by studies that demonstrated the effect of temperature on baseline LH levels in white-crowned sparrows (*Zonotrichia leucophrys*) [13, 29–31] and the effect of temperature on the onset of laying in gulls [22, 25], although we are aware that these assumptions may not hold true for glaucous-winged gulls. For example, studies showed that the gonadotropin cycles of various species of European tits (*Parus* spp.) responded differently to low temperatures [23]. Endocrine activity in wild glaucous-winged gulls would be extremely difficult to monitor without disturbance to the birds which would seriously interfere with their reproductive activity.

The process model (8.8) amounts to a very simplified expression of any actual mechanism for synchrony. It allows for only a narrow range of biological possibilities. It does not include, for example, the chance that females may experience acceleration or delay of LH surges on the first day of the ovulation cycle, but only on the second day. It is a discrete-time model which does not address gradual endocrine changes which could lead to acceleration or delay. However, the model allows for a simple method for exploring the dynamic consequences of hypotheses before the actual endocrine mechanisms are worked out.

Although it seems obvious that circadian entrainment cannot explain every-other-day egg-laying synchrony, the possibility remains that environmental forcing was involved. Given that seven of the 10 data sets from 2006 and 2007 did not show synchrony, however, environmental forcing appears unlikely.

8.8.3 Possible Synchronizing Stimulus

Collections of spontaneously synchronizing oscillators synchronize via some form of "communication." Visual, auditory, and/or olfactory communication could serve as synchronizers of LH oscillators and could involve the presence and behavior of males.

In a related experiment, recorded copulation (Fig. 8.7) calls were broadcast to a small group of nine pairs of glaucous-winged gulls on Protection Island [2]. The group was somewhat isolated acoustically and visually from the main colony. Although the broadcast calls did not increase the overall numbers of head-tossing and copulatory mounts, they did significantly concentrate head-tossing and mounts to a time period immediately following call broadcast. Control periods, by contrast, during which no sound was broadcast,

Figure 8.7: Copulation in glaucous-winged gulls often lasts for more than 1 minute. During this time the female pecks at the breast of the male, and the male utters a rhythmic, high-amplitude call that can be heard throughout the colony. Evidence suggests this rhythmic call serves as a synchronizing signal for the synchronization of every-other-day ovulation and egg laying in females living close together (Photo by J. L. Hayward)

experienced no such concentration of head-tossing and mounts in the period immediately following call broadcast. Playback of non-gull sounds produced no such concentration of head-tossing and mounts. Thus, the copulation call alone seems sufficient to coordinate the timing of head-tossing and mounting in neighboring glaucous-winged gulls. Additional experiments [1] demonstrated that a female oviposition event reduced the likelihood of mounts for that day as compared with the day before and that total numbers of copulations and eggs laid occurred in an out-of-phase rhythm. Results from these experiments support the hypothesis that social facilitation of mounts, in concert with an inhibition of mounting on egg-laying days, synchronizes every-other-day egg laying among female gulls living in close proximity to one another.

The identification of egg-laying synchrony raises the question about function. Does reproductive synchrony confer a selective advantage on participants, or is it merely an epiphenomenon without selective advantage [14, 17]? We provide a tentative answer to this question in Chap. 9.

References

1. Atkins GJ, Hayward JL, Henson SM (2021) How do gulls synchronize every-other-day egg laying? Wilson J Ornith 133:226–235

2. Atkins GJ, Reichert AA, Henson SM, Hayward JL (2017) Copulation call coordinates timing of head-tossing and mounting behaviors in neighboring glaucous-winged gulls (*Larus glaucescens*). Wilson J Ornith 129:560–567

3. Dawson A, Sharp PJ (2007) Photorefractoriness in birds—photoperiodic and non-photoperiodic control. General and Comparative Endocrinology 153:378–384

4. Dennis B, Desharnais RA, Cushing JM, Henson SM, Costantino RF (2001) Estimating chaos and complex dynamics in an insect population. Ecol Monogr 71:277–303

5. Desharnais RA, Costantino RF, Cushing JM, Henson SM, Dennis B, King AA (2006) Experimental support of the scaling rule for demographic stochasticity. Ecol Lett 9:537–547

6. Faaborg J (1988) Ornithology: an ecological approach. Prentice Hall, Englewood Cliffs, NJ

7. Hahn TP, Boswell T, Wingfield JC, Ball GF (1997) Temporal flexibility in avian reproduction: patterns and mechanisms. In: Nolan V, Jr, Ketterson ED, Thompson CF (eds) Current ornithology, vol 14. Plenum, New York, pp 39–80

8. Hahn TP, MacDougall-Shackleton SA (2008) Adaptive specialization, conditional plasticity and phylogenetic history in the reproductive cue response systems of birds. Phil Trans R Soc B 363:267–286

9. Hahn TP, Watts HE, Cornelius JM, Brazeal KR, MacDougall-Shackleton SA (2009) Evolution of environmental cue response mechanisms: adaptive variation in photorefractoriness. General and Comparative Endocrinology 163:193–200

10. Henson SM, Costantino RF, Cushing JM, Desharnais RA, Dennis B, King AA (2001) Lattice effects observed in chaotic dynamics of experimental populations. Science 294:602–605

11. Henson SM, Hayward JL, Curshing JM, Galusha JG (2010) Socially induced synchronization of every-other-day egg laying in a seabird colony. Auk 127:571–580

12. Liu H-K., Nestor KE, Long DW, Bacon WL (2001) Frequency of luteinizing hormone surges and egg production rate in turkey hens. Biol Reprod 64:1769–1775

13. Maney DL, Hahn TP, Schoech SJ, Sharp PJ, Morton ML, Wingfield JC (1999) Effects of ambient temperature on photo-induced prolactin secretion in three subspecies of white-crowned sparrow, *Zonotrichia leucophrys*. General and Comparative Endocrinology 113:445–456

14. McClintock MK (1981) Social control of the ovarian cycle and the function of estrous synchrony. Am Zool 21:243–256

15. McClintock MK (1984a) Estrous synchrony: modulation of ovarian cycle length by female pheromones. Physiol Behav 32:701–705

16. McClintock MK (1984b) Group mating in the domestic rat as a context for sexual selection: consequences for analysis of sexual behavior and neuroendocrine responses. Advances in the Study of Behavior 14:1–50

17. McClintock MK (1998) Whither menstrual synchrony? Annual Review of Sex Research 9:77–95

18. Nicholls TJ, Goldsmith AR, Dawson A (1988) Photorefractoriness in birds and comparison with mammals. Physiol Rev 68:133–176

19. Patterson IJ (1965) Timing and spacing of broods in the black-headed gull, *Larus ridibundus*. Ibis 107:433–459

20. Pikovsky A, Rosenblum M, Kurths J (2003) Synchronization: a universal concept in nonlinear sciences. Cambridge University, Cambridge

21. Sandler AG, Megna LC, Hayward JL, Henson SM, Tkachuck CM, Tkachuck RD (2016) Every-other-day clutch-initiation synchrony in ring-billed gulls (*Larus delawarensis*). Wilson J Ornith 128:760–765

22. Schreiber EA (1980) Climate and weather effects on seabirds. In: Schreiber EA, Burger J (eds) Biology of marine birds. CRC Press, Boca Raton, FL, pp 179–216

23. Silverin B, Viebke PA (1994) Low temperatures affect the photoperiodically induced LH and testicular cycles differently in closely related species of tits (*Parus* spp.). Horm Behav 28:199–206

24. Strogatz S (2003) Sync: the emerging science of spontaneous order. Hyperion, New York

25. Verbeek NAM (1979) Timing of primary molt and egg-laying in glaucous-winged gulls. Wilson Bulletin 91:420–425

26. Vermeer K (1963) The breeding ecology of the glaucous-winged gull (*Larus glaucescens*). In: Occasional Papers of the British Columbia Provincial Museum No. 13

27. Wiebe KL, Martin K (1995) Ecological and physiological effects on egg laying intervals in ptarmigan. Condor 97:708–717

28. Wingfield JC (2008) Organization of vertebrate annual cycles: implications for control mechanisms. Phil Trans R Soc B 363:425–441

29. Wingfield JC, Hahn TP, Maney DL, Schoech SJ, Wada M, Morton ML (2003) Effects of temperature on photoperiodically induced reproductive development, circulating plasma luteinizing hormone and thyroid hormones, body mass, fat deposition and molt in mountain white-crowned sparrows, *Zonotrichia leucophrys oriantha*. General and Comparative Endocrinology 131:143–158

30. Wingfield JC, Hahn TP, Wada M, Astheimer LB, Schoech S (1996) Interrelationship of day length and temperature on the control of gonadal development, body mass, and fat score in white-crowned sparrows, *Zonotrichia leucophrys gambelii*. General and Comparative Endocrinology 101:242–255

31. Wingfield JC, Hahn TP, Wada M, Schoech SJ (1997) Effects of day length and temperature on gonadal development, body mass, and fat depots in white-crowned sparrows, *Zonotrichia leucophrys pugetensis*. General and Comparative Endocrinology 107:44–62

32. Zador SG, Piatt JF (2007) Simulating the effects of predation and egg-harvest at a gull colony. In: Piatt JF, Gende SM (eds) Proceedings of the fourth glacier bay science symposium, October 26–28, 2004. U.S. Geological Survey Scientific Investigations Report 1007–5047, pp 188–192

Chapter 9

Synchronized Egg-Laying as an Adaptive Response to Egg Cannibalism in Gulls

In Chap. 7 we described the occurrence of egg cannibalism by gulls, and in Chap. 8 we reported that under certain conditions female gulls lay their eggs on a synchronous, every-other-day schedule. Here we use logistic regression, chi-square analysis, and Monte Carlo simulations to show that every-other-day egg-laying synchrony functions as an adaptive response that reduces the chance that a given egg will be cannibalized during years of high sea surface temperature. This research, initially published in The Wilson Journal of Ornithology *in 2020, was authored by Sumiko Weir, who evaluated this topic for her undergraduate honors thesis, and coauthored by senior researchers Shandelle Henson, James Hayward, and Gordon Atkins and by students Ashley Polski, WayAnne Watson, and Amanda Sandler [64].*

9.1 Introduction

Climate extremes associated with El Niño-Southern Oscillation (ENSO) occur on an approximately decadal time scale and are expected to increase in frequency with greenhouse warming [14]. ENSO events are associated with higher sea surface temperatures (SST), weakened upwellings, lower thermoclines, and concomitant decreased plankton productivity in the California Current system. Forage fish follow plankton to deeper water, thus decreasing the food supply for surface-feeding seabirds, leading to reproductive failures and population declines [1–3, 8, 37, 39, 50, 51, 54]. The Salish Sea, composed of the marine waters of southwestern British Columbia and northwestern Washington which lie to

J. M. Cushing et al., *Modeling Behavior and Population Dynamics*, Interdisciplinary Applied Mathematics 57, https://doi.org/10.1007/978-3-031-34283-7_9

the east and south of Vancouver Island, is connected to the California Current upwelling system and constitutes a natural laboratory for studying the effects of climate on marine organisms. Resident seabirds serve as sentinels of environmental quality [11, 30, 32].

Two studies at Protection Island National Wildlife Refuge, USA, near the east end of the Strait of Juan de Fuca in the Salish Sea, examined the effect of local SST on feeding and reproductive tactics in a colony of approximately 2,000 pairs of glaucous-winged gulls (*Larus glaucescens*). One of the studies, based on data from 2006 to 2011, showed that cannibalism by adult gulls on other colonists' eggs increased and hatching success decreased with rises in SST. In particular, a 0.1°C rise in SST was associated with a 10% increase in the odds that an egg would be cannibalized and a 10% decrease in the odds that an egg would hatch [22] (see Chap. 7).

The second study, based on data from 2006 to 2008, demonstrated the existence of every-other-day clutch-initiation synchrony in sufficiently dense areas of the colony [25] (see Chap. 8). Clutch initiations at the Protection Island colony are approximately normally distributed in time over a 4-week interval beginning during the last half of May. Clutch initiation occurs when a breeding female lays her first egg (A-egg), after which she lays subsequent eggs (B- and C-eggs) approximately every other day until completing a two- or three-egg clutch. Every-other-day clutch-initiation synchrony thus leads to an every-other-day pulse of egg-laying on the colony. Evidence suggests that the high-amplitude copulation call by male gulls, in combination with an apparent inhibition of copulation on days of egg-laying, leads to the synchronization process [6, 7].

Both cannibalism and reproductive synchrony are posited to confer selective advantages. During years of food shortage, cannibalism may act as a "lifeboat" mechanism in a variety of taxa, redirecting reproductive effort to times of higher resource levels [18, 23]. One gull egg contains about half the energy required per day by an adult gull, and we suggested that egg cannibalism might be adaptive during years of high SST and its associated low resource availability [22]. Yearly reproductive synchrony, in which egg-laying occurs in an annual pulse within a time window set by photoperiod [42], may be advantageous because of external predator satiation (the Fraser Darling Hypothesis [16]). In the Protection Island study on egg-laying synchrony, we similarly suggested that the every-other-day clutch-initiation synchrony observed at Protection Island could confer an advantage in the presence of egg cannibalism [25, 26]. This is because if clutch initiations are synchronized on an every-other-day schedule, each egg is less likely to be cannibalized on days when more eggs are laid. Moreover, mathematical models showed that the clutch-initiation synchronization process would require longer egg-laying seasons [25], suggesting the possibility of a trade-off between protection from external predation (high yearly synchrony and short egg-laying season) and protection from conspecific predation (every-other-day clutch-initiation synchrony but longer egg-laying season). If true, clutch-initiation synchrony would likely be most pronounced during years of high SST.

In this chapter we pose five predictions for the Protection Island colony tested to further investigate the alternative feeding and reproductive tactics:

(1) The degree of clutch-initiation synchrony increases with SST.

(2) The length of the clutch-initiation season increases with SST.

(3) The A-egg in a clutch is more likely to be cannibalized than the B- or C-egg.

(4) The A-egg in a clutch is most likely to be cannibalized within the first 24 hours after it is laid.

(5) The odds that the A-egg is cannibalized within the first 24 hours decreases with increasing additional clutches initiated that day. That is, clutch-initiation synchrony confers a reproductive advantage in the presence of cannibalism.

9.2 Methods and Results

9.2.1 Location and Sample Plots

At the Violet Point breeding colony ($48°07'40''N, 122°55'3''W$) at Protection Island National Wildlife Refuge, Washington, USA, we monitored all glaucous-winged gull nests in sample plots centered on dense socially cohesive groups. During 2006–2008 we focused on nests in five sample plots, Plots A–E (Fig. 6.3, Chap. 6). Throughout this time, the northwest boundary of the colony moved further west; thus, in 2009–2011 we added a sixth study area northwest of Plot C which we called Plot C-auxiliary (Caux; Fig. 6.3, Chap. 6). We did not collect relevant data in 2012–2013. During this time Plots A and B became mostly depopulated, so we added another new study area Plot L to the southeast of Plot D (Fig. 6.3, Chap. 6). In 2014–2015 we continued the study in Plots C, Caux, D, E, and L.

9.2.2 Nest Monitoring

Nest monitoring methods are detailed in [22], but we repeat the main points here for convenience. Each afternoon throughout the laying season, late May to mid-July, 2006–2011 and 2014–2015, we monitored the sample plots on foot. At the time of clutch initiation, each nest was marked with a wooden stake, positioned to avoid interference with movement of territory occupants. Each staked nest was checked daily for new eggs, which were marked with permanent ink at their apices and in order of their laying. We monitored each nest until the fate of each egg was recorded as cannibalized, eagle predated (see below), addled, died during pipping, hatched, or some other fate (punctured, nest-flooded, or rolled out of nest). For each nest, we determined the distance from its center to the center of the nest of its "nearest neighbor" [43] using a tape measure or laser rule (2006–2010) or a Trimble GPS and ArcGIS Desktop 10 with the "pointdistances" tool in Geospatial Modeling Environment software (2011, 2014, and 2015).

9.2.3 Identification of Causes of Egg Loss

During our more than 30 years of observations of the Protection Island colony, only two species of gull egg predators were observed:

(1) Bald eagles (*Haliaeetus leucocephalus*) tend to feed on all of the eggs in a clutch at the nest site, leaving behind fragments of eggshell.
(2) Glaucous-winged gull cannibals which typically steal only one egg from a nest and carry it away to their own territory, leaving no eggshell at the predated territory.

Thus, for each egg lost during the laying season, we recorded the cause of egg loss based on the number of eggs taken and the presence or absence of eggshell fragments at the predated nest site [22].

9.2.4 Measuring Local SST

The reproductive success of gulls, which are capital breeders, depends partially on food availability before egg-laying [11, 12, 29, 36]. A time lag occurs between oceanic conditions and the impact of these conditions on oceanic food webs. Consequently, gulls and other seabirds are affected by oceanic conditions ahead of the egg-laying season [20, 51, 63]. We determined an average local sea surface temperature (SST) for the 9 months (September–May) prior to each egg-laying season using data from the National Oceanic and Atmospheric Administration (NOAA), Port Townsend, Washington, buoy (PTWW1), located 12 km east of the study site [22].

Descriptive data from the 8 years of study are provided in Table 9.1.

9.2.5 Test of Prediction 1: The Degree of Clutch-Initiation Synchrony Increases with SST

We measured the degree of clutch-initiation synchrony after [25] (see also Chap. 8). If E_i is the number of clutches initiated in a sample plot of the colony on day i, with $E_0 = E_{D+1} = 0$ bracketing the beginning and end of clutch initiation, then the degree of daily synchrony was defined to be

$$X = \frac{\sum_{i=0}^{D} |E_{i+1} - E_i|}{2\sum_{i=1}^{D} E_i}.$$

The numerator provides the total variation of the time series; the denominator measures the total variation for any perfectly synchronous time series with the same total number of eggs. The quantity $X = 1$ indicates perfect synchrony, and $X < 1$ indicates a departure from perfect synchrony. A large X can occur by chance. To test whether an observed X

Table 9.1: For each of the 8 breeding seasons, first clutch initiation dates, mean clutch initiation dates, and sea surface temperatures (SST) during the 9 months (September–May) prior to the egg-laying and chick-rearing seasons for nests in the study plots are shown. Also, number of territories with clutches, number of eggs, number of eggs cannibalized, and percent of total eggs cannibalized 1–30 June are shown for the study plots. Data for 2006–2011 are from Hayward et al. (2014), with minor corrections

Year of breeding season	Date of first clutch-initiation	Mean clutch-initiation date	Mean Sep-May SST(°C) prior to breeding season	Number of territories in study plots	Number of eggs in study plot nests	Number of eggs cannibal-ized in study plots	% of total egg can-nibalized in study plot
2006	25 May	13 June	9.12	199	465	93	20.0
2007	24 May	10 June	9.18	201	478	200	41.8
2008	1 June	16 June	8.61	226	560	94	16.8
2009	25 May	7 June	8.68	248	650	92	14.2
2010	25 May	10 June	9.33	267	639	142	22.2
2011	24 May[a]	7 June	8.78	231	606	98	16.2
2014	20 May	8 June	9.01	279	566	121	21.4
2015	20 May	5 June	10.15	289	721	121	16.8
TOTALS	–	–	–	1,940	4,685	961	–

[a] Extrapolated. First day on which a nest check was done was 27 May, when one 2-egg nest was present, and a C-egg was laid on 28 May. Assuming the usual 2-day interval between ovipositions, the first egg was probably laid on 24 May

for a given plot and year was significantly greater than expected by chance, the observed X was compared to the distribution of X values generated by 10^6 Monte Carlo simulations of random clutch initiations, which consisted of the observed number of clutches chosen from a truncated normal distribution which spanned D days with the observed sample mean and standard deviation from that plot. The associated P value is the fraction of the null distribution lying to the right of the observed X value [25].

In the current study we compared the degree of synchrony, which depends on the number of clutches, across sample plots with different numbers of clutches. Thus, we standardized each observed X value to an equivalent standard Z value associated with a hypothetical standard 68-clutch plot. We used 68 clutches for historical reasons because that was the number of clutches in Plot C in 2006 when synchrony was first observed. Specifically, for each observed X value associated with clutch initiations spanning D days, we constructed a standard null distribution from Monte Carlo simulations of 68 random clutch initiations occurring according to a truncated normal distribution over D days with the same sample mean and standard deviation observed in the plot for which X was computed. We then transformed the observed X value to the standard Z that preserved its P value. The Z values were correlated with SST.

We found that significant levels ($\alpha = 0.05$) of clutch-initiation synchrony coincided with years of high SST (Table 9.2; $\chi^2 = 4.18$, d.f. $= 1$, $P < 0.041$). In addition, the degree of clutch-initiation synchrony increased with SST (Fig. 9.1a, $r = 0.600$). Squares correspond to years in which the degree of clutch-initiation synchrony was significant.

Table 9.2: Years for which significant clutch-initiation synchrony occurred in at least on sample plot. Average SST was determined for the 9 months (Sep–May) prior to each 3-month egg-laying and chick-rearing season. Over the 8-year study, local SST ranged from 8.61 to 10.15 °C

Average SST	Daily synchrony	No. daily synchrony
SST $\geq$ 9.01 °C	2006, 2007, 2014, 2015	2010
SST $\leq$ 8.78 °C	None	2008, 2009, 2011

9.2.6 Test of Prediction 2: The Length of the Clutch-Initiation Season Increases with SST

We combined the clutch-initiation data from all sample plots each year to construct an overall temporal distribution of clutch initiations. From this distribution we quantified the length of the clutch-initiation season in two ways. We first computed the number of days spanned by the first and last clutch initiations and correlated this with SST. Second, we computed the sample standard deviation of the distribution of clutch initiations and correlated this with SST.

We found that the length of the clutch-initiation season (the number of days between the first and last clutch initiation) increased with SST (Fig. 9.1b; $r = 0.588$) and the standard deviation of the temporal distribution of clutch initiations also increased with SST (Fig. 9.1c; $r = 0.575$). Furthermore, the day of the year of the first clutch initiation decreased with increased SST (Fig. 9.1d; $r = -0.605$).

9.2.7 Test of Prediction 3: The A-Egg in a Clutch Is More Likely to be Cannibalized than the B- or C-Egg

We tallied the number of A-, B-, and C-eggs cannibalized for the 8 years of data. We used a 2×3 chi-squared test to compare numbers of A-, B-, and C-eggs cannibalized and a 2×2 chi-squared test to compare the number of A-eggs cannibalized with combined numbers of B- and C-eggs cannibalized.

We found that the A-egg in a clutch was much more likely to be cannibalized than the B- or C-egg (Table 9.3—comparing A-, B-, and C-eggs, $\chi^2 = 44.41$, d.f. $= 2$, $P < 0.001$; Table 9.4—comparing A-eggs with B- and C-eggs combined, $\chi^2 = 41.61$, d.f. $= 1$, $P < 0.001$).

Table 9.3: Comparison of numbers of cannibalized A-, B-, and C-eggs in relation to non-cannibalized counterparts during the 8-year study

	A-eggs	B-eggs	C-eggs	Total
Cannibalized	483	299	179	961
Not cannibalized	1,394	1,340	990	3,724
Total	1,877	1,639	1,169	4,685

Table 9.4: Comparison of numbers of A-eggs and non-A-eggs cannibalized in relation to number of non-cannibalized eggs during the 8-year study

	A-eggs	Non-A-eggs	Total
Cannibalized eggs	483	478	961
Not cannibalized eggs	1,394	2,330	3,724
Total	1,877	2,808	4,685

9.2.8 Test of Prediction 4: The A-Egg in a Clutch Is Most Likely to be Cannibalized Within the First 24 Hours After It Is Laid

We binned the A-eggs lost to cannibalism by the number of days they survived. We used a chi-squared test to compare the number of A-eggs lost in the first 24 hours to the expected values calculated under the assumption that A-egg loss occurred uniformly over time.

We found that the first egg in a clutch was most likely to be cannibalized within the first 24 hours after it was laid ($N = 136$; $\chi^2 = 1,075.8$, d.f. $= 1$, $P < 0.001$).

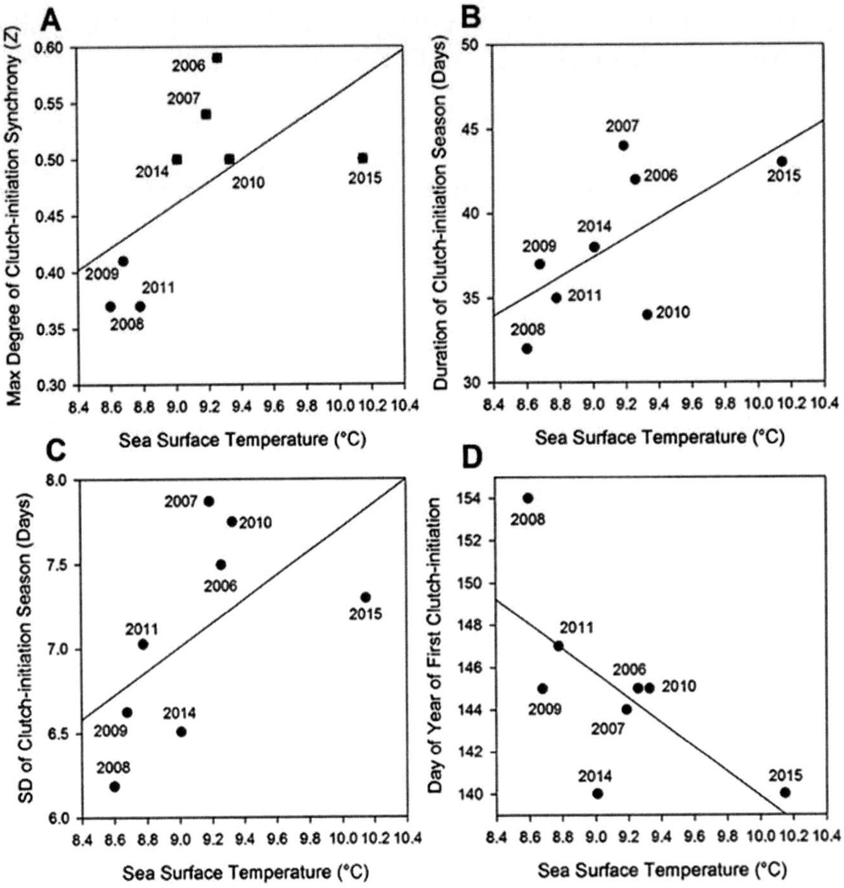

Figure 9.1: Characteristics of clutch initiation of glaucous-winged gulls at Protection Island National Wildlife Refuge, Washington, as functions of SST. (**a**) For each year, the maximum degree of clutch-initiation synchrony observed across all sample plots, plotted against SST. Squares indicate years in which that degree of clutch-initiation synchrony was significant. There is an increasing trend. (**b**) Number of days between the first and last clutch initiations trends positively with SST. (**c**) Standard deviation of temporal distribution of clutch initiations trends positively correlated with SST. (**d**) Day of year on which the first clutch initiation occurred trends negatively with SST (Originally published in Weir et al., "Every-other-day clutch-initiation synchrony as an adaptive response to egg cannibalism in glaucous-winged gulls," *Wilson Journal of Ornithology* 132:575–586. Copyright © The Wilson Ornithological Society 2020. Used by permission.)

9.2.9 Test of Prediction 5: The Odds That the A-Egg Is Cannibalized Within the First 24 Hours Decreases with Increasing Additional Clutches Initiated That Day

For every day on which a first egg was cannibalized over the eight study years, we recorded the total number of A-eggs laid that same day and the number of those A-eggs that were lost to cannibalism that same day. We used logistic regression (with Statistics Toolbox in MATLAB, The MathWorks, Inc., Natick, Massachusetts) to regress the log-odds that an A-egg was cannibalized against the number of A-eggs laid (clutches initiated) that day.

We found that the odds that the A-egg was cannibalized within the first 24 hours decreased 19% for each additional five clutches initiated that day ($P < 0.001$; OR $= 0.81$). Thus, clutch-initiation synchrony conferred an advantage in the presence of cannibalism.

9.3 Discussion

Our results suggest that at the Protection Island glaucous-winged gull colony:

(1) Clutch-initiation synchrony increases in years of higher SST compared with years of lower SST.
(2) In years of lower SST, egg-laying begins earlier and continues for a longer period of time.
(3) An A-egg in a nest has a higher chance to be cannibalized than either the B- or C-egg.
(4) Cannibalization of an A-egg is most likely during the first 24 hours after it is laid.
(5) The likelihood of A-egg cannibalization during its first 24 hours decreases with an increase in the number of clutches initiated that day, thus lowering the chance that a given egg will be cannibalized and conferring an adaptive advantage.

The linkages between SST and the behavioral tactics we examined are unlikely to be direct but are likely mediated through the food web. Nutrient levels are high when ocean water is cool and well-mixed, and these conditions allow plankton to thrive. When ocean water is warm and highly stratified, nutrient levels are low and plankton numbers become depleted [49]. Under these conditions, numbers of Pacific herring (*Clupea pallasii*), a primary food source for glaucous-winged gulls [10, 57, 58], also decline. Herring spawning stock biomass in the California Current Ecosystem decreased from 1980 to 2013, including all six major spawning regions in Washington State [56]. The Olympic/Discovery Bay region, which includes Protection Island, exhibited a 15% decline per year, much higher than the other five regions. In addition, populations of two other forage fish species in the California Current Ecosystem, northern anchovy (*Engraulis mordax*) and Pacific sardine (*Sardinops sagax*), had nearly collapsed by 2016 [27, 35].

Unpredictability of the environment can lead to behavioral plasticity, the heritable capacity of individuals to respond to different environments in different ways, and in which early learning and memory may play a role [9]. For example, *Daphnia magna* typically avoid light, but some genotypes exhibit an enhanced capacity to move toward or away from light when sensing chemicals secreted by predatory fish [17]; western fence lizards (*Sceloporus occidentalis*), when moved from their natal environments to one that is foreign, are able to alter their microhabitat selection [5]; and when water temperatures are low, female western gulls (*L. occidentalis*) spend more time incubating eggs than males, but when water temperatures are high, females and males spend about the same time incubating [46]. Results reported from our study suggest that ENSO-related climate variability and resulting resource fluctuations select for behavioral plasticity in glaucous-winged gulls, and this allows individuals to switch between alternative life history tactics depending on annual environmental cues such as local SST (Fig. 9.2). The same conclusion was reached following a study of western gulls [4]. Foraging plasticity in gulls apparently allows them to survive and reproduce at times when more specialized feeders, such as alcids and terns, find this difficult or impossible.

Previous studies on the diets of glaucous-winged gulls on Protection Island have been based on data from regurgitated boli. A study of 58 boli from this colony found that 43.0% contained fish remains, 37.9% mollusk remains, 27.7% arthropod remains, 19.0% egg components, and 12.1% bones and bone fragments, and most boli contained plant and inorganic material [34]. A later study showed that the odds of egg material or chick parts in 108 boli from cannibals were significantly higher than in 63 boli from non-cannibals, but the odds of finding fish remains in cannibal boli were significantly lower than those for non-cannibal boli [48]. Consequently, it appears that cannibals substituted eggs for fish as a protein source, especially those cannibals that consumed more than one egg each day. It is important to recognize, however, that data from boli are biased toward food items with hard parts [13].

Seabirds in temperate regions lay eggs in an annual pulse within a time window set by photoperiod [42]. Reproductive synchrony on a yearly time scale such as this is hypothesized to induce external predator satiation [16, 31]. Our study suggests that every-other-day clutch-initiation synchrony confers the same advantage in the face of egg cannibalization [26]. Mathematical modeling suggests that this latter type of synchrony is associated with longer egg-laying seasons [26], a conclusion consistent with the data trend exhibited in Fig. 9.1b. Based on our data, we hypothesize a trade-off between protection from external predation (a high level of annual synchrony with a short laying season [16]) and protection from conspecific predation (every-other-day clutch-initiation synchrony with a longer egg-laying season, this study) (Fig. 9.2). In glaucous-winged gulls this trade-off appears to be mediated by alternating feeding tactics. Our data, however, suggest that the colony as a whole did not exhibit a well-defined, dichotomous switch in tactics but instead responded in a more continuous fashion (with noise) as individuals changed tactics.

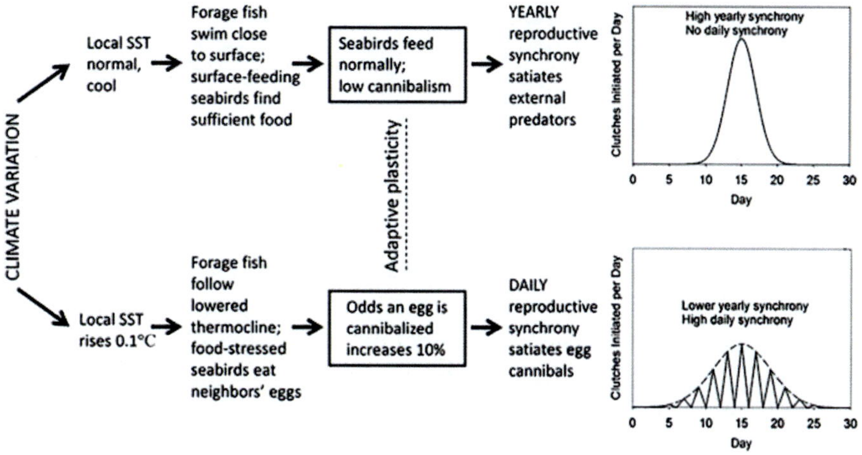

Figure 9.2: Conceptual model featuring alternative feeding and reproductive tactics of glaucous-winged gulls at Protection Island National Wildlife Refuge, Washington, predicated on local SST. Colony residents tended to switch between two sets of feeding and reproductive tactics depending on annual SST. In years of high SST, they exhibited high egg cannibalism, every-other-day clutch-initiation synchrony, and longer egg-laying seasons. In years of low SST, they exhibited lower egg cannibalism, no significant clutch-initiation synchrony, and shorter egg-laying seasons (Originally published in Weir et al., "Every-other-day clutch-initiation synchrony as an adaptive response to egg cannibalism in glaucous-winged gulls," *Wilson Journal of Ornithology* 132:575–586. Copyright © The Wilson Ornithological Society 2020. Used by permission.)

Behavioral tactics chosen during times of poor resource availability may delay essential functions until "better times," advantages that derive from behavioral plasticity. Glaucous-winged gulls breeding on Mandarte Island, British Columbia, for example, delayed the onset of egg-laying in response to cool weather [58]; glaucous-winged gulls nesting on Squab Island, Alaska, engaged in high levels of chick cannibalism when forced to feed on blue mussels (*Mytilus edulis*) [40]; and herring gulls (*L. argentatus*) nesting in Terschelling, the Netherlands, engaged in high rates of intraspecific chick predation when population density was high [53]. Indeed, individuals of many taxa resort to cannibalism in response to scarce food resources, a tactic that may function as a "lifeboat" mechanism by redirecting reproductive efforts to times when resources become more available [18, 19, 23, 47]. Gulls seem to be preadapted for switching feeding tactics given their generalized feeding habits [4].

Neither tactic illustrated in Fig. 9.2 is likely to be adaptive in every breeding season. When resource availability is low, cannibalism may enhance adult survivability and simply delay their reproductive output until better times; but when resource availability is high, cannibalism could be maladaptive, given that egg cannibals exhibit lower reproductive success [48]. In seasons with low resource availability, every-other-day clutch-initiation synchrony increases the chance that an egg avoids being cannibalized; but the extended period of egg-laying during those seasons may expose colony residents to higher levels of intraspecific predation, as from bald eagles.

The chance that egg cannibalism could contribute to colony extirpation raises an interesting question: could individual fitness and population fitness be in conflict? This possibly is suggested by the outcome of group selection experiments with flour beetles (*Tribolium castaneum*) [61, 62]. Another experiment with flour beetles uncovered an example of evolutionary suicide, which resulted from an increase in the frequency of a mutant allele that increased the rates of both reproduction and cannibalism yet led to population extinction (R. A. Desharnais, unpublished data). On Protection Island, an increase in the fitness of egg cannibals could play a role in colony extirpation. In such a case, however, nesting gulls would likely shift their activities to other colonies. Such a scenario would be compounded by the direct effect of bald eagle predation on gull eggs, chicks, and adults [22, 24, 28] and by habitat reduction due to encroaching vegetation [15].

Do our results have implications for the possible impact of long-term climate warming on gull behavior? Between 1950 and 2000, the mean sea surface temperature (SST) in the Strait of Juan de Fuca increased about 1 °C [55]. Until at least 2007 [30], this trend continued at a rate higher than the rate of global warming [52]. In a previous study on cannibalism on the Protection Island colony [22], a 0.1 °C rise in SST was associated with a 10% increase in the odds that an egg would be cannibalized ($\beta = 0.9148$; $\mathrm{OR} = exp[0.9148 \text{ x } 0.1] = 1.10; 1.10 - 1 = 0.10 = 10\%$). Thus, by exponential extrapolation, a 1 °C rise would be associated with a 150% increase in the odds of cannibalism ($\mathrm{OR} = \exp[0.9148 \text{ x } 1] = 2.50; 2.50 - 1 = 1.50 = 150\%$). This extrapolation extends beyond the scope of the Protection Island data set. However, at a glaucous-winged gull colony on nearby Mandarte Island, British Columbia, only 5.0% (83 of 1,652) of sample glaucous-winged gull eggs were lost or eaten in 1961–1962 [60], and 5.6% (26 of 466) of sample eggs were lost or eaten in 1979–1980 [59]. By comparison, from 2006 to 2011, 28.5% (969 of 3,398) of sample eggs on Protection Island were lost or eaten, and 21.4% of the total (727 of 3,398) were cannibalized [22]. Although these data are from a different colony studied decades before our study, they suggest the possibility that the trends observed for ENSO fluctuations during 2006–2014 might be extrapolated to account for the effects of long-term warming. It should be noted, however, that bald eagle numbers have increased markedly in the region since the studies on Mandarte Island, and these predators now prey extensively on seabird eggs, chicks, and adults now that their more traditional prey species have experienced dramatic declines [15, 21, 24, 28, 65] (see also Chap. 10).

We must mention an additional caution of using the effects of short-term fluctuations in SST as proxies for the effects of long-term warming. The mean September–May SST around Protection Island before the 2016 breeding season was 10.02 °C (NOAA Port Townsend, Washington, buoy PTWW1), almost 1 °C higher than during most years of our reproductive success and cannibalism study (Table 9.1) [22]. Yet we found few cannibal territories in

2016, although we did not do a colony-wide nest count that year. Environmental factors such as oxygen concentration, acidity, and temperature in the water surrounding Protection Island are strongly affected by water flowing in from the Pacific Ocean [30]. In 2014, the Blob, a mass of warm water and independent of ENSO-influenced upwelling, formed in the North Pacific and persisted until the summer of 2016 [33, 38, 44, 45]. The presence of the Blob led to complex responses by the marine ecosystem [44, 45], including sufficient food resources for surface-feeding gulls (SMH and JLH, personal observation), yet starvation conditions for deep-water-feeding rhinoceros auklets (*Cerorhinca monocerata*) in the region (S. F. Pearson, personal communication).

In conclusion, *in years of higher SST, glaucous-winged gulls nesting on Protection Island engaged in higher levels of egg cannibalism, higher levels of clutch-initiation synchrony, and earlier and longer-lasting egg-laying seasons than in years of lower SST. The chance that a given egg was cannibalized decreased under conditions of clutch-initiation synchrony.* Our results suggest that *short-term resource fluctuations linked with ENSO events may select for behavioral plasticity in gulls, allowing these long-lived birds to switch between alternative life history tactics.* The implications for longer-term global warming remain unknown.

References

1. Ainley DG (1994) Seasonal and annual patterns in the marine environment near the Farallones. In: Ainley DG (ed) Seabirds of the Farallon Islands. Stanford University Press, Stanford, pp 23–50

2. Ainley DG, Strong CS, Penniman TM, Boekelheide RJ (1994) The feeding ecology of Farallon seabirds. In: Ainley DG (ed) Seabirds of the Farallon Islands. Stanford University Press, Stanford, pp 51–127

3. Ainley DG, Sydeman WJ, Norton J (1995) Upper trophic level predators indicate interannual negative and positive anomalies in the California Current food web. Mar Ecol Prog Ser 118:69–79

4. Annett CA, Pierotti R (1999) Long-term reproductive output in western gulls: consequences of alternative tactics in diet choice. Ecology 80:288–297

5. Asbury DA, Adolph SC (2007) Behavioural plasticity in an ecological generalist: microhabitat use by western fence lizards. Evol Ecol Res 9:801–815

6. Atkins GJ, Reichert AA, Henson SM, Hayward JL (2017) Copulation call coordinates timing of head-tossing and mounting behaviors in neighboring glaucous-winged gulls (*Larus glaucescens*). Wilson J Ornith 129:560–567

7. Atkins GJ, Hayward JL, Henson SM (2021) How do gulls synchronize every-other-day egg laying? Wilson J Ornith 133:226–235

8. Barber RT, Chavez FP (1983) Biological consequences of El Niño. Science 222:1203–1210

9. Bhat A, Greulich MM, Martins EP (2015) Behavioral plasticity in response to environmental manipulation among zebrafish (*Danio rerio*) populations. PLoS ONE 10:e0125097

10. Bishop MA, Green SP (2001) Predation on Pacific herring (*Clupea pallasi*) spawn by birds in Prince Williams Sound, Alaska. Fish Oceanogr 10(Supplment 1):149–158

11. Blight LK (2011) Egg production in a coastal seabird, the glaucous-winged gull (*Larus glaucescens*), declines during the last century. PLoS ONE 6:e22027

12. Boersma PD (1998) Population trends of the Galápagos penguin: impacts of El Niño and La Niña. Condor 100:245–253

13. Brown KM, Ewins PJ (1996) Technique-dependent biases in determination of diet composition: an example with ring-billed gulls. Condor 98:34–41

14. Cai W, Borlace S, Lengaigne M, van Rensch P, Collins M, Vecchi G, Timmermann A, Santoso A, McPhaden MJ, Wu L, England MH, Wang G, Guilyardi E, Jin F-F (2014) Increasing frequency of extreme El Niño events due to greenhouse warming. Nat Clim Change 4:111–116

15. Cowles DL, Galusha JL, Hayward JL (2012) Negative interspecies interactions in a glaucous-winged gull colony on Protection Island, Washington. Northwest Nat 93:89–100

16. Darling FF (1938) Bird flocks and the breeding cycle. Cambridge University Press, Cambridge

17. De Meester L (1996) Evolutionary potential and local genetic differentiation in a phenotypically plastic trait of a cyclical parthenogen, *Daphnia magna*. Evolution 50:1293–1298

18. Elgar MA, Crespi BJ (1992) Ecology and evolution of cannibalism. J Evol Biol 7:1–12

19. Fox LR (1975) Cannibalism in natural populations. Ann Rev Ecol Syst 6:87–106

20. Grémillet D, Boulinier T (2009) Spatial ecology and conservation of seabirds facing global climate change: a review. Mar Ecol Prog Ser 391:121–137

21. Hayward JL, Galusha JG, Henson SM (2010) Foraging-related activity of Bald Eagles at a Washington seabird colony and seal rookery. J Raptor Res 44:19–29

22. Hayward JL, Weldon LM, Henson SM, Megna LC, Payne BG, Moncrieff AE (2014) Egg cannibalism in a gull colony increases with sea surface temperature. Condor Ornith Appl 116:62–73

23. Henson SM (1997) Cannibalism can be beneficial even when its mean yield is less than one. Theor Popul Biol 51:109–117

24. Henson SM, Desharnais RA, Funasaki ET, Galusha JG, Watson JW, Hayward JL (2019) Predator-prey dynamics of bald eagles and glaucous-winged gulls at Protection Island, Washington, USA. Ecol Evol 9:3850–3867

25. Henson SM, Hayward JL, Cushing JM, Galusha JG (2010) Socially induced synchronization of every-other-day egg laying in a seabird colony. Auk 127:571–580

26. Henson SM, Cushing JM, Hayward JL (2011) Socially induced ovulation synchrony and its effect on seabird population dynamics. J Biol Dyn 5:495–516

27. Hill KT, Crone PR, Dorval E, Macewicz BJ (2016) Assessment of the Pacific sardine resource in 2016 for U.S.A. management in 2016–17. NOAA Technical Memorandum NMFS. NOAA-TM-NMFS-SWFSC-562. NOAA. La Jolla, CA: NMFS Southwest Fisheries Science Center

28. Hipfner JM, Blight LK, Lowe RW, Wilhelm SI, Robertson GJ, Barrett RT, Ander-Nilssen T, Good TP (2012) Unintended consequences: how the recovery of sea eagle *Haliaeetus* spp. populations in the northern hemisphere is affecting seabirds. Mar Ornith 40:39–52

29. Hodder J, Graybill MR (1985) Reproduction and survival of seabirds in Oregon during the 1982–1983 El Niño. Condor 87:535–541

30. Irvine JR, Crawford WR (2013) State of physical, biological, and selected fishery resources of Pacific Canadian marine ecosystems in 2012. Research Document 2013/032. Canadian Science Advisory Secretariat, Fisheries and Oceans Canada, Pacific Region

31. James R, Bennett PG, Krause J (2004) Geometry for mutualistic and selfish herds: the limited domain of danger. J Theoret Biol 228:107–113

32. Kershner J, Samhouri JF, James CA, Levin PS (2011) Selecting indicator portfolios for marine species and food webs: a Puget Sound case study. PLoS ONE 6:e25248

33. Kintisch E (2015) "The Blob" invades Pacific, flummoxing climate experts. Science 348:17–18

34. Lindborg VA, Ledbetter JF, Walat JM, Moffett C (2012) Plastic consumption and diet of glaucous-winged gulls (*Larus glaucescens*). Mar Poll Bull 64:2351–2356

35. MacCall AD, Sydeman WJ, Davison PC, Thayer JA (2016) Recent collapse of northern anchovy biomass off California. Fish Res 175:87–94

36. Marra PP, Hobson KA, Holmes RT (1998) Linking winter and summer events in a migratory bird by using stable-carbon isotopes. Science 282:1884–1886

37. McGowan JA, Cayan DR, Dorman LM (1998) Climate-ocean variability and ecosystem response in the Northeast Pacific. Science 281:210–217

38. Moore S, Wold R, Stark K, Bos J, Williams P, Dzinbal K, Krembs C, Newton J (eds) (2016) Puget Sound marine waters 2015 overview. Northwest Fisheries Science Center, National Oceanographic and Atmospheric Administration, Seattle

39. Murphy RC (1936) Oceanic birds of South America, vol I. Macmillan, New York

40. Murphy EC, Day RH, Oakley KL, Hoover AA (1984) Dietary changes and poor reproductive performance in glaucous-winged gulls. Auk 101:532–541

41. Newsome AE (1964) Anoestrus in the red kangaroo *Megaleia rufa* (Demarest). Aust J Zool 12:9–17

42. Nicholls TJ, Goldsmith AR, Dawson A (1988) Photorefractoriness in birds and comparison with mammals. Physiol Rev 68:133–176

43. Patterson IJ (1965) Timing and spacing of broods in the black-headed gull *Larus ridibundus*. Ibis 107:433–450

44. Peterson W, Bond N, Robert M (2016) The Blob (part three): going, going, gone? PICES Press 24(1):46–48

45. Peterson W, Bond N, Robert M (2016) The Blob is gone but has morphed into a strongly positive PDO/SST pattern. PICES Press 24(2):46–48

46. Pierotti R (1981) Male and female parental roles in the western gull under different environmental conditions. Auk 98:532–549

47. Polis GA (1981) The evolution and dynamics of intraspecific predation. Ann Rev Ecol Syst 12:225–251

48. Polski AA, Osborn KJ, Hayward JL, Joo E, Mitchell AT, Sandler AG, Henson SM (2021) Egg cannibalism as a foraging tactic by less fit glaucous-winged gulls (*Larus glaucescens*). Wilson J Ornith 133:552–567

49. Richardson, AJ (2008) In hot water: zooplankton and climate change. ICES J Mar Sci 65:279–295

50. Schreiber RW, Schreiber EA (1984) Central Pacific seabirds and the El Niño Southern Oscillation: 1982 to 1983 perspectives. Science 225:713–716

51. Smith RS, Weldon LM, Hayward JL, Henson SM (2017) Time lags associated with effects of oceanic conditions on seabird breeding in the Salish Sea region of the northern California Current system. Mar Ornith 45:39–42

52. Snover AK, Mote PW, Binder LW, Hamlet AF, Mantua NJ (2005) Uncertain future: climate change and its effects on Puget Sound. Climate Impacts Group, Center for Science in the Earth System, Joint Institute for the Study of the Atmosphere and Oceans, University of Washington, Seattle

53. Spaans AL, de Wit AAN, van Vlaardingen MA (1987) Effects of increased population size in herring gulls on breeding success and other parameters. Stud Avian Biol 10:58–65

54. Stenseth NC, Mysterud A, Ottersen G, Hurrell JW, Chan KS, Lima M (2002) Ecological effects of climate fluctuations. Science 297:1292–1296

55. Strom A, Francis RC, Mantua NJ, Miles EL, Peterson DL (2004) North Pacific climate recorded in growth rings of geoduck clams: a new tool for paleoenvironmental reconstruction. Geophys Res Lett 31:L06206

56. Thompson SA, Sydeman WJ, Thayer JA, Weinstein A, Krieger KL, Hay D (2017) Trends in the Pacific herring (*Clupea pallasii*) metapopulation in the California Current Ecosystem. Calif Cooperative Oceanic Fish Invest 58:77–94

57. Trapp JL (1979) Variation in summer diet of glaucous-winged gulls in the western Aleutian islands: an ecological interpretation. Wilson Bull 91:412–419

58. Verbeek NAM (1979) Timing of primary molt and egg-laying in glaucous-winged gulls. Wilson Bull 91:420–425

59. Verbeek NAM (1986) Aspects of the breeding biology of an expanded population of glaucous-winged gulls in British Columbia. J Field Ornith 57:22–33

60. Vermeer K (1963) The breeding ecology of the glaucous-winged gull (*Larus glaucescens*) on Mandarte Island, BC. Occasional Papers of the British Columbia Provincial Museum, No. 13

61. Wade MJ (1977) An experimental study of group selection. Evolution 31:134–153

62. Wade MJ (1979) The primary characteristics of *Tribolium* populations group-selected for increased and decreased population size. Evolution 33:749–764

63. Walther G, Post RE, Convey P, Menzel A, Parmesan C, Beebee TJC, Fromentin J-M, Hoegh-Guldberg O, Bairlein F (2002) Ecological responses to recent climate change. Nature 416:389–395

64. Weir SK, Henson SM, Hayward JL, Atkins GJ, Polski AA, Watson W, Sandler AG (2020) Every-other-day clutch initiation synchrony as an adaptive response to egg cannibalism in glaucous-winged gulls (*Larus glaucescens*). Wilson J Ornith 132:575–586

65. White AF, Heath JP, Gisborne B (2006) Seasonal timing of bald eagle attendance and influence on activity budgets of glaucous-winged gulls in Barkley Sound, British Columbia. Waterbirds 29:497–500

Chapter 10

Predator-Prey Dynamics of Bald Eagles and Glaucous-Winged Gulls

In this chapter we use a modification of the classic Lotka-Volterra predator-prey ordinary differential equation model to test whether rebounding bald eagle numbers were primarily responsible for declining numbers of gulls nesting on Protection Island, Washington. We ask if changes in eagle numbers are primarily responsible for changes in numbers of gulls nesting on the island. The results of this research were initially published in the journal Ecology and Evolution *in 2019 by Shandelle Henson, Robert Desharnais, Eric Funasaki, Joseph Galusha, James Watson, and James Hayward [41].*

10.1 Introduction

Bald eagle (*Haliaeetus leucocephalus*) populations, once widespread in most of North America, experienced significant declines during the first half of the twentieth century. Concerns over these declines led to the outlawing of pesticides such as DDT, a reduction in the use of lead shot by hunters, and increased protection of these birds through a variety of other means [43, 76]. These efforts led to a major population rebound which resulted in one of the great twentieth-century conservation success stories [56]. The recovery of this species was especially evident in North America's Pacific Northwest, which provides ideal habitat for these raptors [26, 69, 75, 76].

One consequence of bald eagle recovery has been a sharp decline in seabird populations, already suffering from habitat destruction, gill netting, and overfishing [5, 8]. Populations of salmon and other fish taken by bald eagles have experienced serious declines, forcing increasing numbers of eagles to focus on other food sources such as seabirds (Fig. 10.1), which have always been at least a part of their diets [26, 67]. This shift has led to serious

J. M. Cushing et al., *Modeling Behavior and Population Dynamics*, Interdisciplinary Applied Mathematics 57, https://doi.org/10.1007/978-3-031-34283-7_10

declines in seabird populations, including local extirpations resulting from increased eagle predation pressure [26, 43, 58, 59, 70, 73, 77]. Resident eagles and the rates at which they consume prey seem to be more responsible than overwintering eagles for changes in marine food-web structure [34].

The effect of bald eagle predation on seabirds can be both direct and indirect [43, 59]. Predation on adults, juveniles, and eggs constitutes the most obvious direct effect (Fig. 10.1) [20, 36, 37], but another direct effect is the energy expended by stressed nesting and feeding seabirds in the presence of eagles [40, 59]. An indirect effect occurs when eagle disturbances displace breeding seabirds from their nests (Fig. 10.2), thus exposing their eggs and young to other predators [37]. Another indirect effect occurs when changes in distribution patterns of seabirds occur in response to eagle predation [55].

Figure 10.1: Bald eagles prey on seabirds of many species. (**a**) An adult bald eagle on Protection Island National Wildlife Refuge, Washington, has just killed an adult glaucous-winged gull on which it is standing. (**b**) A juvenile bald eagle on Protection Island feeds on an adult rhinoceros auklet (Photos by J. L. Hayward)

Breeding populations of glaucous-winged gulls (*Larus glaucescens*) in the Georgia Basin of the Salish Sea of Washington and British Columbia rose markedly during the first eight decades of the twentieth century. By 2010, however, populations of this species had declined to about half of what they were at their highest level [8, 70]. A study that included more southerly regions of the Salish Sea reported similar declines from 1975 to 2007 [10]. Since at least the 1940s [62] and probably before, Protection Island, located at the southeast end of the Strait of Juan de Fuca in the Salish Sea, has served as a breeding center for marine birds. By the early 1960s, a large glaucous-winged gull colony was established there [63]. Currently, glaucous-winged gulls, rhinoceros auklets (*Cerorhinca monocerata*), pigeon guillemots (*Cepphus columba*), and harbor seals (*Phoca vitulina*) breed there in large numbers. Resident and visiting bald eagles feed on gull adults, chicks, and eggs, on seal afterbirths and dead seals, and on adult rhinoceros auklets and pigeon guillemots [18, 35, 37].

Figure 10.2: Fish prey of bald eagles have dwindled in recent years, and consequently eagles have often refocused their attention to seabird colonies. Just the sight of an eagle flying over the colony causes a major colony disturbance during which the gulls temporarily abandon their nests. Gulls that fail to fly during these disturbances may become eagle prey (Photo by J. L. Hayward)

Few disturbances of the Protection Island gull colony by bald eagles were noted during the 1980s, and between 1980 and 1993 gull nest numbers increased from 3,797 to 5,189 (37%, Table 10.1). During the 1990s, however, a significant increase in numbers of eagle disturbances occurred [31, 37] with a concurrent decrease in numbers of nesting gulls (Table 1 in [18]), although the decrease occurred slightly later than decreases generally noted for the Salish Sea [8]. Systematic counts of gull nests on the upper plateau of Protection Island were not made, but some areas once containing gull nests were now vacated by these birds (J. L. Hayward, unpublished observations).

In view of the reduction in size of the Protection Island gull colony, two questions arise:

(1) Is bald eagle activity responsible for at least some of the decline?
(2) Is this colony, along with other gull colonies in the region, simply declining to historic levels?

Here we use mathematical modeling to determine if dynamic trends in numbers of gulls nesting on the Protection Island gull colony can be explained by the dynamics of numbers of active bald eagle nests in Washington State. We also used modeling methodology to learn if a stable coexistence of gulls and eagles on the island is possible.

Table 10.1: Observed numbers of glaucous-winged gull nests and bald eagles on Protection Island, Washington (PI), and numbers of occupied bald eagle territories in Washington State (WA)

Year	Gull nests (PI)	Occupied eagle territories (WA)	Observed eagles (PI)
1980	3,796	105	
1981		126	
1982	4,068	138	
1983		168	
1984	4,726	206	
1985		231	
1986		250	
1987	4,958	268	
1988		309	
1989	5,045	369	
1990		403	
1991	4,551	445	
1992		468	
1993	5,189	493	9
1994		547	8
1995		558	19
1996		594	16
1997	4,278	582	16
1998		666	24
1999			26
2000			17
2001		673	12
2002	2,472		23
2003			
2004	2,925		
2005		840	38
2006	2,281		
2007			
2008	2,830		
2009	3,018		
2010	2,495		
2011	2,364		
2012	2,093		
2013	1,850		
2014	1,589		
2015	1,832		
2016	2,512		

10.2 Mathematical Model

10.2.1 Classic Lotka-Volterra Predator-Prey Model

A classic paper on the relationship between population cycles of Canada lynxes (*Lynx canadensis*) and snowshoe hares (*Lepus americanus*) incorporated a century of records on numbers of pelts of both species purchased from trappers by the Hudson Bay Company [28]. The cycles highlighted in that study were modeled by the Lotka-Volterra predator-prey ordinary differential equation model [39, 49, 74]:

$$\begin{aligned} G' &= aG - \alpha GE \\ E' &= -bE + \beta GE. \end{aligned} \tag{10.1}$$

In this model, the "prime" denotes the derivative with respect to time, G and E are numbers of prey (gulls) and predators (eagles), $a > 0$ is the per capita growth rate of the prey population in the absence of predators, $b > 0$ is the per capita rate of decline of the predator population in the absence of prey, $\alpha > 0$ is the probability per unit time that a given prey individual will be taken by a given predator, and $\beta > 0$ is the conversion rate of prey into predators. This classic Lotka-Volterra model (10.1) has two important deficiencies:

(1) The prey population exhibits unlimited exponential growth in the absence of predators.
(2) The predator population declines exponentially to extinction in the absence of prey.

Neither of these conditions occur in most ecosystems because self-limitation eventually limits prey population growth; moreover, predators usually can switch prey and avoid starvation, which certainly is what happens with opportunistic foragers such as bald eagles [11].

10.2.2 Eagle-Gull Predator-Prey Model

In [64] a modified Lotka-Volterra predator-prey model, specifically the model

$$\begin{aligned} G' &= rG - \frac{r}{K}G^2 - \alpha GE \\ E' &= sE - \frac{s}{C}E^2 + \beta GE \ , \end{aligned} \tag{10.2}$$

was used to examine the interaction between the Protection Island gull population and bald eagle activity. Here G and E are numbers of gull and eagle pairs, respectively (as opposed to individual animals in the original Lotka-Volterra model). The inherent per pair growth rates for gulls and eagle pairs, at small population sizes, are r and s, respectively, and $r/K > 0$ and $s/C > 0$ are the rates of self-limitation. The parameter $\alpha > 0$ denotes the predation rate of eagles on gulls, and β is the conversion rate of gulls into eagle births. In the absence of the other species (when $\alpha = \beta = 0$), each population exhibits logistic growth with carrying capacities $K > 0$ for gulls and $C > 0$ for eagles.

Sustained predator-prey cycles are not predicted by model (10.2) but only equilibrium dynamics. Analysis of the equilibria of model (10.2) and their stability are given in [41]; here we summarize the possibilities.

Model (10.2) has four equilibrium states:

(1) An equilibrium $(0, 0)$ in which both species are absent

(2) An equilibrium $(K, 0)$ in which eagles are absent and gulls are at carrying capacity

(3) An equilibrium $(0, C)$ in which gulls are absent and eagles are at carrying capacity

(4) A coexistence equilibrium $(\overline{G}, \overline{E})$ with gull and eagle numbers given by

$$\overline{G} = \frac{sK(r - \alpha C)}{\alpha\beta KC + rs} \text{ and } \overline{E} = \frac{rC(s + \beta K)}{\alpha\beta KC + rs}. \tag{10.3}$$

There are two primary dynamic alternatives:

(i) If $r > aC$, then both $\overline{G}$ and $\overline{E}$ are positive in Eq. (10.3); thus, the coexistence state (10.3) is biologically feasible. The equilibria $(0, 0)$, $(K, 0)$, and $(0, C)$ are unstable. The coexistence equilibrium $(\overline{G}, \overline{E})$ is stable; specifically, it is either a stable spiral or a stable node. This means that gulls and eagles either approach the coexistence equilibrium via damped predator-prey oscillations (Fig. 10.3b), or they approach equilibrium in a non-oscillatory manner. In the latter case, early transient dynamics may resemble predator-prey oscillations, despite the fact they do not persist (Fig. 10.3c).

(ii) If $r < \alpha C$, the coexistence equilibrium (10.3) is not biologically possible because the equilibrium for the number of gulls $\overline{G}$ is negative. The equilibria $(0, 0)$ and $(K, 0)$ are still unstable. But the equilibrium $(0, C)$ in which gulls are absent and eagles are at carrying capacity C is now stable. That is, gulls approach extinction, but eagles approach their carrying capacity C. Early transient dynamics may resemble predator-prey oscillations before gulls eventually go extinct (Fig. 10.3d).

These alternative possibilities lead to the following biological interpretations. Note that the quantity αC is the rate at which gulls are predated by C eagle pairs at equilibrium. *If the inherent population growth rate of gull pairs is greater than the rate at which C eagle pairs can predate gulls, then the gull and eagle populations both survive and approach a positive* *coexistence equilibrium. If the inherent population growth rate of gull pairs is smaller than the rate at which C eagle pair can predate gulls, then the gulls go extinct and the eagles approach their carrying capacity C.*

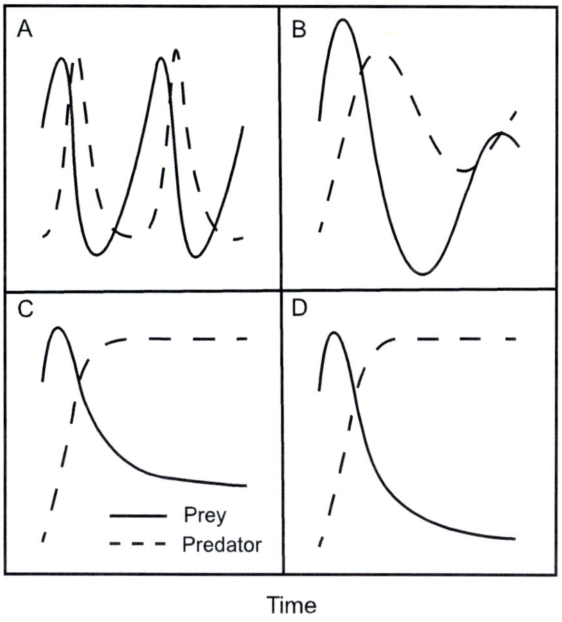

Figure 10.3: Predator-prey dynamics. (**a**) Classic predator-prey cycles. (**b**) Coexistence approached through damped oscillations. (**c**) Coexistence approached in a non-oscillatory fashion. (**d**) Extinction of prey (Originally published in Henson et al., "Predator-prey dynamics of bald eagles and glaucous-winged gulls at Protection Island, Washington, USA," Ecology and Evolution 9:3850–3867. Copyright © 2019, The Authors)

10.3 Methods

10.3.1 Gull Nest Counts from Violet Point, Protection Island

Glaucous-winged gull nest count data were collected between 1980 and 2016 on the Violet Point colony, Protection Island, Washington ($48°07'40''N, 122°55'3''W$). Protection Island is located at the southeastern end of the Strait of Juan de Fuca in the Salish Sea. The colony is occupied mostly by glaucous-winged gull x western gull (*L. occidentalis*) hybrids [6, 7], but given that most of the hybrids resemble glaucous-winged gulls more than western gulls [54, 57], we refer to them all as glaucous-winged gulls.

Counts of gull nests from 1980 to 2014 were made in squad fashion [32]. In 2015 and 2016, nests were mapped using ArcGIS Desktop 10 with data collected with a Trimble 6000 Series GPS.

10.3.2 Data on Occupied Eagle Territories in Washington State

Numbers of occupied bald eagle breeding territories from 1980 to 1998 were obtained from a previously published paper [75], whereas data for 2001 and 2005 were obtained from the Wildlife Resource Data System, Washington Department of Fish and Wildlife, Olympia, Washington (Table 10.1).

10.3.3 Counts of Nesting and Non-nesting Bald Eagles on Protection Island

Annual maximum numbers of adult and subadult bald eagle simultaneously observed from 1993 to 2002 (Table 10.1) were obtained from a previously published paper [37]. Counts for 2005 were from previously unpublished observations (J. L. Hayward).

10.3.4 Use of Occupied Bald Eagle Territories in Washington as a Proxy for Eagles on Protection Island

Numbers of eagles counted on Protection Island correlated strongly with numbers of occupied eagle territories in Washington ($\rho = 0.86$, $P = 0.006$, $N = 8$) for the 8 years for which these data sets overlapped (1993–1998, 2001, 2005; Table 10.1). The following relationship was produced by Poisson regression:

$$\ln(\text{PI Eagles}) = 0.5074 + 0.003701\,(\text{WA Eagle Territories}). \tag{10.4}$$

The regression line exhibited a significant slope coefficient ($P = 0.0057$), but a nonsignificant intercept ($P = 0.42$). A small amount of overdispersion was indicated by the estimated dispersion parameter of 1.23.

Because data on numbers of bald eagles observed on Protection Island were quite noisy, for modeling purposes we used statewide numbers in our analysis.

10.3.5 Model Parameterization

A number of factors must be considered when fitting theoretical models to ecological time series data. Both demographic and environmental stochasticity contribute to "process noise" in the data [23]. Moreover, inaccuracies in population density estimates can result in measurement error [12]. Methods are available to deal with these factors (e.g., [24]), but the data demands for these methods can be prohibitive. The data in our study, for example, include gaps in time series, missing data for one species some years, different sample sizes for gull and eagle numbers, and different magnitudes in numbers for the two species. Thus, application of techniques based on autoregressive time series that underlie many of the methods cannot be applied [9]. Parameter estimation was therefore challenging for this project, although sophisticated techniques are available to deal with deficiencies in ecological time series data (e.g., [15, 16]). Our aim in this study was to arrive at a good fit of the model to the data and to evaluate the implications of model predictions for wildlife management.

Model (10.2) parameter estimation was carried out as follows: first, we scaled state variables G and E, and the data, by dividing each one by the observed standard deviations σ_g and σ_e, such that $\widehat{G} = G/\sigma_g$ and $\widehat{E} = E/\sigma_e$; this operation scaled the data to comparable magnitudes. The derivatives, then, are $\widehat{G}' = G\prime/\sigma_g$ and $\widehat{E} = E\prime/\sigma_e$. Thus, model (10.2) can be rewritten in terms of the scaled variables as follows:

$$\begin{aligned} \widehat{G}' &= r\widehat{G} - \frac{r\sigma_g}{K}\widehat{G}^2 - \alpha\sigma_e\widehat{G}\widehat{E} \\ \widehat{E}' &= s\widehat{E} - \frac{s\sigma_e}{C}\widehat{E}^2 + \beta\sigma_g\widehat{G}\widehat{E}. \end{aligned} \tag{10.5}$$

In order to fit model (10.5) to the scaled data, the ode45 differential equation solver in Matlab was used to produce predicted model trajectories from 1980 to 2016. The scaled initial conditions $\widehat{g}_0$ and $\widehat{e}_0$ were treated as parameters to be estimated. Given a vector of parameter estimates $\theta = (\widehat{g}_0, \widehat{e}_0, r, K, \alpha, s, C, \beta)$, residuals on the log scale were computed to account for environmental stochasticity, which is approximately additive on the log scale [19, 22]:

$$\begin{aligned} \gamma_t &= \ln\left(\widehat{G}_t\right) - \ln\left(\widehat{g}_t(\theta)\right) \\ \varepsilon_t &= \ln\left(\widehat{E}_t\right) - \ln\left(\widehat{e}_t(\theta)\right), \end{aligned} \tag{10.6}$$

where $\widehat{g}_t(\theta)$ and $\widehat{e}_t(\theta)$ are predicted values obtained by numerically integrating model (10.5) from 1980 to 2016 data, using parameters θ, the parameters $\widehat{g}_0$ and $\widehat{e}_0$ as initial conditions, and the standard deviations σ_g and σ_e for gulls and eagles, respectively. We obtained best fit parameters $\overline{\theta}$ by minimizing the sum of the root mean squares (RMS) of the residuals

$$RMS(\theta) = \sqrt{\frac{\sum_{t=1}^{n_g} \gamma_t^2}{n_g}} + \sqrt{\frac{\sum_{t=1}^{n_e} \varepsilon_t^2}{n_e}} \tag{10.7}$$

as a function of θ, where n_g and n_ϵ are the numbers of residuals for gulls and eagles, respectively, using the fminsearch downhill search algorithm in Matlab.

Four considerations are involved in the fitting method described above.

(1) The two time series are unpaired, because in some years estimates are available for only one of the two species. As a result, we are unable to view the data for each year as a standard bivariate observation which would allow for a traditional sum of squared errors; moreover, we cannot use one-step predictions in determining residuals.

(2) Different numbers of observations were made for the two species; thus, the sum of the squared residuals must be scaled by the number of observations.

Otherwise, the parameter estimates would bias the species with more observations.
(3) Parameters cannot be estimated using separate RMS values because the equations are coupled, and the fit of one species influences the fit of the other species.
(4) The magnitudes of observations differ for the two species. Consequently, we scaled the data by the standard deviations such that the two terms in Eq. (10.7) would be commensurate and the parameter estimates would not be biased in favor of fitting the species with the higher numbers.

Diagnostic analyses were performed on the eagle and gull residuals to check for independence and normality. Details are given in [41].

10.3.6 Goodness-of-Fit

A generalized R^2 was used to evaluate the goodness-of-fit of the scaled model (10.5) to the scaled data:

$$R^2 = 1 - \frac{RMS_M}{RMS_T} \tag{10.8}$$

where RMS_M equals the fitted root mean square using model (10.5) as the predictor, and using Eqs. (10.6) to compute the residuals, and RMS_T is the sum of the root mean squares using the data mean as the predictor and computing the residuals in Eq. (10.7) for RMS_T in the following way:

$$\begin{aligned} \gamma_t &= \ln\left(\widehat{G}_t\right) - \text{mean}\left[\ln\left(\widehat{G}_t\right)\right] \\ \varepsilon_t &= \ln\left(\widehat{E}_t\right) - \text{mean}\left[\ln\left(\widehat{E}_t\right)\right]. \end{aligned} \tag{10.9}$$

The adjusted goodness-of-fit R_A^2 also was determined by

$$R_A^2 = 1 - \frac{n_g + n_e - 2}{n_g + n_e - p - 2}(1 - R^2), \tag{10.10}$$

where $p = 8$ equals the number of estimated model parameters. In general, $R_A^2 < R^2$ given that it takes into account the number of estimated parameters and penalizes the goodness-of-fit as p increases. See [41].

10.3.7 Confidence Intervals for Parameters

Bootstrapping techniques can be utilized to obtain confidence intervals for the estimated parameters, once a deterministic model has been fitted to population time series data [21, 30]. With replacement we randomly sampled from the model residuals $\gamma_1, \gamma_2, \ldots, \gamma_{n_g}$

and $\varepsilon_1, \varepsilon_2, \ldots, \varepsilon_{n_e}$ to form sets of surrogate residuals $\gamma_1^*, \gamma_2^*, \ldots, \gamma_{n_g}^*$ and $\varepsilon_1^*, \varepsilon_2^*, \ldots, \varepsilon_{n_e}^*$. When sampling, the time order of the residuals was ignored [21, 30].

Surrogate data were created from surrogate residuals:

$$\widehat{G}_t^* = \widehat{g}_t\left(\widetilde{\theta}\right)\exp(\gamma_t^*) \qquad (10.11)$$
$$\widehat{E}_t^* = \widehat{e}_t\left(\widetilde{\theta}\right)\exp(\varepsilon_t^*).$$

We estimated point parameters for each surrogate data set using the methods explained above. Using an independent random sampling of the original residuals for each iteration, we repeated this process $n_s = 2{,}000$ times. If the fminsearch algorithm did not converge on a solution within 1,000 functional evaluations, we repeated these steps for a new set of surrogate data. This occurred at a rate of 21.6%. A lack of convergence in some of the bootstrap realizations was because the time series are quite short so that the overall number of residuals is small. This frequently leads to sets of resampled residuals that negatively impact the rate of convergence for the minimization algorithm. We verified the fact that 2,000 repetitions were sufficient and that the nonconvergent bootstrap realizations were not a problem by independently repeating the anlyses several times with only trivial variations in the results.

A set of bootstrapped parameter estimates $\theta_1^*, \theta_2^*, \ldots, \theta_{nt}^*$ was generated which should reflect the variation seen in the best-fit parameters, assuming model (10.5) is valid and the observed residuals from the model are random effects with no autocorrelation or cross-correlation.

Ranking the parameter estimates for the surrogate data sets and determining the 2.5th and 97.5th percentiles allowed us to obtain the 95% confidence intervals for the point parameter estimates.

10.4 Results

The relationship $r > \alpha C$ holds for the fitted point estimates of the parameters (Table 10.2). This indicates a stable coexistence equilibrium. Parameter β is essentially zero. Consequently, even without dependence of the gull population, the eagle population is predicted to grow. Our model predicts that despite an estimated carrying capacity of $K = 7{,}395$, gull nests on the Protection Island colony will stabilize at 1,072 (Table 10.2). The number of bald eagle territories in Washington is predicted to equilibrate at 823 territories, equal to the predicted carrying capacity C (Table 10.2). The goodness-of-fit indicates that the model accounts for at least 77% of the variability in the data ($R^2 = 0.819$ and $R_{\mathrm{A}}^2 = 0.772$). Given the predicted equilibrium of 823 eagle territories in the state, Eq. (10.4) predicts an equilibrium of 35 eagle visitors to Protection Island. Figure 10.4 shows fitted model predictions for the years 1980–2080.

Table 10.2: Point estimates for the parameters in model (10.2) and the initial conditions, with 95% confidence intervals. Estimates for the equilibria, with 95% confidence intervals

	Estimate	95% CI
Parameter		
G_0	3663	$(2847, 4422)$
E_0	106.7	$(101.8, 112.1)$
r	0.2327	$(0.1044, 06.642)$
K	7395	$(6186, 10444)$
α	0.0002417	$(0.0001460, 0.0005857)$
s	0.1809	$(0.1682, 0.1935)$
C	823.1	$(768.8, 889.0)$
β	1.876×10^{-23}	$(2.652x10^{-63}, 2.082 \times 10^{-13})$
Equilibrium		
G (gull nests)	1072	$(0, 1981)$
E (eagle territories)	823.1	$(768.8, 889.0)$
PI Eagles	34.94	(28.58, 44.59)

The model residuals for eagles and gulls in our analysis (Eq. 10.6) show no evidence of significant violations of model assumptions. Figure 10.5 shows time series and normal quantile-quantile plots of the model residuals. Further details are given in [41].

With the exception of β, which was always close to zero, a scatterplot matrix of the bootstrapped parameter estimates is shown in Fig. 10.6. The diagonal plot are histograms showing the distribution of the 2,000 estimates for each parameter. Unusual properties for the distributions, such as high skewness or multiple modes, are not depicted by these histograms. Pairwise relationships between the parameters for the 2,000 estimated parameter vectors are shown by the off-center scatterplots. Strong or unusual dependencies between parameter estimates, such as for parameters r and a, are revealed by these plots, suggesting that large estimates of gull population growth rates comport with large estimates of eagle predation rates on gulls and vice versa.

The vector of parameter estimates predicted gull colony extinction in 13.4% of cases. Predicted equilibria were derived for each of the 2,000 bootstrapped parameter vectors and plotted for both species (Fig. 10.7). The scatterplot of the 2,000 estimates shows r versus α with the coexistence cutoff condition $r = \alpha C$ for the point estimate of C shown as a dashed line (Fig. 10.8). Estimates below that line lead to colony extinction. The dotted lines are $r = \alpha C$ using the lower and upper 95% CI bounds for parameter C. The area between the dotted lines might be considered an uncertainty parameter region for coexistence.

We cannot predict a time to extirpation with our deterministic model, but for those cases (267 of 2,000) for which the bootstrapped parameter vectors predict extinction, we

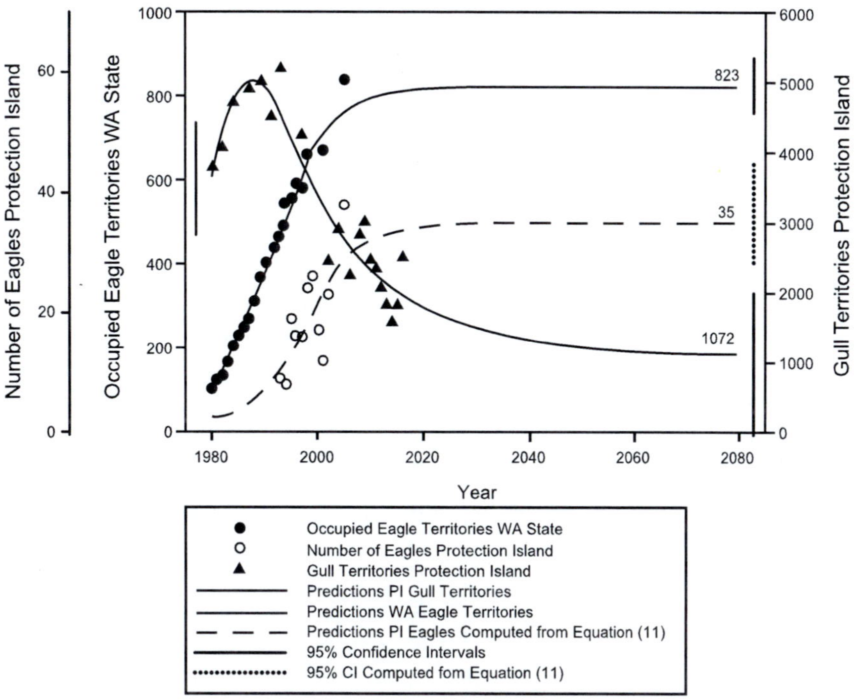

Figure 10.4: Observed data and model predictions. Observed data (symbols) and predictions of models (10.2) and (10.4) from 1980 to 2080 (curves) are numbers of occupied bald eagle territories in Washington State (solid circles, light solid curve); numbers of bald eagles observed at Protection Island, Washington (open circles, dashed curve); and numbers of glaucous-winged gull nests at the Violet Point colony on Protection Island (triangles, dark solid curve). The 95% confidence intervals for fitted initial conditions and predicted equilibria are marked with vertical lines on the left- and right-hand sides of the graph, respectively (Originally published in Henson et al., "Predator-prey dynamics of bald eagles and glaucous-winged gulls at Protection Island, Washington, USA," Ecology and Evolution 9:3850–3867. Copyright © 2019, The Authors)

can estimate the amount of time it takes for numbers of gulls to fall below a threshold. *By arbitrarily defining the threshold as 10% of the estimated carrying capacity for gulls, the estimated mean year within a 95% confidence interval (2022, 2059) in which extirpation of the gull colony occurs is 2039; for a 5% threshold of K within a 95% confidence interval (2040, 2122), the mean is 2073.*

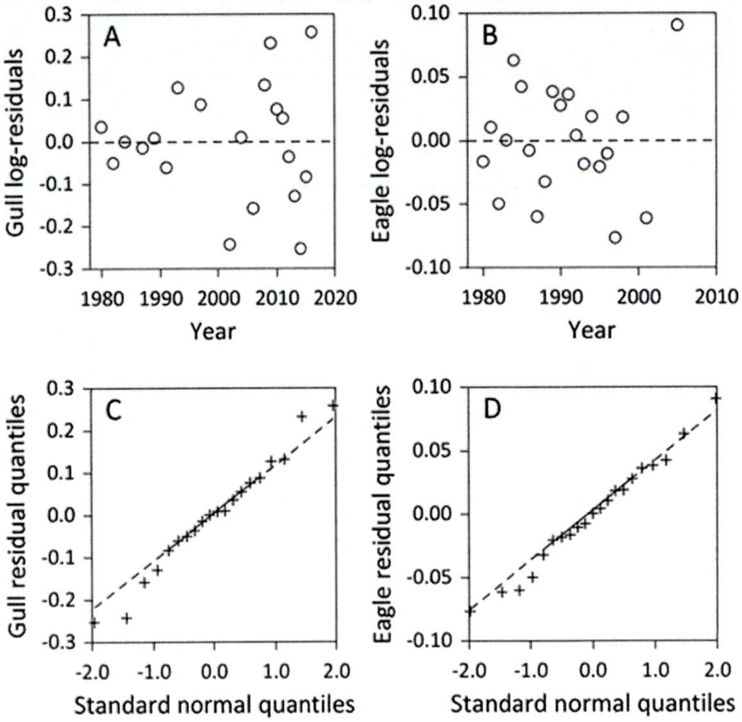

Figure 10.5: Time series plots of model residuals for (**a**) glaucous-winged gulls and (**b**) bald eagles. Normal quantile-quantile plots of model residuals for (**c**) glaucous-winged gulls and (**d**) bald eagles (Originally published in Henson et al., "Predator-prey dynamics of bald eagles and glaucous-winged gulls at Protection Island, Washington, USA," Ecology and Evolution 9:3850–3867. Copyright © 2019, The Authors)

10.5 Discussion

Our model demonstrates a strong dynamic interaction between the bald eagle population in Washington State and the numbers of glaucous-winged gull nests in the Protection Island colony. Given the point estimates of the parameters, this interaction reveals a Lotka-Volterra-type dynamic that predicts long-term coexistence and equilibrium for the two species. Predator-prey oscillations as depicted in the original Lotka-Volterra model [49, 50, 74], however, are not predicted. It is worth noting, though, that the model predicts extirpation of the gull colony for some parameters within the 95% confidence intervals about the point estimates.

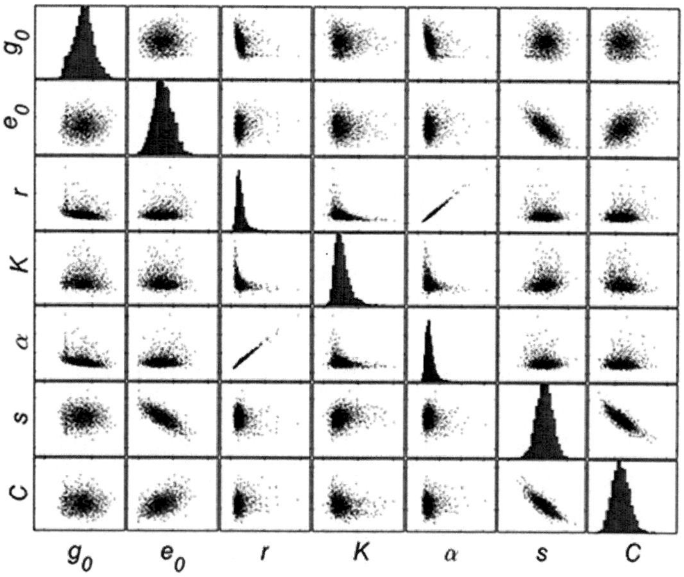

Figure 10.6: Scatterplot matrix of 2,000 bootstrapped parameter estimates. Beta was always approximately zero (Originally published in Henson et al., "Predator-prey dynamics of bald eagles and glaucous-winged gulls at Protection Island, Washington, USA," Ecology and Evolution 9:3850–3867. Copyright © 2019, The Authors)

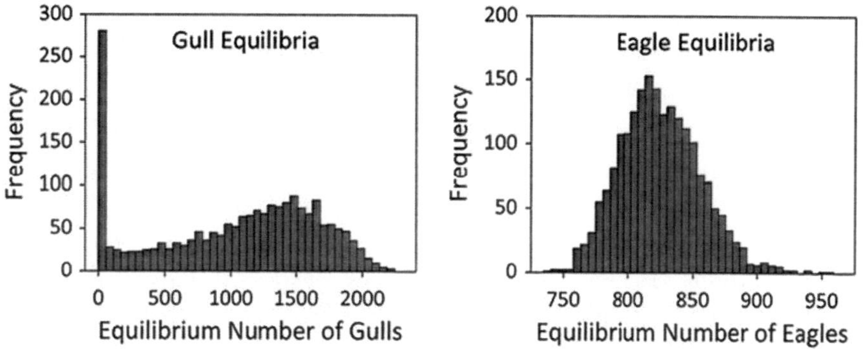

Figure 10.7: Histograms of the equilibria for both species, based on 2,000 bootstrapped parameter estimates (Originally published in Henson et al., "Predator-prey dynamics of bald eagles and glaucous-winged gulls at Protection Island, Washington, USA," Ecology and Evolution 9:3850–3867. Copyright © 2019, The Authors)

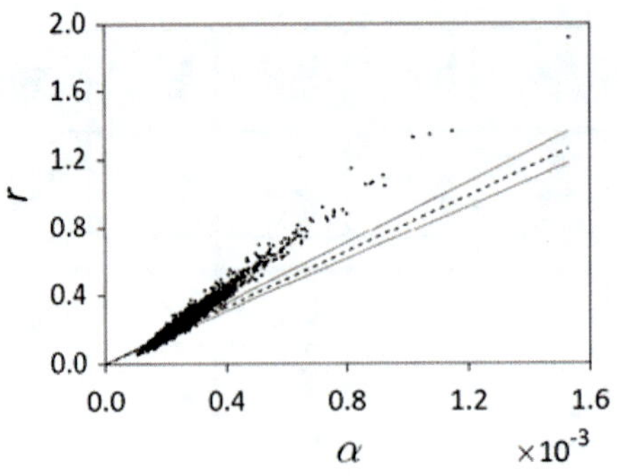

Figure 10.8: Scatterplot of the 2,000 estimates of r versus α with the coexistence cutoff condition $r = \alpha C$ for the point estimate of C appearing as a dashed line. Estimates below that line lead to gull colony extinction. The dotted lines are $r - \alpha C$ using the lower and upper 95% CI bounds for parameter C, The area between the dotted lines could be considered an "uncertainly parameter region" for coexistence

The estimate of carrying capacity for bald eagle territories in Washington State (823) is currently realized or exceeded; as of 2018 there were more than 1,000 nesting territories, although not all were active in the same years. In 2001, 923 territories were checked, of which 705 were active, and in 2005, 1,158 territories were checked, of which 893 were active (J. W. Watson, unpublished data).

Bald eagles have nested on Protection Island since at least the 1920s [17], and one or two eaglets were hatched during many years since the early 1980s [37]. The island attracts numerous transient eagles during the breeding season when gull chicks are hatching and harbor seal afterbirths and dead seal pups are readily available. Eagle numbers typically peak during the second week in July [37]. Resident eagles sometimes chase transient eagles back to the mainland (J. L. Hayward, unpublished observations).

As bald eagle populations have recovered, the impact of these predators on seabirds has become increasingly obvious given that some marine fish populations on which eagles depended have declined [2, 3, 68, 71]. Bald eagles have been responsible for declines in local populations of common murres (*Uria aalge)* [42, 59], double-crested cormorants (*Nannopterum auritum*) [14, 33], pelagic cormorants (*Urile pelagicus*) [13, 14, 33], great blue herons (*Ardea herodias*) [73], western grebes (*Aechmophorus occidentalis*) [10], and glaucous-winged gulls [37, 70]. In some cases the declines have been dramatic. A 131-km stretch along the Oregon coast, for example, once supported more than 380,000 breeding pairs of common murres, but successful reproduction by these birds along this stretch was virtually absent at the time of our study. Murres harassed by bald eagles usually abandon the breeding process [43]. In Northern Europe, white-tailed eagles (*Haliaeetus albicilla*), whose populations also have rebounded, have exerted a similar effect on regional seabirds [43].

The reduction in numbers of gulls nesting on Protection Island mirrored declines and the ultimate extirpation of double-crested and pelagic cormorants on the island. Hundreds of double-crested cormorants thrived on the island in two colonies for many years, and a small pelagic cormorant colony occurred for several years during our study. Eagle disturbances, however, became more frequent causing the cormorants to flee their nests. All three cormorant colonies had been vacated by 2007 (J. L. Hayward, unpublished observations). Rhinoceros auklets (*Cerorhinca monocerata*) continued to breed in large numbers on the island during our study [61], despite the fact the most common prey remains beneath an active bald eagle nest on the island were of this species [37]. The auklets appear to be taken by eagles mostly during predawn when they are leaving their nest burrows to forage.

At the point parameter estimates, our model predicts coexistence between the eagles and gulls, but the possibility of extirpation occurs within 95% confidence intervals. The model predicted extirpation in 13.4% of the 2,000 bootstrapped parameter vectors. If we restrict our prediction to the 1,788 cases for which the elements of the bootstrapped parameter triplets (r, α, C) are included within their 95% confidence intervals, in 11% of the cases gull colony extirpation was predicted. The estimated mean year in which extirpation occurs is 2039 with a 95% confidence interval of (2022, 2059), if we define the threshold for extirpation as 10% of the carrying capacity for gulls. In the case of a 5% threshold, the mean year for extirpation is 2073 with a 95% confidence interval (2040, 2122). A large gull colony on Colville Island, located 33 km north of Protection Island, increased from 1,273 breeding pairs in 1963 to 1,808 pairs in 1975 [1, 72]. But by the year 2000, only approximately 20 pairs nested in the colony, and more recently no nesting gulls were observed there (J. L. Hayward, unpublished data). Although it is unknown if bald eagles caused the decline and eventual extirpation of this colony, disturbance and predation by eagles were documented there during the 1970s [36].

The Protection Island ecosystem would be impacted by extirpation of the gull colony. First, extirpation of the gulls would affect the island's vegetation. Gulls alter patterns of vegetation by trampling, collecting nest material, digging nest scrapes, depositing feces, and regurgitating undigestible food components [27, 47, 66]. The decomposition of adult and juvenile corpses contributes nutrients to the soil [29, 48]. Consequently, extirpation of the gulls would result in significant vegetation changes and changes in the organisms that depend on that vegetation [66]. There also would be a loss of nutrients by runoff to the waters encircling Protection Island [44, 46, 52]. Extirpation would also eliminate a food source for both resident and transient bald eagles; in response, eagles might relocate to other feeding grounds [51].

Gull populations are impacted by factors other than bald eagles. Gull populations grew substantially during the twentieth century due in part to the availability of food from landfills [1, 25, 45, 70], and closure of landfills is thought to be responsible for gull population declines [60]. The closure of the Coupeville Landfill in 1992 [4], a popular feeding locality for Protection Island gulls [65], was followed by a 10-year decline in gull nest counts at the colony (Fig. 10.4). Forage fish declines [8, 53, 71] may also have played a role. Dune grass cover, which restricts gull nesting except along the edges, increased on the colony from 2.5 ha in 1980 to 6.6 ha by 2009 [18], although ample nest habitat remained, suggesting this factor was not limiting. Increasing sea surface temperatures associated with El Niño events, and possibly climate change, are known to increase the rate of egg cannibalism and decrease reproductive success at the Protection Island gull colony [38]. See Chap. 7. We do not know how much these confounding factors, which are not included in our model, contributed to the decline in gull numbers. *Nonetheless, it is significant that the model successfully predicted the reduction in gull numbers based only on eagle dynamics. This is evidence that eagle dynamics constitute one of the most important factors contributing to the decline of nesting gulls on the island.*

10.6 Conclusion

Our results indicate that the population dynamics of glaucous-winged gulls nesting on Protection Island from 1980 to 2016 can be predicted on the basis of numbers of active bald eagle territories in Washington State (generalized $R^2 = 0.82$). *Thus, our hypothesis that the dynamics of the gull population on Protection Island are largely due to the decline and subsequent recovery of bald eagles in the region is supported. Moreover, we predicted that, with 95% confidence, gull and eagle populations on Protection Island will continue to coexist, although there is a small possibility the gull colony will disappear.*

Our study highlights the point that management of one species can exert dramatic effects on other species, although the nature of these effects may be uncertain. This has implications for the management not only of gulls but also of other seabirds. Specifically, in order to document and mitigate trends that could lead to colony extirpation, managers should monitor the numbers of seabird nests, as well as predator activity on a colony.

References

1. Amlaner CJ, Hayward JL, Schwab ER, Stout JF (1977) Increases in a population of nesting glaucous-winged gulls disturbed by humans. Murrelet 58:18–20

2. Anderson EM, Bower JL, Nysewander DR, Evenson JR, Lovvorn JR (2009) Changes in avifaunal abundance in a heavily used wintering and migration site in Puget Sound, Washington, during 1966–2007. Mar Ornith 37:19–27

3. Anderson EM, Lovvorn JR, Esler D, Boyd WS, Stick KC (2009) Using predator distributions, diet, and condition to evaluate seasonal foraging sites: sea ducks and herring spawn. Mar Ecol Prog Ser 386:287–302

4. Anonymous (2001) Trash talk: waste division heads it up, moves it out. Whidbey News-Times, 6 June. Accessed online on 2 Jul 2018. http://www.whidbeynewstimes.com/news/trash-talk-waste-division-heads-it-up-moves-it-out/

5. Atkins N, Heneman B (1987) The dangers of gill netting to seabirds. Am Birds 41:1395–1403

6. Bell DA (1996) Genetic differentiation, geographic variation and hybridization in gulls of the *Larus glaucescens-occidentalis* complex. Condor 98:527–546

7. Bell DA (1997) Hybridization and reproductive performance in gulls of the *Larus glaucescens-occidentalis* complex. Condor 99:585–594

8. Blight LK, Drever MC, Arcese P (2015) A century of change in glaucous-winged gull (*Larus glaucescens*) populations in a dynamic coastal environment. Condor 117:108–120

9. Bolker BM, Gardner B, Maunder M, Berg CW, Brooks M, Comita L, Crone E, Cubaynes S, Davies T, de Valpine P, Ford J, Gimenez O, Kéry M, Kim EJ, Lennert-Cody C, Magnusson A, Martell S, Nash J, Nielsen A, Regetz J, Skaug H, Zipkin E (2013) Strategies for fitting nonlinear ecological models in R, AD Model Builder, and BUGS. Methods Ecol Evol 4:501–512

10. Bower JL (2009) Changes in marine bird abundance in the Salish Sea: 1975 to 2007. Mar Ornith 37:9–17

11. Buehler DA (2022) Bald eagle (*Haliaeetus leucocephalus*), version 2.0. In: Rodewald PG, Mlodinow SG (eds) Birds of World. Cornell Lab of Ornithology, Ithaca

12. Carpenter SR, Cottingham KL, Stow CA (1994) Fitting predator-prey models to time series with observation errors. Ecology 75:1254–1264

13. Carter HR, Hebert PN, Clarkson PV (2009) Decline of pelagic cormorants in Barkley Sound, British Columbia. Wildl Afield 4:3–32

14. Chatwin TA, Mather MH, Giesbrecht TD (2002) Changes in pelagic and double-crested cormorant nesting populations in the Strait of Georgia, British Columbia. Northwest Nat 83:109–117

15. Clark JS (2007) Models for ecological data: an introduction. Princeton University Press, Princeton

16. Clark JS, Bjørnstad ON (2004) Population time series: process variability, observation errors, missing values, lags, and hidden states. Ecology 85:3140–3150

17. Cowles DL, Hayward JL (2008) Historical changes in the physical and vegetational characteristics of Protection Island, Washington. Northwest Sci 82:174–184

18. Cowles DL, Galusha JG, Hayward JL (2012) Negative interspecies interactions in a glaucous-winged gull colony on Protection Island, Washington. Northwest Nat 93:89–100

19. Cushing JM, Costantino RF, Dennis B, Desharnais RA, Henson SM (2003) Chaos in ecology: experimental nonlinear dynamics. Academic Press, San Diego

20. DeGange AR, Nelson JW (1982) Bald eagle predation on nocturnal seabirds. J Field Ornith 53:407–409

21. Dennis B, Desharnais RA, Cushing JM, Henson SM, Costantino RF (2001) Estimating chaos and complex dynamics in an insect population. Ecol Monogr 712:77–303

22. Dennis B, Munholland PL, Scott JM (1991) Estimation of growth and extinction parameters for endangered species. Ecol Monogr 61:115–143

23. Dennis B, Ponciano JM, Lele SR, Taper ML, Staples DF (2006) Estimating density dependence, process noise, and observation error. Ecol Monogr 76:323–341

24. de Valpine P, Hastings A (2002) Fitting population models incorporating process noise and observation error. Ecol Monogr 72:57–76

25. Duhem C, Roche PK, Vidal E, Tatoni T (2008) Effects of anthropogenic food resources on yellow-legged gull colony size on Mediterranean islands. Popul Ecol 50:91–100

26. Elliott KH, Elliott JE, Wilson LK, Jones I, Stenerson K (2011) Density-dependence in the survival and reproduction of bald eagles: linkages to chum salmon. J Wildl Manag 75:1688–1699

27. Ellis JC, Fariña JM, Witman JD (2006) Nutrient transfer from sea to land: the case of gulls and cormorants in the Gulf of Maine. J Anim Ecol 75:565–574

28. Elton C, Nicholson M (1942) The ten-year cycle in numbers of the lynx in Canada. J Anim Ecol 11:215–244

29. Emslie SD, Messenger SL (1991) Pellet and bone accumulation at a colony of western gulls (*Larus occidentalis*). J Vertebr Paleontol 11:133–136

30. Falck W, Bjornstad ON, Stenseth NC (1995) Bootstrap estimated uncertainty of the dominant Lyapunov exponent for Holarctic microtine rodents. Proc R Soc London B 261:159–165

31. Galusha JG, Hayward JL (2002) Bald eagle activity at a gull colony and seal rookery on Protection Island, Washington. Northwest Nat 83:23–25

32. Galusha JG, Vorvick B, Opp MR, Vorvick PT (1987) Nesting season censuses of seabirds on Protection Island, Washington. Murrelet 68:103–107

33. Harris ML, Wilson LK, Elliott JE (2005) An assessment of PCBs and OC pesticides in eggs of double-crested (*Phalacrocorax auritus*) and pelagic (*P. pelagicus*) cormorants from the west coast of Canada, 1970 to 2002. Ecotoxicology 14:607–625

34. Harvey CJ, Good TP, Pearson SF (2012) Top–down influence of resident and overwintering bald eagles (*Haliaeetus leucocephalus*) in a model marine ecosystem. Can J Zool 90:903–914

35. Hayward JL (2009) Bald eagle predation on harbor seal pups. Northwest Nat 90:51–53

36. Hayward JL, Gillett WH, Amlaner CJ, Stout JF (1977) Predation on gulls by bald eagles in Washington. Auk 94:375

37. Hayward JL, Galusha JG, Henson SM (2010) Foraging-related activity of bald eagles at a Washington seabird colony and seal rookery. J Raptor Res 44:19–29

38. Hayward JL, Weldon LM, Henson SM, Megna LC, Payne BG, Moncrieff AE (2014) Egg cannibalism in a gull colony increases with sea surface temperature. Condor Ornith Appl 116:62–73

39. Henson SM (2012) Phase plane analysis. In Hastings A, Gross L (eds) Encyclopedia of theoretical ecology. University of California Press, Berkeley, pp 538–545

40. Henson SM, Weldon LM, Hayward JL, Greene DJ, Megna LC, Serem MC (2012) Coping behaviour as an adaptation to stress: post-disturbance preening in colonial seabirds. J Biol Dyn 6:17–37

41. Henson SM, Desharnais RA, Funasaki ET, Galusha JG, Watson JW, Hayward JL (2019) Predator-prey dynamics of bald eagles and glaucous-winged gulls at Protection Island, Washington, USA. Ecol Evol 9:3850–3867

42. Hipfner JM, Morrison KW, Darvill R (2011) Peregrine Falcons enable two species of colonial seabirds to breed successfully by excluding other aerial predators. Waterbirds 34:82–88

43. Hipfner JM, Blight LK, Lowe RW, Wilhelm SI, Robertson GJ, Barrett RT, Anker-Nilssen T, Good TP (2012) Unintended consequences: how the recovery of sea eagle *Haliaeetus* spp. populations in the northern hemisphere is affecting seabirds. Mar Ornith 40:39–52

44. Hutchinson GE (1950) Survey of contemporary knowledge of biogeochemistry, 3. The biogeochemistry of vertebrate excretion. Bull Am Mus Nat Hist 96:3–4

45. Kadlec J, Drury WH (1968) Structure of the New England herring gull population. Ecology 49:645–676

46. Leentvaar P (1967) Observations on guanotrophic environments. Hydrobiologia 29:441–489

47. Lindborg VA, Ledbetter JF, Walat JM, Moffett C (2012) Plastic consumption and diet of glaucous-winged gulls (*Larus glaucescens*). Mar Poll Bull 64:2351–2356

48. Lord WD, Burger JF (1984) Arthropods associated with herring gull (*Larus argentatus*) and great black-backed gull (*Larus marinus*) carrion on islands in the Gulf of Maine. Environ Entomol 13:1261–1268

49. Lotka AJ (1925) Elements of physical biology. Williams and Wilkins, Baltimore

50. Lotka AJ (1932) The growth of mixed populations: two species competing for a common food supply. J Wash Acad Sci 22:461–469

51. McClelland BR, Young LS, McClelland PT, Crenshaw JG, Allen HL, Shea DS (1994) Migration ecology of bald eagles from autumn concentrations in Glacier National Park, Montana. Wildl Monogr 125:1–61

52. McColl JG, Burger J (1976) Chemical inputs by a colony of Franklin's gulls nesting in cattails. Am Midl Nat 96:270–280

53. McKechnie I, Lepofsky D, Moss ML, Butler VL, Orchard TJ, Coupland G, Foster F, Caldwell M, Lertzman K (2014) Archaeological data provide alternative hypotheses on Pacific herring (*Clupea pallasii*) distribution, abundance, and variability. Proc Natl Acad Sci USA 111:E807–E816

54. Megna LC, Moncrieff AE, Hayward JL, Henson SM (2014) Equal reproductive success of phenotypes in the *Larus glaucescens-occidentalis* complex. J Avian Biol 45:410–416

55. Middleton HA, Butler RW, Davidson P (2018) Waterbirds alter their distribution and behavior in the presence of bald eagles (*Haliaeetus leucocephalus*). Northwest Nat 99:21–30

56. Millar JG, Lynch D (2006) U.S.D.I., Endangered and Threatened wildlife and plants: removing the bald eagle in the lower 48 states from the list of endangered and threatened wildlife. U.S.D.I. Fish and Wildlife Service, Fed Regist 71(32):8238–8251

57. Moncrieff AE, Megna LC, Hayward JL, Henson SM (2013) Mating patterns and breeding success in gulls of the *Larus glaucescens-occidentalis* complex, Protection Island, Washington, USA. Northwest Nat 94:67–75

58. Moul IE, Gebauer MB (2002) Status of the double-crested cormorant in British Columbia, B. C. Minist. Water, Land and Air Protection, Biodiversity Branch, Wildl. Working Rep. No. WR-105, Victoria

59. Parrish JK, Marvier M, Paine RT (2001) Direct and indirect effects: interactions between bald eagles and common murres. Ecol Appl 11:1858–1869

60. Payo-Payo A, Oro D, Igual JM, Jover L, Sanpera C, Taveccia G (2015) Population control of an overabundant species achieved through consecutive anthropogenic perturbations. Ecol Appl 25:2228–2239

61. Pearson SF, Hodum PJ, Good TP, Schrimpf M, Knapp SM (2013) A model approach for estimating colony size, trends, and habitat associations of burrow-nesting seabirds. Condor 115:356–365

62. Power EA (1976) Protection Island and the Power family. Unpublished manuscript, Historical Society Archives, Jefferson County

63. Richardson F (1961) Breeding biology of the rhinoceros auklet on Protection Island, Washington. Condor 63:456–473

64. Ricklefs RE (1990) Ecology, 3rd edn. W. H. Freeman and Company, New York

65. Schmidt DF (1986). The numbers, distribution and behavior of glaucous-winged gulls (*Larus glaucescens*) at a sanitary landfill. M.S. Thesis, Walla Walla College

66. Sobey DG, Kenworthy JB (1979) The relationship between herring gulls and the vegetation of their breeding colonies. J Ecol 67:469–496

67. Stalmaster MV (1987) The bald eagle. Universe Books, New York

68. Stick KC, Lindquist A (2009) 2008 Washington State Herring Stock Status Report. Washington Department of Fish and Wildlife, Fish Program, Fish Management Division, Olympia

69. Stinson DW, Watson JW, McAllister KR (2001) Washington state status report for the bald eagle. Washington Dept. Fish and Wildlife, Olympia

70. Sullivan TM, Hazlitt SL, Lemon MJF (2002) Population trends of nesting glaucous-winged gulls, *Larus glaucescens*, in the southern Strait of Georgia, British Columbia. Can. Field Nat 116:603–606

71. Therriault TW, Hay DE, Schweigert JF (2009) Biological overview and trends in pelagic forage fish abundance in the Salish Sea (Strait of Georgia, British Columbia). Mar Ornith 37:3–8

72. Thoresen AC, Galusha JG (1971) A nesting population study of some islands in the Puget Sound area. Murrelet 52:20–23

73. Vennesland RG, Butler RW (2004) Factors influencing great blue heron nesting productivity on the Pacific Coast of Canada from 1998 to 1999. Waterbirds 27:289–296

74. Volterra V (1926) Variazioni e fluttuazioni del numero d'individui in specie animali conviventi. Memoria della Reale Accademia Nazionale dei Lincei 2:31–113

75. Watson JW (2002) Comparative home ranges and food habits of bald eagles nesting in four aquatic habitats in western Washington. Northwest Nat 83:101–108

76. Watson JW, Stinson DW, McAllister KR, Owens TE (2002) Population status of bald eagles breeding in Washington at the end of the 20th century. J Raptor Res 36:161–169

77. White AF, Heath JP, Gisborne B (2006) Seasonal timing of bald eagle attendance and influence on activity budgets of glaucous-winged gulls in Barkley Sound, British Columbia. Waterbirds 29:497–500

Part III

Models of Population Dynamics

In Parts I and II we used models designed with specific species in mind for which we had observational data. The primary purpose was to make predictions of population behavior and dynamics. In the chapters of Part III we instead derive and analyze "proof-of-concept" models. These models are motivated by the field observations described in Parts I and II, but are not designed to predict the dynamics of any specific species or population. Instead they focus on a small number of key mechanisms of interest. The models are typically of low dimension (that is to say, involve a small number of state variables) and are thereby more amenable to rigorous mathematical analysis. The goal is to use such models to probe the dynamic consequences of the mechanisms under consideration, thereby not only suggesting hypotheses that can explain the observations we (and others) have made, but guiding future observational protocols. In the process, these models also serve to sharpen ideas and clarify definitions.

Given the nature of the phenomena we investigate, we use discrete-time (difference equation) models of structured populations. This methodology distinguishes different categories of individuals and focuses on the interactions among them. It is most appropriate when the stages have strongly discrete characteristics, such as juvenile and adult or larva-pupa-adult stages, and life history events occur at well-defined time intervals. Unlike continuous-time structured models, which involve partial differential equations, these models pose no problems with regard to the definition and existence of solutions, numerical simulation, or the validity of basic analytic methods (such as linearization, bifurcations, etc.). The unit of time in our discrete models is generally determined by either a basic time scale in the modeled population (e.g., a maturation period) or the census time interval.

The predictions of a model, for example, its equilibrium states and their stability properties, are determined by the model coefficients or parameters, which represent various vital rates (birth, survival, resource consumption, etc.). A significant change in a model prediction caused by a change in a model parameter is called a bifurcation. By "significant change" one typically means a change in the stability properties of an equilibrium (or some other attractor, such as a periodic state) or a change in the number of attractors. A destabilization of an equilibrium generally signals the creation of a new attractor. Such bifurcations can be of significant interest because the old and new attractors often predict very different dynamics and long-term consequences for a population. For example, the most basic difference concerns survival versus extinction. Others include changes from equilibrium to cyclic fluctuations (crash and boom outbreaks) or other dramatic states such as chaos. The parameter values at which bifurcations occur are called bifurcation points (or sometimes tipping points).

One compelling reason we are interested in bifurcation points in our models is that the location of our field studies is experiencing significant climate change (specifically, increased mean sea surface temperature), which in turn affects certain vital rates of the populations we are modeling. Thus, climate change is represented in our models by a change in model parameters, which can cause bifurcations to occur. A basic question to ask of a model is, then, where are the bifurcation points that indicate a threat to the survival of the population? Which vital rates are involved and are the most important? We can also inquire as to what counter-changes in other parameters might mitigate the threat of extinction. We can use a model to investigate whether or not observed behavioral changes in the population that highly correlate with climate change are likely to be adaptive in the long run.

In Chap. 11 we derive and analyze a population model that structures the population into juveniles and adults only. After some basic analysis of the model, we specialize it to include some central mechanisms involved in adult-on-juvenile cannibalism and use it to explore the hypothesis that in a degraded environment of lower food resources cannibalism can be an evolutionarily stable strategy. The low-dimensional base model used in Chap. 11 is extended in Chap. 12 to explore reproductive synchrony as a response to cannibalism. That model makes the simplifying assumption that juvenile maturation and adult ovulation cycle are on the same time scale, an assumption that is removed in the more complicate model for across-season dynamics studied in Chap. 13.

Chapter 11

Models of Adult-on-Juvenile Cannibalism

In this chapter we set up the modeling framework for a study of cannibalism by means of low-dimensional structured population models that distinguish only between cannibals and victims. Our specific interest is on adult cannibalism of juveniles. In Sect. 11.2 we use matrix modeling methodology [5, 7, 8] to develop and analyze a general discrete-time juvenile-adult matrix model. A Darwinian dynamics (evolutionary game theoretic) version of the model is studied in Sect. 11.4 to investigate whether cannibalism can be an evolutionarily stable trait (or strategy). These and further results can be found in [15, 16, 18].

11.1 Introduction

In Sect. 11.2 we derive a basic discrete-time matrix model for a population structured into only two classes: juveniles and adults. We provide an analysis of the stability properties of the extinction equilibrium and of the dynamics that result from the bifurcation that occurs when it destabilizes. With this foundational analysis at hand, in Sect. 11.3 we specialize the model so as to include a few key mechanisms involved in the cannibalism of juveniles by adults. Although our motivation derives from the egg cannibalism in gull colonies (Chap. 7), the model we study in Sect. 11.3 is not designed to apply specifically to gull populations (or any other specific species). Instead, the approach is to focus on these central mechanisms, in exclusion of other details, and to keep the model low dimensional and thereby more tractable to mathematical analysis. The goal is to attain insights into the role that these

J. M. Cushing et al., *Modeling Behavior and Population Dynamics*, Interdisciplinary Applied Mathematics 57, https://doi.org/10.1007/978-3-031-34283-7_11

mechanisms play in the dynamics of a cannibalistic population. In this way, we can explore certain hypotheses concerning the interplay of cannibalism and various population dynamic outcomes. Specifically, we study circumstances when cannibalism can result in population survival in a degraded environment when the population would otherwise go extinct. In Sect. 11.4 we use the methodology of Darwinian dynamics (evolutionary game theory) to study the evolution of cannibalism as an evolutionarily stable strategy.

11.2 Juvenile-Adult Matrix Models

Consider a biological population, closed to immigration and emigration, whose individuals are classified as either juveniles or adults. Juveniles do not reproduce, of course, and we assume all adults are reproductively active. We consider the population numbers (or densities) of these two classes, denoted by $x_1(t)$ and $x_2(t)$, respectively, at discrete census times $t = 0, 1, 2, \ldots$. The basic modeling procedure for the dynamics of the population is a simple but general, class-specific bookkeeping rule based on the fact that the individuals in a specific class at time $t+1$ are of only two types: those who were not in that class at time t (recruits into the class) and those who were in that class at time t (survivors within the class).

To formulate a model that predicts the juvenile and adult densities at the next census time $t+1$ from their densities at time t, we need to devise formulas that predict the recruitment and survival processes for both classes. For the juvenile-adult models considered in this chapter, we make the following assumptions:

- The population is closed to immigration and emigration.
- Population counts are taken at the start of the breeding season.
- The time unit in the model is equal to the time between breeding Seasons.
- A juvenile in a breeding season is a mature adult at the next breeding season (if it survives).
- Each adult produces $b > 0$ newborns who survive a unit time.
- s_1 $(0 < s_1 < 1)$ is probability a juvenile survives a unit of time.
- s_2 $(0 < s_2 < 1)$ is probability an adult survives a unit of time.

According to these assumptions, the bookkeeping rule applied to both the juvenile and adult classes yields

$$x_1(t+1) = bx_2(t) \tag{11.1}$$

$$x_2(t+1) = s_1x_1(t) + s_2x_2(t). \tag{11.2}$$

These equations form a system of difference equations that we can write using matrix notation as

$$\begin{pmatrix} x_1(t+1) \\ x_2(t+1) \end{pmatrix} = \begin{pmatrix} 0 & b \\ s_1 & s_2 \end{pmatrix} \begin{pmatrix} x_1(t) \\ x_2(t) \end{pmatrix}. \tag{11.3}$$

The vital fertility and survival rates b ands_1 and s_2 can be functions of $x_1(t)$ and/or $x_2(t)$ depending on density-dependent effects that the modeler wishes to include in the model (competition for resources, intraspecific predation, etc.). In that case, Eqs. (11.1)–(11.2) become nonlinear difference equations and Eq. (11.3) is a nonlinear matrix equation.

Equation (11.3) is an example of a nonlinear Leslie matrix equation

$$\mathbf{x}(t+1) = P(\mathbf{x}(t))\,\mathbf{x}(t), \tag{11.4}$$

where

$$\mathbf{x} = \operatorname{col}(x_1, x_2) = \begin{pmatrix} x_1 \\ x_2 \end{pmatrix}$$

and

$$P(\mathbf{x}) = \begin{pmatrix} 0 & p_{12}(\mathbf{x}) \\ p_{21}(\mathbf{x}) & p_{22}(\mathbf{x}) \end{pmatrix}. \tag{11.5}$$

See [5, 7, 8, 26, 27]. The Leslie matrix $P(\mathbf{x})$ is called the population *projection matrix* in that it projects the demographic vector $\mathbf{x}$ from time t to time $t+1$.

We continue our model analysis of a cannibalistic population in Sect. 11.3 by providing submodels for the entries $p_{ij}(\mathbf{x})$ in the projection matrix (11.5). Before doing that, however, we summarize some basic mathematical analysis of the general nonlinear matrix equation (11.4)–(11.5) which will be of use in Sect. 11.3.

Let R^2 denote two-dimensional Euclidean space and let

$$R_+^2 = \left\{\mathbf{x} \in R^2 : x_i > 0\right\}, \quad \bar{R}_+^2 = \left\{\mathbf{x} \in R^2 : x_i \geq 0\right\}, \quad \partial R_+^2 = \bar{R}_+^2 \backslash R_+^2$$

denote the positive cone, its closure (the nonnegative cone), and its boundary, respectively.

We assume the entries in $P(\mathbf{x})$ satisfy the following conditions.

(A1) There exists an open set $\Omega^2 \subseteq R^2$ containing $\bar{R}_+^2$ on which the entries $p_{ij}(\mathbf{x})$ are twice continuously differentiable and satisfy

$$p_{12}(\mathbf{x}) > 0 \quad \text{and} \quad 0 < p_{21}(\mathbf{x}), p_{22}(\mathbf{x}) < 1.$$

Under this assumption the nonnegative projection matrix (11.5) is primitive and has, for each $\mathbf{x} \in \bar{R}_+^2$, the strictly dominant eigenvalue

$$r(\mathbf{x}) \mathrel{\mathring{=}} \frac{1}{2}\left(p_{22}(\mathbf{x}) + \sqrt{p_{22}^2(\mathbf{x}) + 4p_{12}(\mathbf{x})\,p_{21}(\mathbf{x})}\right). \tag{11.6}$$

(See Appendix 11.6.) This eigenvalue is called the *population growth rate* (when the population is at $\mathbf{x}$). We also define the *reproduction number*

$$R(\mathbf{x}) := p_{12}(\mathbf{x}) \frac{p_{21}(\mathbf{x})}{1 - p_{22}(\mathbf{x})}, \tag{11.7}$$

which is the expected number of juveniles produced per juvenile per lifetime (when the population is at $\mathbf{x}$). This quantity is mathematically derived by writing

$$P(\mathbf{x})\,\mathbf{x} = F(\mathbf{x})\,\mathbf{x} + T(\mathbf{x})\,\mathbf{x}\,,$$

where $F(\mathbf{x})\,\mathbf{x}$ and $T(\mathbf{x})\,\mathbf{x}$ are the newborns and survivors, respectively, and

$$F(\mathbf{x}) = \begin{pmatrix} 0 & p_{12}(\mathbf{x}) \\ 0 & 0 \end{pmatrix}, \quad T(\mathbf{x}) = \begin{pmatrix} 0 & 0 \\ p_{21}(\mathbf{x}) & p_{22}(\mathbf{x}) \end{pmatrix}$$

are the fertility and transition matrices, and obtaining the dominant eigenvalue of the next generation matrix [5, 17]

$$F(\mathbf{x})\,(I - T(\mathbf{x}))^{-1} = \begin{pmatrix} p_{12}(\mathbf{x}) \frac{p_{21}(\mathbf{x})}{1-p_{22}(\mathbf{x})} & p_{12}(\mathbf{x}) \frac{1}{1-p_{22}(\mathbf{x})} \\ 0 & 0 \end{pmatrix}.$$

The *inherent* (or *intrinsic*) *population growth rate* and the *inherent reproduction number* are

$$r_0 := r(\mathbf{0}) \quad \text{and} \quad R_0 := R(\mathbf{0})\,,$$

respectively. These can be viewed as the population's growth rate and reproduction number in the absence of density effects. Note that r_0 and R_0 are on the same side of 1 and that $r_0 = 1$ if and only if $R_0 = 1$. (These facts relating r_0 and R_0 hold for general population matrix models [14, 17, 28].)

For notional simplification we will also use a superscript 0 to denote evaluation at $\mathbf{x} = \mathbf{0}$. For example,

$$p_{ij}^0 := p_{ij}(\mathbf{0})\,, \quad \partial_k^0 p_{ij} := \left.\frac{\partial p_{ij}(\mathbf{x})}{\partial x_k}\right|_{\mathbf{x}=\mathbf{0}}, \quad \nabla^0 p_{ij} := \begin{pmatrix} \partial_1^0 p_{ij} & \partial_2^0 p_{ij} \end{pmatrix}.$$

The latter is the (row vector) gradient of the entry p_{ij} with respect to x_1 and x_2 evaluated at $\mathbf{x} = \mathbf{0}$.

Note that $r_0 = 1$ (or $R_0 = 1$) if and only if $1 - p_{22}^0 = p_{12}^0 p_{21}^0$. Left and right eigenvectors $\mathbf{w}_L, \mathbf{w}_R \in R_+^2$ of $P(\mathbf{0})$ associated with eigenvalue $r_0 = 1$ are

$$\mathbf{w}_L^T = \frac{1}{1 + p_{12}^0 p_{21}^0} \begin{pmatrix} p_{21}^0 & 1 \end{pmatrix} \quad \text{and} \quad \mathbf{w}_R = \begin{pmatrix} p_{12}^0 \\ 1 \end{pmatrix}.$$

($\mathbf{w}_L^T$ denotes the transpose of the vector $\mathbf{w}_L$.)

A solution $\mathbf{x}$ of the algebraic equation $\mathbf{x} = P(\mathbf{x})\mathbf{x}$ is an equilibrium of the matrix equation (11.4). Clearly $\mathbf{x} = \mathbf{0}$ is an equilibrium for any projection matrix $P(\mathbf{x})$. The existence of non-extinction equilibria $\mathbf{x} \neq \mathbf{0}$ depends on the properties of the density-dependent projection matrix $P(\mathbf{x})$ and, in particular, the coefficients that appear in its entries (which we call model parameters). The *extinction equilibrium* $\mathbf{x} = \mathbf{0}$ exists for all values of the model parameters. The existence of *positive equilibria* $\mathbf{x} \in R_+^2$ (or *nonnegative and non-extinction equilibria* $\mathbf{x} \in \bar{R}_+^2 \setminus \{\mathbf{0}\}$), however, depends in general on the values of the model parameters appearing in the projection matrix. While this seemingly makes the study of equilibria very model specific, the two parameters r_0 and R_0 can be used to obtain some general conclusions about the equilibria of the matrix equation (11.4). The approach is to treat equilibria as functions of the inherent population growth rate r_0 or of the inherent reproduction number R_0, both of which are related to model parameters through formulas (11.6) and (11.7) evaluated at $\mathbf{x} = \mathbf{0}$.

If $\mathbf{x}$ is an equilibrium of Eq. (11.4) and r_0 is the inherent population growth rate, then we refer to the pair $(r_0, \mathbf{x}) \in R_+ \times R^2$ as an *equilibrium pair.* If $\mathbf{x}$ is a positive (or nonnegative) equilibrium, then we refer to the equilibrium pair $(r_0, \mathbf{x})$ as a *positive (or nonnegative)* equilibrium pair. Note that $(r_0, \mathbf{0})$ is an equilibrium pair for all values of $r_0 \in R_+$, which we call an *extinction equilibrium pair.* If $\mathbf{x}$ is a (locally asymptotically) stable equilibrium, then we say the *equilibrium pair* $(r_0, \mathbf{x})$ *is stable.* Similarly, if $\mathbf{x}$ is an unstable equilibrium, then we say the *equilibrium pair* $(r_0, \mathbf{x})$ *is unstable.* (Similar definitions are made for equilibrium pairs $(R_0, \mathbf{x})$.)

A straightforward application of the linearization principle in stability theory [22] shows that the extinction equilibrium $(r_0, \mathbf{0})$ loses stability as r_0 (equivalently R_0) increases through 1. This follows from that fact that the Jacobian matrix associated with (11.4), when evaluated at $\mathbf{x} = \mathbf{0}$ equals the inherent projection matrix $P(\mathbf{0})$. The expectation from general bifurcation theory is that this destabilization will result in the bifurcation of non-extinction equilibrium pairs $(r_0, \mathbf{x})$, $\mathbf{x} \neq \mathbf{0}$, from $(1, \mathbf{0})$, i.e., the existence of a continuum of non-extinction equilibrium pairs whose closure contains the bifurcation point $(1, \mathbf{0})$ [23].

Theorem 2 gives conditions under which there exists a continuum of *positive* equilibrium pairs $(r_0, \mathbf{x})$, $\mathbf{x} \in R_+^2$, that bifurcates from $(1, \mathbf{0})$ and that the (local asymptotic) stability of the bifurcating equilibrium pairs depends on the *direction of bifurcation.* The direction of bifurcation is *forward* (respectively *backward*) if, in a neighborhood of the bifurcation point $(1, \mathbf{0})$, the bifurcating positive equilibria exist for $r_0 \gtrsim 1$ (respectively $r_0 \lesssim 1$). We say that the *bifurcation is stable or unstable* if in this neighborhood the positive equilibrium pairs are (locally asymptotically) stable or unstable, respectively.

The following theorem is a special case of a more general theorem in [15] (also see [7, 8]). Define the scalar quantity

$$\kappa := -\mathbf{w}_L^T \begin{pmatrix} 0 & \nabla^0 p_{12} \mathbf{w}_R \\ \nabla^0 p_{21} \mathbf{w}_R & \nabla^0 p_{22} \mathbf{w}_R \end{pmatrix} \mathbf{w}_R \,, \tag{11.8}$$

which is a measure of the low population level density effects on the entries of the projection matrix.

Theorem 2 *Consider the nonlinear Leslie matrix equation (11.4)–(11.5) under assumption A1. If* $r_0 < 1$ *then the extinction equilibrium pair* $(r_0, \mathbf{0})$ *is stable and if* $r_0 > 1$ *then it is unstable.*

Assume in addition that $\kappa \neq 0$ *when* $r_0 = 1$. *Then a continuum of positive equilibrium pairs* $(r_0, \mathbf{x})$ *bifurcates from the extinction equilibrium pair* $(1, \mathbf{0})$. *The bifurcation is forward and stable if* $\kappa > 0$ *and backward and unstable if* $\kappa < 0$.

Remark 3 *From the fact that* r_0 *and* R_0 *are on the same side of* 1 *[17, 28], we see that Theorem 2 remains valid when* r_0 *is replaced by* R_0.

Remark 4 *A stronger conclusion holds when* $r_0 > 1$, *namely, the matrix model is uniformly persistent with respect to* ∂R_+^2; *see [7, 24]. This implies not only that the extinction equilibrium is unstable, but that no orbits in the positive cone tend to the extinction equilibrium* 0 *as* $t \to +\infty$.

Theorem 2 is a local bifurcation result in that it guarantees the existence of bifurcating positive equilibria in a neighborhood of the bifurcation point $(1, \mathbf{0})$, i.e., for r_0 (or R_0) near 1 and $\mathbf{x}$ near $\mathbf{0}$. The global existence of the continuum is known to occur in general circumstances, by which it is meant that the continuum of positive equilibrium pairs is unbounded in $R_+ \times R_+^2$ [6, 7, 30]. This means the set of r_0 (or R_0) values from the continuum (the spectrum of the continuum) or the set of equilibria $\mathbf{x}$ from the continuum (the range of the continuum) is unbounded (or both). We return to this issue in Sect. 11.3 when it becomes important in the case of a backward unstable bifurcation and the occurrence of a strong Allee effect.

A calculation using formula (11.8) shows that

$$\kappa = -\frac{p_{21}^0\left[p_{12}^0\partial_1^0 p_{12} + \partial_2^0 p_{12}\right] + p_{12}^0\left[p_{12}^0\partial_1^0 p_{21} + \partial_2^0 p_{21}\right] + \left[p_{12}^0\partial_1^0 p_{22} + \partial_2^0 p_{22}\right]}{1 + p_{12}^0 p_{21}^0} \tag{11.9}$$

when $r_0 = 1$ (or equivalently $R_0 = 1$). We see from Theorem 2 and formula (11.9) that the direction of bifurcation, and hence the stability of the bifurcating positive equilibrium pairs, is determined by the partial derivatives of the matrix entries p_{ij} with respect to the components of $\mathbf{x}$, evaluated at $\mathbf{x} = \mathbf{0}$ and $r_0 = 1$ (or $R_0 = 1$). A negative derivative $\partial_1^0 p_{ij} < 0$ and/or $\partial_2^0 p_{ij} < 0$ means a *negative feedback* mechanism is in play for increased low population densities. On the other hand, a positive derivative $\partial_1^0 p_{ij} > 0$ and/or $\partial_2^0 p_{ij} > 0$ means a *positive feedback* mechanism is in play for increased low population densities (a *component Allee effect* [6]).

Remark 5 *If only negative feedback mechanisms are present in a model (as is often the case), then clearly* $\kappa > 0$ *and a forward (hence stable) bifurcation occurs as* r_0 *(equivalently*

R_0) increases through 1. In fact, it is shown in [7] that if $\partial_1 p_{ij}(\mathbf{x}) \leq 0$ and $\partial_2 p_{ij}(\mathbf{x}) \leq 0$ for $\mathbf{x} \in \Omega^2$, then the extinction equilibrium pair is globally asymptotically stable if $r_0 < 1$ (equivalently $R_0 < 1$), i.e., extinction is assured for all initial population densities. See Fig. 11.1a.

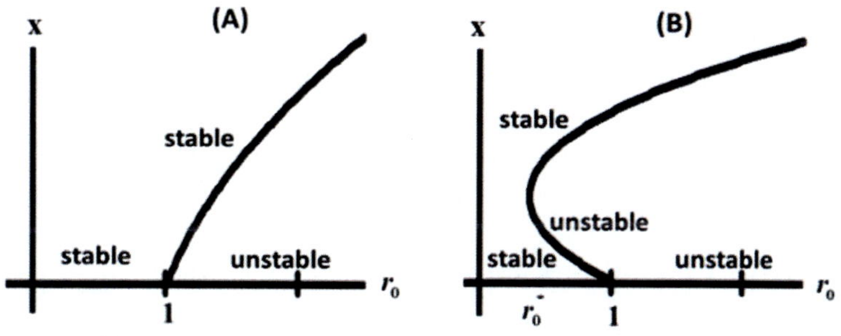

Figure 11.1: Schematic representations of the two basic (transcritical) bifurcations that occur at the extinction equilibrium $\mathbf{x} = \mathbf{0}$ in nonlinear matrix models as r_0 increases through 1 are shown. (**a**) A forward (right or super-critcal) bifurcation of positive equilibria $\mathbf{x}$ occurs if $\kappa > 0$ and the bifurcating positive equilibria are stable (at least in a neighborhood of the bifurcation point). (**b**) A backward (left or subcritical) bifurcation of positive equilibria $\mathbf{x}$ occurs if $\kappa < 0$ and the bifurcating positive equilibria are unstable (at least in a neighborhood of the bifurcation point). Because of assumed negative effects at high densities, the bifurcating continuum of positive equilibria "turns around" at a value $r_0^* < 1$ of r_0. This is generally a saddle-node bifurcation, which usually (but not always) results in a branch of stable positive equilibria as indicated. This creates an interval of r_0 values less than 1 on which a strong Allee effect occurs. In both figures r_0 can be replaced by R_0. Figure reproduced from [18] with permission.

If component Allee effects are present and of sufficient magnitude (compared to the negative feedback mechanisms present at low densities) so that $\kappa < 0$, then a backward and hence unstable bifurcation occurs. A common consequence of a backward bifurcation is the occurrence of what is called a *strong Allee effect*, which is a scenario in which there occur two stable equilibria, one of which is the extinction equilibrium and the other of which is positive. (More generally, a strong Allee effect occurs when a stable extinction equilibrium occurs simultaneously with any type of positive attractor, which could be a cycle or even a chaotic attractor [12].)

There are two basic reasons why a strong Allee effect typically occurs in population models when a backward bifurcation occurs.

First of all, the continuum of positive equilibria that bifurcates from the extinction equilibrium $(1, \mathbf{0})$, as guaranteed by Theorem 2, exists globally in the sense that it connects to ∞ (is unbounded in $R_+ \times R_+^2$) [7, 8].

Secondly, the spectrum (i.e., the set of r_0 values obtained from equilibrium pairs on the continuum) is an unbounded interval in R_+. This is a consequence of the fact that in virtually all population models negative density effects dominate at high population levels. The result is that the continuum "turns around" or "folds over" at a point $r_0^c < 1$, as illustrated in Fig. 11.1b. It is generally true that a change of stability occurs at such a turning or "tipping" point (known as a saddle-node bifurcation), as also illustrated in Fig. 11.1b. Since the extinction equilibrium is stable for $r_0 < 1$, this creates the simultaneous existence of the stable extinction equilibrium with a positive equilibrium, at least for r_0 near the tipping point r_0^c.

An ecological consequence of the strong Allee effect is that there is the possibility of the population's survival in a severely deteriorated environment, by which we mean an environment in which $r_0 < 1$ (and hence the extinction equilibrium is stable). This survival possibility comes with two caveats, however.

(C_1) Firstly, a population must stay out of the basin of attraction of the extinction equilibrium. (This basin is called the Allee basin and its boundary is called the Allee threshold.) This means population densities must remain sufficiently high.

(C_2) Secondly, the population will experience a sudden collapse to extinction (from a seemingly high and healthy equilibrium level) should the environment deteriorate so much that r_0 decreases below the tipping (saddle-node bifurcation) point r_0^c.

These consequences of a backward bifurcation do not necessarily occur for nonlinear matrix models in general. But in population models, they are the expectation when a backward bifurcation occurs. To confirm their occurrence requires further investigation of the specific model of interest, either by mathematical analysis or numerical simulation.

11.3 A Model of Adult-on-Juvenile Cannibalism

In this section we consider a model of adult cannibalism of juveniles studied in [18]. This model is derived from the general juvenile-adult model (11.4) with projection matrix (11.5) by prescribing features of the projection matrix entries $p_{ij}(x)$ that reflect the following trade-off assumptions:

(a) Environmental resource uptake and cannibalism activity are negatively correlated.

(b) Cannibalism decreases the juvenile survival rate and increases the adult survival rate.

Also included in this model are:

(c) A victim saturation effect which postulates that the per juvenile probability of being cannibalized decreases as the density of juveniles increases
(d) A standard adult fertility density self-regulation assumption

Assumptions (a), (b), and (c) are motivated by the seabird cannibalism studies in Chap. 7. In particular, assumption (a) is motivated by the empirical result that increased sea surface temperature (and the concomitant decreased food supply) is associated with higher rates of egg cannibalism in gulls. Assumption (b) is motivated by the "lifeboat" hypothesis that gulls cannibalize eggs when food supply is low in order to survive and redirect reproductive efforts to "normal" years. Assumption (c) is motivated by a version of the "Fraser Darling" effect in which a larger number of eggs mean that each egg has a reduced chance of being cannibalized due to predator (cannibal) satiation.

We begin by assuming, in the absence of cannibalism, that juvenile and adult survival probabilities entries p_{21} and p_{22} and the adult fertility entry p_{12} in the projection matrix $P(\mathbf{x})$ of the matrix equations (11.4)–(11.5) are increasing functions of the amount of environmental (non-cannibalistic) food resource $\rho \geq 0$ available in the habitat, i.e.,

$$p_{12} = b(\rho), \quad p_{21} = s_1(\rho), \quad p_{22} = s_2(\rho),$$

where $b(\rho) > 0$ and the fractions $s_i(\rho)$ are all increasing functions of ρ. When cannibalism is present, the juvenile survival entry p_{21} is modified by the probability that a juvenile survives cannibalism, so that

$$p_{21}(x_1, x_2, \rho) = s_1(\rho)(1 - \pi(x_1, x_2, \rho) x_2).$$

Here the fraction $\pi(x_1, x_2, \rho) x_2$ is the probability a juvenile is cannibalized in the presence of x_2 adults and x_1 juveniles and $1 - \pi(x_1, x_2, \rho) x_2$ is the probability a juvenile survives under these circumstances. In specifying a mathematical expression for $\pi(x_1, x_2, \rho)$, we need to make sure that $\pi(x_1, x_2, \rho) x_2 \leq 1$. The basic mathematical features of $\pi(x_1, x_2, \rho) x_2$ are that it decreases as the environmental resource ρ increases (cf. (a)), increases as the density x_2 of adults increases (cf. (b)), and decreases as juvenile density x_1 increases (cf. (c)).

With regard to adult survival probability $p_{22}(x_1, x_2, \rho)$, we modify $s_2(\rho)$ by a factor σ that reflects an increase due to cannibalistic resources

$$p_{22}(x_1, x_2, \rho) = s_2(\rho)\,\sigma(w)$$

where $w := \pi(x_1, x_2, \rho) x_1$ is the number of juveniles cannibalized per adult. We assume σ is an increasing function of w (cf. (b)), but in such a way that $s_2(\rho)\,\sigma(w) \leq 1$.

Finally, we assume

$$p_{12}(x_1, x_2) = b(\rho)\,\varphi(x_2)$$

where φ is a decreasing function of x_2 (cf. (d)).

In summary, putting all this together we have the projection matrix

$$P(\mathbf{x}) = \begin{pmatrix} 0 & b(\rho)\,\varphi(x_2) \\ s_1(\rho)\,(1 - \pi(x_1, x_2, \rho)\,x_2) & s_2(\rho)\,\sigma(\pi(x_1, x_2, \rho)\,x_1) \end{pmatrix} \tag{11.10}$$

in the juvenile-adult model (11.4). To apply Theorem 2 to this model, we need to ensure assumption A1 holds. To do this, we assume the following.

- **Inherent vital rates:** $b(\rho)$ and $s_i(\rho)$ are increasing continuous functions of ρ on an open interval Ω containing $\bar{R}_+$ that satisfy $b(0) = 0$ (i.e., reproduction ceases in the absence of the resource ρ) and $b(\rho) \geq 0$, $0 < s_i(\rho) \leq 1$ for $\rho \geq 0$.
- **Population regulation in the absence of cannibalism:** $\varphi(x_2)$ is a continuously differentiable and non-increasing function of x_2 on Ω that satisfies

$$\varphi(0) = 1, \;\; \varphi'(0) < 0, \;\; \varphi(x_2) \geq 0 \text{ for } x_2 \geq 0\;.$$

We assume $\varphi(x_2)\,x_2$ is bounded as a function of $x_2 \geq 0$, i.e., there is a constant $\varphi_0 > 0$ such that $\varphi(x_2)\,x_2 \leq \varphi_0$ for $x_2 \in \bar{R}_+$.

- **Cannibalism interactions:**

(1) $\pi(x_1, x_2, \rho)\,x_2$ is a continuous function on $\Omega^2 \times \Omega$ that is twice continuously differentiable in x_1 and x_2 and satisfies the following conditions for $\rho \geq 0$ and $x_i \geq 0$:

$$0 \leq \pi(x_1, x_2, \rho)\,x_2 \leq 1$$
$\pi(x_1, x_2, \rho)\,x_2$ is increasing in $x_2$and decreasing in both x_1 and ρ.

(2) $\sigma(w)$ is twice continuously differentiable on Ω, is non-decreasing for $w \geq 0$, and satisfies

$$\sigma(0) = 1, \quad \sigma'(0) > 0, \quad 0 < s_2(\rho)\,\sigma(w) \leq \sigma_0(\rho) < 1$$

for $\rho, w \geq 0$ and some constant $\sigma_0(\rho)$.

In our application of Theorem 2 we utilize R_0 as the bifurcation parameter. We denote the dependence of R_0 on the environmental resource availability ρ by writing $R_0(\rho)$, which from (11.7) (evaluated at $\mathbf{x} = \mathbf{0}$) is

$$R_0(\rho) := b(\rho)\,\frac{s_1(\rho)}{1 - s_2(\rho)}.$$

The destabilization of the extinction equilibrium and the bifurcation of positive equilibrium pairs occurs at $R_0(\rho) = 1$, at which the formula for the diagnostic quantity κ becomes

$$\kappa = \frac{1}{s_1(\rho)} \frac{1 - s_2(\rho)}{2 - s_2(\rho)} \left(\left[s_1(\rho) - s_2(\rho)\, \sigma'(0) \right] \pi(0,0,\rho) - s_1(\rho)\, \varphi'(0) \right).$$

The sign of κ is determined by the factor

$$\left[s_1(\rho) - s_2(\rho)\, \sigma'(0) \right] \pi(0,0,\rho) - s_1(\rho)\, \varphi'(0), \tag{11.11}$$

whose sign is ambiguous under the monotonicity assumptions made on φ and σ. Although the last term is positive in this expression, the sign of the first term is the same as that of the bracketed term

$$s_1(\rho) - s_2(\rho)\, \sigma'(0)$$

in which the magnitude of $\sigma'(0) > 0$ measures the impact that an increase in cannibalism has on adult survival. This term, and hence κ, is positive if $\sigma'(0) > 0$ is sufficiently large.

We begin by noting that if cannibalism is absent (i.e., if $\pi(x_1, x_2, \rho) \equiv \sigma(w) \equiv 0$), then $\kappa > 0$, and, as a result of Theorem 2, the bifurcation at $R_0(\rho) = 1$ is forward and stable. Moreover, in this case we see by Remark 5 that $R_0(\rho) < 1$ implies the extinction equilibrium $\mathbf{x} = \mathbf{0}$ is globally asymptotically stable. Thus, there is no chance of survival of a non-cannibalistic population if $R_0(\rho) < 1$, which is true if ρ is too small (since $R_0(0) = 0$). Thus, in the absence of cannibalism in this model, an environment that is so severely degraded that $R_0(\rho) < 1$ results in the (global) extinction of the population.

A major conclusion in Sect. 11.2 is that survival when $R_0(\rho) < 1$ is possible if a backward bifurcation occurs and a strong Allee effect is thereby created. A backward bifurcation is present in the model if $\kappa < 0$, i.e., if the term (11.11) is negative. One way to view this criterion is that $\sigma'(0) > 0$ is sufficiently large relative to $\varphi'(0) < 0$, that is to say, if the positive feedback on adult survival obtained from juvenile cannibalism is sufficiently large. In this event, long-term population survival is possible (with caveats C_1 and C_2) in a degraded environment when $R_0(\rho) < 1$ whereas if the positive feedback from cannibalism were low (or absent), extinction would occur because the bifurcation is forward and $R_0(\rho) < 1$.

To illustrate this phenomenon we look at an example in which the components of the model are specified more explicitly, using rational functions of types commonly used in population dynamics (such as Holling type II and Leslie-Holt functionals). We assume that the uptake rate of the environmental resource ρ also is described by a saturating, Holling II type functional

$$\eta(\rho) := \frac{c_4 \rho}{1 + c_4 \rho}, \quad c_4 > 0 \tag{11.12}$$

and assume that inherent adult fertility and the inherent survival of adults and juveniles are each proportional to this consumed amount of environmental resource. We model the three entries p_{12}, p_{21}, and p_{22} in the projection matrix (11.5) as follows. All indicated coefficients are nonnegative.

Juvenile survival p_{21}. Inherent juvenile survivorship is assumed proportional to the resource uptake rate $\eta(\rho)$:

$$s_1(\rho) = s_1^0 \eta(\rho) \quad 0 < s_1^0 < 1. \tag{11.13}$$

Here s_1^0 is the maximum possible inherent juvenile survival probability. We model the probability that a juvenile is cannibalized by the expression

$$\pi(x_1, x_2, \rho) x_2 = \frac{1}{1 + c_4 \rho} \frac{1}{1 + c_1 x_1} \frac{v x_2}{1 + v x_2}. \tag{11.14}$$

The first factor in this expression assumes juvenile cannibalism decreases if the environmental resource level ρ increases. The second factor accounts for a victim saturation effect, i.e., that the probability an individual juvenile is cannibalized decreases when the density of juveniles increases. The third factor describes a cannibalism saturation effect, as modeled by a Holling II type functional of cannibal density x_2.

The derivative (sensitivity) of $\pi(x_1, x_2, \rho) x_2$ with respect to x_2 evaluated at $\mathbf{x} = \mathbf{0}$, namely,

$$\frac{1}{1 + c_4 \rho} v,$$

is a measure of adult cannibal aggressiveness (at low population densities) in that it measures the increase in the probability a juvenile is cannibalized that results from an increase in the density of cannibals. Note that this cannibalism aggressiveness is inversely related to the amount ρ of available environmental resource and that the coefficient v is the maximal aggressiveness, which occurs when the resource ρ vanishes. We will refer to v as the *cannibalism aggressiveness* coefficient.

Finally, putting all of these components together, we see that the juvenile survival entry in the projection matrix (11.5) is

$$p_{21}(x_1, x_2) = s_1^0 \eta(\rho) \left(1 - \frac{1}{1 + c_4 \rho} \frac{1}{1 + c_1 x_1} \frac{v x_2}{1 + v x_2}\right). \tag{11.15}$$

Adult Fertility p_{12}. We assume that inherent adult fertility is proportional to the resource uptake rate (11.12) and that the constant of proportionality is a decreasing function of the cannibalism aggressiveness v (cf. (a)). Specifically, we take it to be $e^{-c_5 v}$ so that if β is the number of juveniles produced per adult per unit resource then

$$b(\rho) = \beta e^{-c_5 v} \eta(\rho). \tag{11.16}$$

We also assume adult fertility is regulated by the density of adults (cf. (d)) according to the discrete logistic (or Beverton-Holt) functional

$$\varphi(x_2) = \frac{1}{1 + c_2 x_2}. \tag{11.17}$$

In summary, the adult fertility entry in the projection matrix (11.5) is

$$p_{12}(x_1, x_2) = \beta e^{-c_5 v} \eta(\rho) \frac{1}{1 + c_2 x_2}. \tag{11.18}$$

Adult Survival p_{22}. We assume that inherent adult survival is proportional to the resource uptake rate (11.12) and that the constant of proportionality is an increasing function of v (cf. (a)):

$$s_2(\rho, v) = \left[s_2^0 + \left(s_2^m - s_2^0\right)\left(1 - e^{-c_6 v}\right)\right] \eta(\rho), \quad 0 < s_2^0 < s_2^m < 1. \tag{11.19}$$

The bracketed factor is an increasing function of v that ranges from s_2^0 when cannibalism activity is absent ($v = 0$) to a maximum of s_2^m as the cannibalism aggressiveness v increases without bound. Adult survival is also a function of the number of juveniles cannibalized per adult $w = \pi(x_1, x_2, \rho) x_1$. We model this by modifying the inherent survival probability by a factor $\sigma(w)$ that is an increasing function of w. In this example, we use

$$\sigma(w) = \frac{(1 + \gamma)(1 + c_3 w)}{1 + \gamma + c_3 w}. \tag{11.20}$$

This factor monotonically ranges from $\sigma(0) = 1$ when no juveniles are cannibalized to a maximum of $1 + \gamma$. In order to keep the survival probability p_{22} less than 1 we must assume

$$s_2^m (1 + \gamma) \leq 1.$$

In summary, the adult survival entry in the projection matrix (11.5) is

$$p_{22}(x_1, x_2) = \left[s_2^0 + \left(s_2^m - s_2^0\right)\left(1 - e^{-c_6 v}\right)\right] \eta(\rho) \sigma(\pi(x_1, x_2, \rho) x_1). \tag{11.21}$$

The population matrix model (11.4)–(11.5) that results from the projection matrix entries (11.15), (11.18), and (11.21) contains no cannibalistic interactions if $v = 0$. As $v > 0$ increases, less effort is put into environmental resource gathering and more into juvenile cannibalism, resulting in a lower inherent adult fertility rate but higher inherent survival rate. There is also a trade-off between environmental resource availability ρ and cannibalism activity.

A summary of the model parameters appears in Table 11.1.

Table 11.1: Listed are the parameters appearing in the model (11.4) whose projection matrix (11.5) has entries (11.15), (11.18), and (11.21).

Parameter	Biological interpretation
v	Cannibalism aggressiveness
ρ	Environmental resource availability
c_1	Victim saturation effectiveness against cannibalism
c_2	Coefficient of adult density regulation of fertility
c_3	Coefficient that measures increases of adult survival as a function of juveniles cannibalized
c_4	Inverse of the half-saturation level of the Holling II type resource uptake rate
c_5	Coefficient that measures the decrease in environmental resource uptake due to an increase in cannibalism intensity
c_6	Coefficient that measures the increase in inherent adult survival due to an increase in cannibalism intensity
β	Maximum adult fertility in absence of cannibalism
γ	Maximum relative gain in adult survival from cannibalistic resources
s_1^0, s_2^0	Maximum inherent juvenile and adult survival probabilities
$s_2^m(1+\gamma)$	Maximal adult survival probability in the presence of cannibalism

By Theorem 2 we know that the extinction equilibrium is stable if $R_0(\rho) < 1$, that a continuum of positive equilibria bifurcates from the destabilized extinction equilibrium as $R_0(\rho)$ increases through 1, and that the bifurcating positive equilibria are stable if the bifurcation is forward and unstable if it is backward.

Our goal is to address this question: *if our model population is threatened with extinction because environmental change has reduced resource availability ρ to such an extent that its inherent reproduction number $R_0(\rho)$ is less than 1, can extinction nonetheless be avoided, and long-term survival ensured, by means of cannibalism activity on juveniles (albeit with caveats C_1 and C_2)?* Mathematically, our approach is to determine if cannibalism can introduce a backward bifurcation and, as a result, a strong Allee effect.

We begin by considering the model population when $R_0(\rho) < 1$ and cannibalism is absent (i.e., $v = 0$) :

$$\begin{pmatrix} x_1(t+1) \\ x_2(t+1) \end{pmatrix} = \begin{pmatrix} 0 & \beta\eta(\rho)\frac{1}{1+c_2x_2(t)} \\ s_1^0\eta(\rho) & s_2^0\eta(\rho) \end{pmatrix} \begin{pmatrix} x_1(t) \\ x_2(t) \end{pmatrix}.$$

In this case, the only nonlinearity in the projection matrix is a decreasing function of x_2 and it immediately follows that $\kappa > 0$ (cf. Remark 5). By Theorem 2 the destabilization of the extinction equilibrium as

$$R_0(\rho) = \beta\eta(\rho)\frac{s_1^0\eta(\rho)}{1 - s_2^0\eta(\rho)} \tag{11.22}$$

increases through 1 results in a forward bifurcation of positive equilibria. According to Remark 5, the extinction equilibrium is *globally* attracting when $R_0(\rho) < 1$. This can be explicitly seen as follows. Consider the *linear* matrix equation

$$\begin{pmatrix} y_1(t+1) \\ y_2(t+1) \end{pmatrix} = \begin{pmatrix} 0 & \beta e^{-c_5 v}\eta(\rho) \\ s_1^0\eta(\rho) & s_2^0\eta(\rho) \end{pmatrix} \begin{pmatrix} y_1(t) \\ y_2(t) \end{pmatrix} \tag{11.23}$$

with the same initial conditions $y_1(0) = x_1(0) \geq 0$ and $y_2(0) = x_2(0) \geq 0$ as the nonlinear matrix model. It is straightforward to see that

$$\begin{aligned} 0 &\leq x_1(1) \leq \beta e^{-c_5 v}\eta(\rho) x_2(0) = y_1(1) \\ 0 &\leq x_2(1) = s_1^0\eta(\rho) y_1(0) + s_2^0\eta(\rho) y_2(0) = y_2(1) \end{aligned}$$

and by an induction argument that

$$\begin{aligned} 0 &\leq x_1(t) \leq y_1(t) \\ 0 &\leq x_2(t) \leq y_2(t) \end{aligned}$$

for all $t = 0, 1, 2, 3, \ldots$. If $R_0(\rho) < 1$ then the dominant eigenvalue $r_0(\rho)$ of the linear equation is also less than 1, which implies that $\mathbf{y}(t) = \mathrm{col}(y_1(t), y_2(t))$, and, by these inequalities, $\mathbf{x}(t)$ both tend to the origin $\mathbf{0}$ as $t \to +\infty$. We conclude that if $R_0(\rho) < 1$ in the matrix model (11.4) with projection matrix (11.10) and entries (11.15), (11.18), and (11.21) and with low (or absent) cannibalism aggressiveness v, then $\lim_{t\to+\infty} \mathbf{x}(t) = \mathbf{0}$ for all initial conditions $\mathbf{x}(0) \in R_+^2$. Since the resource uptake rate satisfies $\eta(0) = 0$, it follows from formula (11.22) that $R_0(\rho) < 1$ for small $\rho > 0$.

That is to say, with low (or absent) cannibalism aggressiveness the matrix model (11.4) with projection matrix (11.10) and entries (11.15), (11.18), and (11.21) predicts global extinction in a degraded environment. By a degraded environment we mean one whose resource availability ρ is so low that $R_0(\rho) < 1$. To explore whether or not a higher level of cannibalism ($v >> 0$) can allow survival in such a degraded environment, we first determine whether or not a backward bifurcation can occur and, if so, whether this induces a strong Allee effect, as discussed in Sect. 11.2.

Recall that a backward bifurcation occurs if κ is negative. A calculation shows

$$\kappa = \frac{1}{s_1(\rho)}\frac{1 - s_2(\rho, v)}{2 - s_2(\rho, v)}\left[s_1(\rho)c_2 + \left(s_1(\rho) - s_2(\rho, v)\frac{\gamma}{1+\gamma}c_3\right)\frac{1}{1 + c_4\rho}v\right],$$

from which we see that a backward bifurcation occurs if the bracketed factor is negative:

$$s_1(\rho)\, c_2 + \left(s_1(\rho) - s_2(\rho, v) \frac{\gamma}{1+\gamma} c_3 \right) \frac{1}{1 + c_4 \rho} v < 0. \tag{11.24}$$

From this inequality one can determine the key mechanisms in this model that promote a backward bifurcation. The parenthetical factor in (11.24) must be negative for this to occur and that occurs if the benefit to adult survival is high, i.e., if c_3 is large enough. In that case, inequality (11.24) and a backward bifurcation occur, if the cannibalism aggressiveness v is sufficiently high. On the other hand, working against a backward bifurcation is a high level of food resource ρ, a high resource uptake rate c_4, and strong density regulation of fertility by adult density (large c_2).

A backward bifurcation alone is not a guarantee of a strong Allee effect. To do that, the existence of a stable equilibrium needs to be established when $R_0(\rho) < 1$. When a backward bifurcation occurs, the local bifurcation given by Theorem 2 provides the existence of unstable positive equilibria in a neighborhood of the bifurcation point $(1, \mathbf{0})$ and consequently if a strong Allee occurs a stable positive equilibrium must be sought outside this neighborhood. The first question to ask concerns the existence of positive equilibria outside this neighborhood. This question is answered by global bifurcation results in [7] (which are a result of the celebrated Rabinowitz global bifurcation theorem in nonlinear functional analysis [23, 31]).

Specifically, after introducing $R_0(\rho)$ explicitly into the model by using

$$\beta = \frac{1}{\eta(\rho)} \frac{1 - s_2^0 \eta(\rho)}{s_1^0 \eta(\rho)} R_0(\rho)$$

in entry p_{12} of the projection matrix and then noting that the equilibrium equation $\mathbf{x} = A(\mathbf{x})\,\mathbf{x}$, together with the Perron-Frobenius Theorem, implies $r(\mathbf{x}) = 1$ and hence $R(\mathbf{x}) = 1$ at any equilibrium $\mathbf{x} \in R_+^2$, we have from

$$R(\mathbf{x}) = \frac{R_0(\rho)}{1 + c_2 x_2} \frac{\left(1 - \left[s_2^0 + \left(s_2^m - s_2^0\right)\left(1 - e^{-c_6 v}\right)\right] \eta(\rho)\right) \left(1 - \frac{1}{1+c_4\rho} \frac{1}{1+c_1 x_1} \frac{v x_2}{1+v x_2}\right)}{1 - \left[s_2^0 + \left(s_2^m - s_2^0\right)\left(1 - e^{-c_6 v}\right)\right] \eta(\rho)\, \sigma\left(\pi\left(x_1, x_2, \rho\right) x_1\right)}$$

that

$$\lim_{\|\mathbf{x}\| \to \infty} R(\mathbf{x}) = 0.$$

It follows by Theorem 1.2.8(i) in [7] that the spectrum of $R_0(\rho)$ values from the continuum is a positive half line (that contains 1 in its closure).

If, in addition, the bifurcation is backward, it further follows that the spectrum is an interval $R_0(\rho) \geq R_0^*(\rho)$ for some threshold $R_0^*(\rho) < 1$. In terms of $r_0(\rho)$, this results in a bifurcation diagram as shown in Fig. 11.1b and the existence of multiple positive equilibria for an interval or $r_0(\rho)$ values less than 1. To conclude that a strong Allee effect occurs, however, it remains to show that one of these positive equilibria when $r_0(\rho) < 1$ is stable. As suggested by Fig. 11.1b the continuum of unstable positive equilibria typically stabilizes at the tipping point r_0^c (although this does not necessarily happen in general). Unfortunately, we have no formal proof this occurs in the cannibalism model under consideration. Simulations of the model, however, support the presence of a strong Allee effect when a backward bifurcation occurs. Selected sample simulations shown in Fig. 11.2 illustrate this.

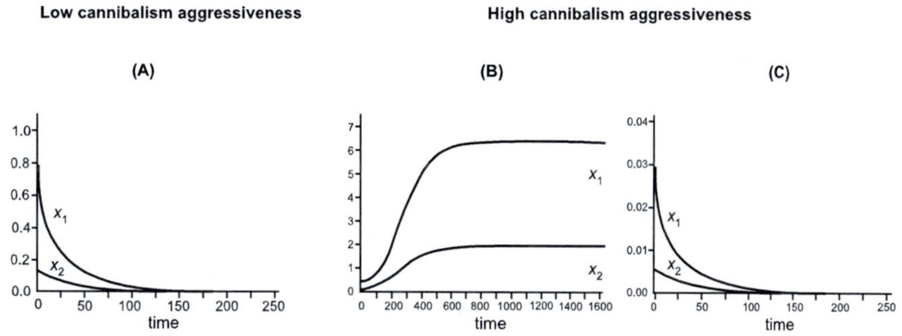

Figure 11.2: Sample time series generated by the juvenile-adult model (11.4)–(11.10) with expressions (11.14)–(11.13) and parameter values $\beta = 4$, $s_1^0 = s_2^0 = 0.25$, $s_2^m = 0.50$, $c_1 = c_2 = c_4 = c_5 = 0.01$, $c_3 = 10$, $c_6 = 0.04$, $\gamma = 0.9$ and $\rho = 600$. (**a**) With a low level of cannibalism aggressiveness $v = 0.01$ the population with initial condition $\operatorname{col}(x_1(0), x_2(0)) = \operatorname{col}(1, 0)$ goes extinct, as is predicted by $R_0(\rho) = 0.936305 < 1$ and the occurrence of a forward bifurcation $\kappa = 0.004659 > 0$. (**b**) With a higher level of cannibalism aggressiveness $v = 3$, the population survives because a backward bifurcation has created a strong Allee effect ($\kappa = -1.821064 < 0$). This survival is initial condition dependent, however, as the plot (**c**) with initial condition $\operatorname{col}(x_1, x_2) = \operatorname{col}(0.4, 0)$ illustrates

Figure 11.2a shows extinction at a low environmental resource level when $R_0(\rho) < 1$ for a model population with a low level of cannibalism aggressiveness. Figure 11.2b shows survival in the same environment when cannibalism aggressiveness is high with, however, the caveat that survival is initial condition dependent as illustrated in Fig. 11.2c. The model explanation for this reversal of outcome is that the high level of cannibalism aggressiveness causes a backward bifurcation which results in a multiple attractor situation (a strong Allee effect).

In summary, *the adult-on-juvenile cannibalism model (11.4) with projection matrix (11.10) and entries (11.15), (11.18), and (11.21) provides theoretical support for the hypothesis that cannibalism can be a means for long-term survival in a degraded environment (when in its absence extinction would occur). In this proof-of-concept model, this conclusion comes, however, with the two caveats C_1 and C_2: survival is initial condition dependent and there is a dangerous "tipping point" threshold, below which a degraded environment results in a sudden catastrophic extinction event.*

The avoidance of extinction in a deteriorated environment by these means is accomplished by an increase in the level of cannibalism aggressiveness in the population, i.e., by an increase in the coefficient v in the model equations. How can such a change occur? Is it possible that such an increase can be attained by means of natural selection? And if so, is the resulting level of cannibalism aggressiveness an evolutionarily stable trait (strategy)? For example, is it possible to move from (A) to (B)–(C) in Fig. 11.2 by means of Darwinian principles? We take up these questions in the next section.

11.4 Cannibalism as an Evolutionarily Stable Strategy

The equations used to model populations appearing in the previous chapters are mathematically called (time) autonomous, which means the predictions of future population densities do not depend explicitly on time (but only implicitly through changes in population densities). This is a reflection of the fact that the model parameters (e.g., fertility and survival rates, resource consumption rates, cannibalism rates, etc.) remain constant through time. There are, of course, numerous reasons why these parameters might not remain constant through time, an important one of which are adaptive changes due to Darwinian principles. In this section we derive and investigate an evolutionary version of the juvenile-adult models in Sects. 11.2 and 11.3. The goal is to determine in what ways the conclusions about survival and long-term dynamics obtained from those models are affected when the coefficients in the models are subject to evolutionary change.

As an example, suppose that with a certain set of parameter values a model makes a favorable prediction—say, avoidance of extinction and long-term survival. Can a model population that initially has a different set of parameter values attain these favorable values through evolutionary adaptation? In Sect. 11.3 we saw how a cannibalistic population model can predict long-term survival in a deteriorated environment when in the absence of cannibalism it predicts extinction. Can a model population without those cannibalistic characteristics attain them by evolutionary adaptation, or simply put, can cannibalism be an evolutionarily stable strategy or trait (an ESS)?

For this project we use the methodology of Darwinian dynamics (or evolutionary game theory) [29, 32], which provides a procedure by which to include the dependence of parameters appearing in a population model

$$\mathbf{x}(t+1) = P(\mathbf{x}(t))\,\mathbf{x}(t) \tag{11.25}$$

on a (continuous) phenotypic trait possessed by individuals that is subject to the axioms of Darwinian evolution: heritability, variability, and differential fitness. A trait v with a heritable component is assumed distributed throughout the population at all times and the (per capita) entries in the projection matrix of a population model are allowed to be dependent on v

$$P(\mathbf{x}, v) = (p_{ij}(\mathbf{x}, v)).$$

This method postulates that the coupled population and evolution dynamics are described by the equations

$$\begin{aligned} \mathbf{x}(t+1) &= P(\mathbf{x}(t), u(t))\,\mathbf{x}(t) \\ u(t+1) &= u(t) + \theta^2 \frac{\partial \ln r(\mathbf{x}(t), u(t))}{\partial v} \end{aligned} \tag{11.26}$$

where the trait has mean $u(t)$ and variance θ^2 [29, 32] and where

$$\frac{\partial \ln r(\mathbf{x}, u)}{\partial v} := \left. \frac{\partial \ln r(\mathbf{x}, v)}{\partial v} \right|_{v=u}.$$

The equation for $u(t)$ in (11.26) states that the change in the population mean trait $u(t)$ is proportional to the fitness gradient; it is called the canonical equation of evolutionary dynamics and often goes by the name of Lande's equation or Fisher's equation (of additive genetic variance) [1–3, 20, 21, 25, 29, 32].

Fitness in the model (11.26) is the (exponential) growth rate $\ln r(\mathbf{x}, v)$ where $r(\mathbf{x}, v)$ is the spectral radius of $P(\mathbf{x}, v)$. For a justification that $\ln r(\mathbf{x}, v)$ is the appropriate measure of fitness for a structured population model, see [4]. When $P(\mathbf{x}, v)$ is nonnegative and irreducible (as it is in the vast majority of population dynamic applications), $r(\mathbf{x}, v)$ is the dominant (Perron) eigenvalue of $P(\mathbf{x}, v)$; it is a real, positive, and simple eigenvalue. In many applications (but not all), $r(\mathbf{x}, v)$ is in fact strictly dominant, i.e., the projection matrix is primitive (cf. Appendix 11.6).

Suppose

$$\operatorname{col}(\mathbf{x}(t), u(t)) = \begin{pmatrix} \mathbf{x}(t) \\ u(t) \end{pmatrix}$$

is a solution of the Darwinian equations (11.26). For a fixed t, $\ln r(\mathbf{x}(t), v)$ is, as a function of v, called the *fitness (or adaptive) landscape*. The trait equation in (11.26) implies $u(t)$ moves uphill on the fitness landscape at each t. Note however that the landscape is not fixed and changes over time.

The $\mathbf{x}$ and u components of an equilibrium $\operatorname{col}(\mathbf{x}, u)$ satisfy the algebraic equations

$$\begin{aligned} \mathbf{x} &= P(\mathbf{x}, u)\,\mathbf{x} \\ 0 &= \frac{\partial}{\partial v} \ln r(\mathbf{x}, u) \end{aligned}. \tag{11.27}$$

The second equation implies u is located at a critical point on the fitness landscape at equilibrium.

The Darwinian model (11.26) is a special case of Darwinian models considered in [9, 10, 16, 19], and [30], where theorems dealing with the existence and stability of equilibria are proved (in particular, extinction equilibria and bifurcating survival equilibria as occurs in non-evolutionary models). We will utilize some of these theorems in Sect. 11.4.1.

An important concept in evolutionary theory is an evolutionarily stable strategy or trait (ESS). The trait component u of an equilibrium $\text{col}(\mathbf{x}, u)$ of a "resident" population is an ESS if a small density of a similar population with a different mean trait u (a mutant) cannot successfully invade and eliminate the resident population. One way to attack this problem would be to create a competitive model by coupling two models of the form (11.26) and determine which population is eliminated in the classic competitive exclusion manner. An alternative (and simpler) way is to use the ESS maximum principle [32]. This principle states that

> u is an ESS if and only if $\text{col}(\mathbf{x}, u)$ is a (locally asymptotically) stable equilibrium of the Darwinian model (11.26) and $v = u$ is located at a global maximum of the adaptive landscape $\ln r(\mathbf{x}, v)$.

11.4.1 Darwinian Juvenile-Adult Matrix Models

The goal in this section is to obtain model support for the hypothesis that cannibalism can be an evolutionarily stable strategy for a population confronted with deteriorated environmental resources. We do this with a Darwinian (EGT) version of the juvenile-adult Leslie model (11.4) with projection matrix (11.10), which we saw in Sect. 11.3 allows for the possibility of long-term survival of a population in a deteriorated environment ($R_0^* < 1$), provided the benefit to adult survival from cannibalism is sufficiently high. Our goal here is to investigate whether, in a population with a low level of cannibalism, a Darwinian dynamics version of the model gives conditions under which a high enough level of cannibalism could be reached by evolutionary adaptation and under which this level of cannibalism is an ESS.

We begin with a Darwinian extension of Theorem 2 to Eq. (11.26) with the projection matrix

$$P(\mathbf{x}, v) = \begin{pmatrix} 0 & p_{12}(\mathbf{x}, v) \\ p_{21}(\mathbf{x}, v) & p_{22}(\mathbf{x}, v) \end{pmatrix}. \tag{11.28}$$

The needed mathematical assumptions on the entries $p_{ij}(\mathbf{x}, v)$ are the following:

> (A2) There exist an open set $\Omega^2 \subseteq R^2$ containing $\bar{R}_+^2$ and an open set $\Omega \subseteq R$ for which the entries $p_{ij}(\mathbf{x}, v)$ are twice continuously differentiable and satisfy the inequalities
>
> $$p_{12}(\mathbf{x}, v) > 0 \text{ and } 0 < p_{21}(\mathbf{x}, v),\ p_{22}(\mathbf{x}, v) < 1$$
>
> on $\Omega^2 \times \Omega$.

Under assumption A2, $P(\mathbf{x}, v)$ is a nonnegative and primitive matrix and therefore has a strictly dominant, positive eigenvalue

$$r(\mathbf{x}, v) := \frac{1}{2}\left(p_{22}(\mathbf{x}, v) + \sqrt{p_{22}^2(\mathbf{x}, v) + 4p_{12}(\mathbf{x}, v)\, p_{21}(\mathbf{x}, v)}\right).$$

From the second equilibrium equation in (11.27), we see that if $\mathrm{col}(\mathbf{x}_e, u_e)$ is an equilibrium, then

$$0 = \frac{\partial}{\partial v} \ln r\left(\mathbf{x}_e, u_e\right), \tag{11.29}$$

which means $v = u_e$ is a critical point on the fitness landscape $\ln r\left(\mathbf{x}_e, v\right)$ at equilibrium. Moreover, from the first equation we see that if $(\mathbf{x}_e, u_e)$, $x_e \in \bar{R}^2_+$, is a nonnegative equilibrium then 1 is an eigenvalue of the primitive matrix $P\left(\mathbf{x}_e, u_e\right)$ and hence $r\left(\mathbf{x}_e, u_e\right) = 1$ by Perron-Frobenius theory.

Since (11.29) implies $0 = \frac{\partial}{\partial v} r\left(\mathbf{x}_e, u_e\right)$ it follows that the extremal properties of $\ln r\left(\mathbf{x}_e, v\right)$ are the same as those of $r\left(\mathbf{x}_e, v\right)$, which by the results in [11], are the same as those of the reproduction number (see (11.7))

$$R\left(\mathbf{x}_e, v\right) := p_{12}\left(\mathbf{x}_e, v\right) \frac{p_{21}\left(\mathbf{x}_e, v\right)}{1 - p_{22}\left(\mathbf{x}_e, v\right)}. \tag{11.30}$$

Clearly $x = 0$ satisfies the first equilibrium equation in (11.27). Thus, there exists an *extinction equilibrium* $\mathrm{col}(\mathbf{x}_e, u_e) = \mathrm{col}\left(\mathbf{0}, u^*\right)$ if and only if u^* satisfies the equation

$$0 = \frac{\partial}{\partial v} \ln r\left(\mathbf{0}, u^*\right) \tag{11.31}$$

or equivalently

$$0 = \frac{\partial}{\partial v} R\left(\mathbf{0}, u^*\right).$$

(A3) Assume there exists a critical trait, i.e., a solution u^* of Eq. (11.31).

Define

$$r_0^* := r\left(\mathbf{0}, u^*\right), \quad R_0^* := R\left(\mathbf{0}, u^*\right).$$

To study the stability of an extinction equilibrium $\mathrm{col}\left(\mathbf{0}, u^*\right)$ by means of the linearization principle, we calculate the Jacobian associated with the model equations (11.26) with projection matrix (11.28) and evaluate it at $\mathrm{col}\left(\mathbf{x}, u\right) = \mathrm{col}\left(\mathbf{0}, u^*\right)$ to obtain the 3×3 matrix

$$\begin{pmatrix} P\left(\mathbf{0}, u^*\right) & 0 \\ * & 1 + \frac{1}{r(\mathbf{0}, u^*)} \theta^2 \frac{\partial^2}{\partial v^2} r\left(\mathbf{0}, u^*\right) \end{pmatrix}$$

where

$$\frac{\partial^2}{\partial v^2} r\left(\mathbf{0}, u^*\right) := \left. \frac{\partial^2}{\partial v^2} r\left(\mathbf{0}, v\right) \right|_{v = u^*}$$

and the asterisk denotes an unneeded term. The eigenvalues of this block diagonal matrix are the eigenvalues of the projection matrix $P(\mathbf{0}, u^*)$ and

$$\lambda = 1 + \frac{1}{r(\mathbf{0}, u^*)}\theta^2 \frac{\partial^2}{\partial v^2} r(\mathbf{0}, u^*).$$

We want to explore the possibility of a strong Allee effect, caused by the positive density effect on adult survival due to juvenile cannibalism, and its implications with regard to survival in a deteriorated environment. The goal is to see if that scenario is attainable by evolutionary dynamics in such a way that cannibalism is an ESS. Toward that end we are motivated by the results in Sect. 11.3 for the non-evolutionary model that showed the occurrence of a strong Allee effect arising from a backward bifurcation which leads us to consider the case when the extinction equilibrium $\text{col}(\mathbf{x},u) = \text{col}(\mathbf{0},u^*)$ is stable for $r_0^* < 1$ and a backward bifurcation of positive equilibria occurs at $r_0^* = 1$. For this to occur we need to assume that $|\lambda| < 1$ which requires

$$\frac{\partial^2}{\partial v^2} r(\mathbf{0}, u^*) < 0 \tag{11.32}$$

and that the speed of evolution θ^2 is not too fast, specifically, that

$$\theta^2 < \theta_c^2 := -2\left(\frac{\partial^2}{\partial v^2} r(\mathbf{0}, u^*)\right)^{-1}$$

when $r_0^* = 1$.

Following the notation in Sect. 11.3, if $\text{col}(\mathbf{x}_e, u_e)$ is an equilibrium associated with r_0, then we call $[r_0, \text{col}(\mathbf{x}_e, u_e)]$ an equilibrium pair. We say $[r_0^*, \text{col}(\mathbf{x}_e, u_e)]$ is a positive equilibrium pair if $\mathbf{x}_e \in R_+^2$ and a stable (unstable) equilibrium pair if $\text{col}(\mathbf{x}_e, u_e)$ is a stable (unstable) equilibrium of (11.26). The following theorem describes the resulting bifurcation of positive equilibrium pairs that occurs as r_0^* increases through 1; it is an extension of Theorem 2. For Darwinian models, the quantity that determines the direction of bifurcation is

$$\kappa^* := -\mathbf{w}_L^T \left(\nabla^0 p_{ij} \mathbf{w}_R\right) \mathbf{w}_R$$

where

$$\mathbf{w}_L^T = \frac{1}{1 + p_{12}^0 p_{21}^0} \begin{pmatrix} p_{21}^0 & 1 \end{pmatrix} \quad \text{and} \quad \mathbf{w}_R = \begin{pmatrix} p_{12}^0 \\ 1 \end{pmatrix}.$$

Here the superscript "0" indicates evaluation at $\text{col}(\mathbf{x}, u) = \text{col}(\mathbf{0}, u^*)$ and $r_0^* = 1$, for example, $p_{12}^0 := p_{12}(\mathbf{0}, u^*)$, etc.

From Theorems 2 and 3 in [19] for general Darwinian model equations, we get the following results. (Also see [19].)

Theorem 6 *Consider the Darwinian model (11.26) with projection matrix (11.28) under assumptions A2 and A3. Assume further that the inequalities (11.32) and $\theta^2 < \theta_c^2$ hold.*

(a) The extinction equilibrium pair $[r_0^*, \text{col}(\mathbf{0}, u^*)]$ is stable if $r_0^* < 1$ and unstable if $r_0^* > 1$.

(b) Assume $\kappa^* \neq 0$ when $r_0^* = 1$. Then a continuum of positive equilibrium pairs $[r_0^*, \operatorname{col}(\mathbf{x}, u)]$ bifurcates from the extinction equilibrium pair $[1, \operatorname{col}(\mathbf{0}, u^*)]$. The bifurcation is forward and stable if $\kappa^* > 0$ and backward and unstable if $\kappa^* < 0$.

Remark 7 *Theorem 6 remains valid when $r_0^* = r(\mathbf{0}, u^*)$ is replaced by $R_0^* = R(\mathbf{0}, u^*)$. This is because the results in [11] imply that inequality (11.32) can be replaced by*

$$\frac{\partial^2}{\partial v^2} R(\mathbf{0}, u^*) < 0$$

at $r_0^* = 1$.

11.4.2 A Darwinian Model for Adult-on-Juvenile Cannibalism

In this section we consider a Darwinian version of the adult-on-juvenile cannibalism model derived and studied in Sect. 11.3 in which the cannibalism aggressiveness of an individual, namely, $v \in R_+$, is subject to evolutionary adaptation. The entries p_{ij} in the projection matrix (11.28) (whose constituents are derived in detail in Sect. 11.3) are

$$\begin{aligned} p_{12}(\mathbf{x}, v) &= b(\rho, v) \tfrac{1}{1+c_2 x_2} \\ p_{21}(\mathbf{x}, v) &= s_1(\rho)(1 - \pi(\mathbf{x}, v, \rho) x_2) \\ p_{22}(\mathbf{x}, v) &= s_2(\rho, v) \sigma(\pi(\mathbf{x}, v, \rho) x_1) \end{aligned} \tag{11.33}$$

where

$$b(\rho, v) = \beta e^{-c_5 v} \eta(\rho)$$

$$s_1(\rho) = s_1^0 \eta(\rho) \quad \text{and} \quad s_2(\rho, v) = \left[s_2^0 + \left(s_2^m - s_2^0\right)\left(1 - e^{-c_6 v}\right)\right] \eta(\rho)$$

$$\begin{aligned} \eta(\rho) &= \frac{c_4 \rho}{1 + c_4 \rho} \quad \text{and} \quad \sigma(w) = \frac{(1+\gamma)(1 + c_3 w)}{1 + \gamma + c_3 w} \\ \pi(\mathbf{x}, v, \rho) &= \frac{1}{1 + c_4 \rho} \frac{1}{1 + c_1 x_1} \frac{v}{1 + v x_2} \end{aligned}$$

and all coefficients are positive and satisfy

$$0 < s_2^0 < s_2^m < 1 \quad \text{and} \quad s_2^m (1 + \gamma) \leq 1.$$

The inherent reproduction number

$$\begin{aligned} R(\mathbf{0}, v) &= b(\rho) \frac{s_1(\rho)}{1 - s_2(\rho, v)} \\ &= \beta \frac{s_1^0 e^{-c_5 v} \eta^2(\rho)}{1 - \left[s_2^0 + \left(s_2^m - s_2^0\right)\left(1 - e^{-c_6 v}\right)\right] \eta(\rho)} \end{aligned}$$

has a critical trait $v = u^* > 0$ if (and only if)

$$\frac{1}{c_5}(c_6 - c_5) > \frac{1 - \eta(\rho) s_2^0}{\eta(\rho)\left(s_2^m - s_2^0\right)} \tag{11.34}$$

in which case u^* is unique and given by the formula

$$u^* = \frac{1}{c_6} \ln\left(\frac{\eta(\rho)\left(s_2^m - s_2^0\right)}{1 - \eta(\rho) s_2^m} \frac{1}{c_5}(c_6 - c_5)\right). \tag{11.35}$$

A calculation shows (since (11.34) implies $c_6 - c_5 > 0$) that

$$\frac{\partial^2}{\partial v^2} R(\mathbf{0}, u^*) = \frac{-\beta s_1^0 \eta^3(\rho)\left(s_2^m - s_2^0\right) c_6 (c_6 - c_5)}{\left(1 - \eta(\rho)\left[s_2^0 + \left(s_2^m - s_2^0\right)\left(1 - e^{-c_6 u^*}\right)\right]\right)^2} e^{-(c_5 + c_6) u^*} < 0.$$

It follows that the requirements for the application of Theorem 6 hold (Remark 7). Thus, the extinction equilibrium loses stability as R_0^* increases through 1 and a bifurcation of positive equilibria occurs whose direction and stability depend on the sign of

$$\kappa^* = \frac{1}{s_1(\rho)} \frac{1 - s_2(\rho, u^*)}{2 - s_2(\rho, u^*)} \left[s_1(\rho) c_2 + \left(s_1(\rho) - s_2(\rho, u^*) \frac{\gamma}{1+\gamma} c_3\right) \frac{1}{1 + c_4 \rho} u^*\right] \tag{11.36}$$

and the speed of evolution θ^2 is not too fast.

We are interested in conditions under which a population with a low mean cannibalism aggressiveness will, when threatened with extinction due to a decreased level of environmental resource availability ρ, survive by evolving a higher mean cannibalism aggressiveness. Using the modeling approach taken here, we address this possibility in the following way. Suppose, in the absence of evolution ($\theta^2 = 0$), extinction is predicted because the mean cannibalism aggressiveness $u(0)$ and environmental resource ρ are so low that a forward bifurcation occurs ($\kappa > 0$) and $R_0 < 1$, but when evolution does occur a backward bifurcation occurs ($\kappa^* < 0$) and survival results even though $R_0^* < 1$; moreover, the mean cannibalism aggressiveness at equilibrium is an ESS.

To see conditions under which this scenario is possible, we note that when cannibalism is absent

$$\kappa = \frac{1}{s_1(\rho)} \frac{1 - s_2(\rho, 0)}{2 - s_2(\rho, 0)} s_1(\rho) c_2 > 0$$

and hence $\kappa > 0$ when a small level of cannibalism is present. If evolutionary does not occur ($\theta^2 = 0$), then $u(t) = u(0)$ remains fixed and small for all t. On the other hand, if evolution occurs ($\theta^2 > 0$) and $u(t)$ evolves, then from (11.36) we see that the requirement (11.34) will hold if c_6 is sufficiently large and κ^* will be negative if c_3 is sufficiently large. With these conditions in place, a population that initially has a low level of cannibalism aggressiveness (and will go extinct in a deteriorated environment) will adapt and evolve a higher level of cannibalism aggressiveness and attain (initial condition dependent) survival. Note that the requirements that c_3 and c_6 are large mean that the effect of cannibalism is highly beneficial to adult survival. An example of this scenario is shown in Fig. 11.3. Figure 11.4 shows that the evolved higher level of cannibalism aggressiveness lies at a global maximum of the fitness landscape at equilibrium and is therefore an ESS.

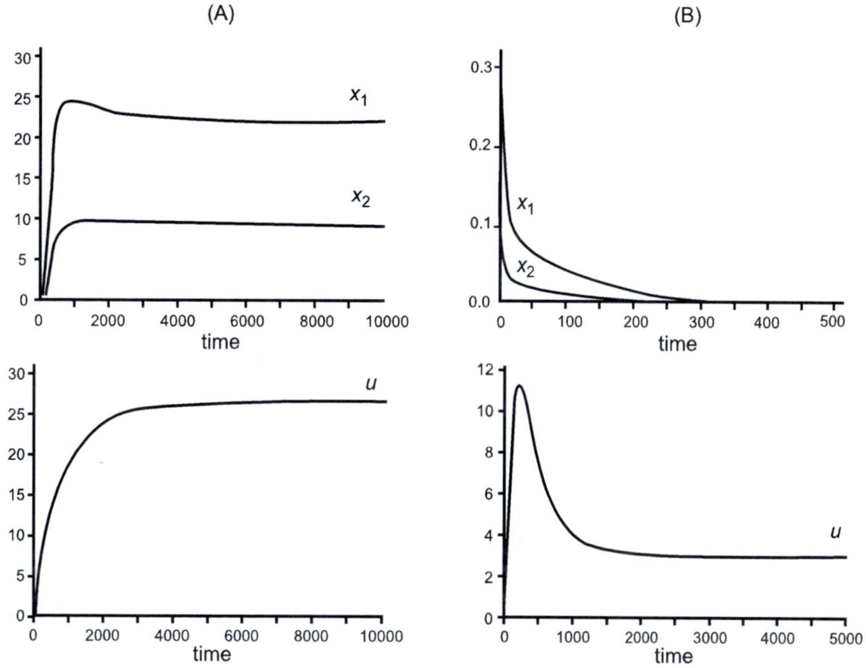

Figure 11.3: Sample time series generated by the juvenile-adult model (11.4)–(11.10) with expressions (11.14)–(11.13) and parameter values the same as in Fig. 11.2. The initial cannibalism aggressiveness is $u(0) = 0.1$, as in Fig. 11.2a when the population goes extinct in the absence of evolution ($\theta^2 = 0$). The speed of evolution here is $\theta^2 = 10$. Furthermore, $\kappa^* = -1.783 < 0$, indicating a backward bifurcation, and $R_0^* = 0.936$. (**a**) With initial population conditions $\mathrm{col}(x_1(0), x_2(0)) = \mathrm{col}(1, 0)$ we see survival with the mean cannibalism aggressiveness evolving to $u_e = 26.545$. (**b**) With initial population conditions $\mathrm{col}(x_1(0), x_2(0)) = \mathrm{col}(0.3, 0)$ we see extinction with $u(t)$ approaching the critical trait $u^* = 2.945$.

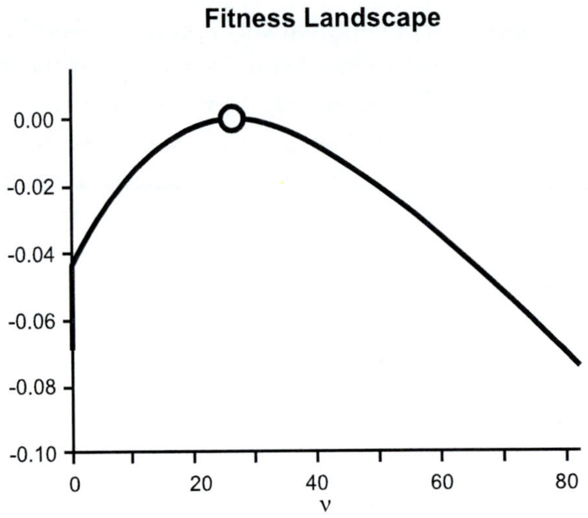

Figure 11.4: The fitness landscape $\ln r(\mathbf{x}_e, v)$ at the equilibrium $\mathbf{x}_e = \mathrm{col}(22.357636, 9.293799)$ attained in Fig. 11.3a. The open circle at the global maximum is located at the equilibrium trait $u_e = 26.545369$

11.5 Conclusion

It is generally true that when the extinction equilibrium in a population model destabilizes because a parameter has changed, a bifurcation results that creates survival equilibria whose stability depends on the direction of bifurcation, at least in a neighborhood of the bifurcation point. A forward bifurcation yields stable survival equilibria, while a backward bifurcation yields unstable survival equilibria. The latter case occurs if sufficiently significant positive correlations occur between fitness components in the model and increased low-level population density (Allee components). In this case, what typically occurs is the formation of a strong Allee effect, which is the simultaneous occurrence of a stable extinction state and a survival attractor (often a stable equilibrium) [12]. (This is due to the common assumption in models that increased high level densities have negative consequences on fitness components in the model.)

These features are present in the general juvenile-adult model studied in Sect. 11.2 and the adult-on-juvenile cannibalism model derived from it in Sect. 11.3. Cannibalism of juveniles by adults results in a negative effect on juvenile survival, of course, but also a positive feedback on adult survival. If, in addition to another positive feedback in the model due to the victim saturation effect, the positive effect on adult survival is significant enough, then a backward bifurcation and strong Allee effect occurs in the model. The result of this is that in a degraded environment—one that results in R_0 less than 1 and hence a stable extinction equilibrium—there exists a stable survival equilibrium and hence survival is still possible. This initial condition-dependent survival possibility disappears, however, if the environment degrades to the extent that R_0 drops below a "tipping point" threshold when the survival equilibrium suddenly and catastrophically disappears, resulting in extinction.

This same scenario holds for the Darwinian version of the model in Sect. 11.4, which shows that the higher-level cannibalism needed to accomplish survival in the event of low environmental resource availability can be attained by natural selection.

11.6 Appendix

An $m \times m$ nonnegative matrix P is a matrix all of whose entries are nonnegative. A positive matrix is one all of whose entries are positive. A nonnegative matrix is called irreducible if its associated cycle graph contains a path from every node to every other node, i.e., in a population model if there is a path from every class x_i in $\mathbf{x}$ to every other class in $\mathbf{x}$. This occurs if and only if $(I + P)^{m-1}$ is positive (where I is the $m \times m$ identity matrix). By Perron-Frobenius theory, a nonnegative, irreducible matrix has a (geometrically) simple and dominant eigenvalue $r > 0$ (i.e., $r \geq |\lambda|$ for all other eigenvalues λ). If r is strictly dominant (i.e., if $r > |\lambda|$ for all other eigenvalues λ), then P is called primitive. A necessary and sufficient condition for P to be primitive is that P^n is positive for some positive integer n.

References

1. Abrams PA (2001) Modelling the adaptive dynamics of traits involved in inter- and intraspecific interactions: an assessment of three methods. Ecol Lett 4:166–175

2. Abrams PA (2006) Adaptive change in the resource-exploitation traits of a generalist consumer: the evolution and coexistence of generalists and specialists. Evolution 60:427–439

3. Abrams PA, Matsuda H, Harada Y (1993) Evolutionarily unstable fitness maxima and stable fitness minima of continuous traits. Evol Ecol 7:465–487

4. Barfield J, Holt RD, Gomulkiewicz R (2011) Evolution in stage-structured populations. Am Nat 177(4):397–408

5. Caswell H (2001) Matrix Population Models: Construction, Analysis and Interpretation, 2nd edn. Sinauer Associates, Inc. Publishers, Sunderland

6. Courchamp F, Berec L, Gascoigne J (2008) Allee Effects in Ecology and Conservation. Oxford University Press, Oxford

7. Cushing JM (1998) An Introduction to Structured Population Dynamics. CBMS-NSF Regional Conference Series in Applied Mathematics, vol 71. SIAM, Philadelphia

8. Cushing JM (2009) Matrix models and population dynamics. In: Lewis M, Chaplain AJ, Keener JP, Maini PK (eds) Mathematical Biology, IAS/Park City Mathematics Series, vol 14. American Mathematical Society, Providence, pp 47–150

9. Cushing JM (2010) On the dynamics of a class of Darwinian matrix models. Nonlin Dyn Syst Theory 10(2):103–116

10. Cushing JM (2010) A bifurcation theorem for Darwinian matrix models. Nonlin Stud 17(1):1–13

11. Cushing JM (2011) On the relationship between r and R_0 and its role in the bifurcation of equilibria of Darwinian matrix models. J Biol Dyn 5:277–297

12. Cushing JM (2014) Backward bifurcations and strong Allee effects in matrix models for the dynamics of structured populations. J Biol Dyn 8:57–73

13. Cushing JM (2021) A bifurcation theorem for Darwinian matrix models with an application to the evolution of reproductive life history strategies. J Biol Dyn 15(Sup1):S190–S213. https://doi.org/10.1080/17513758.2020.1858196

14. Cushing JM, Diekmann O (2016) The many guises of R_0 (a didactic note). J Theor Biol 404:295–302

15. Cushing JM, Farrell AP (2019) A bifurcation theorem for nonlinear matrix models of population dynamics. J Diff Equ Appl 26:25–44. http://dx.doi.org/10.1080/10236198.2019.1699916

16. Cushing JM, MacCracken-Stump S (2010) Darwinian dynamics of a juvenile-adult model. Math Biosci Eng 10(4):1017–1044

17. Cushing JM, Zhou Y (1994) The net reproductive value and stability in structured population models. Nat Resour Model 8:1–37

18. Cushing JM, Henson SM, Hayward JL (2015) An evolutionary game theoretic model of cannibalism. Nat Resour Model 28(4):497–521. https://doi.org/10.1111/nrm.12079

19. Cushing JM, Martins F, Pinto AA, Veprauskas A (2017) A bifurcation theorem for evolutionary matrix models with multiple traits. J Math Biol 75(1):491–520. https://doi.org/10.1007/s00285-016-1091-4

20. Dercole F, Rinaldi S (2008) Analysis of Evolutionary Processes: The Adaptive Dynamics Approach and Its Applications. Princeton University Press, Princeton

21. Dieckmann U, Law R (1996) The dynamical theory of coevolution: a derivation from stochastic ecological processes. J Math Biol 34:569–612

22. Elaydi SN (2005) An Introduction to Difference Equations, 3rd edn. Springer, New York

23. Kielhöfer H (2004) Bifurcation Theory: An Introduction with Applications to PDEs. Applied Mathematical Sciences 156, Springer, New York

24. Kon R, Saito Y, Takeuchi Y (2004) Permanence of single-species stage-structured models. J Math Biol 48:515–528

25. Lande R (1976) Natural selection and random genetic drift in phenotypic evolution. Evolution 30:314–334

26. Leslie PH (1945) On the use of matrices in certain population mathematics. Biometrika 33:183–212

27. Leslie PH (1948) Some further notes on the use of matrices in population mathematics. Biometrika 35:213–245

28. Li C-K, Schneider H (2002) Applications of Perron-Frobenius theory to population dynamics. J Math Biol 44:450–462

29. McGill BM, Brown JS (2007) Evolutionary game theory and adaptive dynamics of continuous traits. Ann Rev Ecol Evol Syst 38:403–435

30. Meissen EP, Salau KR, Cushing JM (2016) A global bifurcation theorem for Darwinian matrix models. J Diff Equ Appl 22:1114–1136. https://doi.org/10.1080/10236198.2016.1177522

31. Rabinowitz PH (1971) Some global results for nonlinear eigenvalue problems. J Funct Anal 7:487–513

32. Vincent TL, Brown JS (2005) Evolutionary Game Theory, Natural Selection, and Darwinian Dynamics. Cambridge University Press, Cambridge

Chapter 12

A Model of Cannibalism and Reproductive Synchrony

Motivated by the empirical findings in Chaps. 7–9 that egg-laying synchrony is adaptive in the presence of egg cannibalism in gulls, we now extend the proof-of-concept cannibalism model in Chap. 11 to include the possibility of reproductive synchrony. The results reported here were first published in the journal articles [12] and [13] and in the Ph.D. dissertation of Amy Veprauskas [11].

12.1 Introduction

Two behavioral tactics observed in gulls (larids) in response to high sea surface temperature (SST) are increased egg cannibalism and every-other-day egg-laying synchrony. Cannibalism is hypothesized to be adaptive in years of high SST, for example, during El Niño events, because of the concomitant reduced food supply at the sea surface. Synchrony was shown to be adaptive in the presence of cannibalism in the sense that an egg laid synchronously had less chance of being cannibalized (see Chaps. 7–9). Henson et al. [9] hypothesized that every-other-day ovulation synchrony in gulls may confer a selective advantage by means of cannibal satiation. In this chapter we derive and analyze a discrete-time, structured population dynamic model that is designed to investigate this life history phenomenon. The goal is to investigate whether reproductive synchrony can be a consequence of adult-on-juvenile cannibalism (a type of "Fraser Darling effect" [7]).

We begin our investigation of the relationship between cannibalism and reproductive synchrony with a proof-of-concept model in which cannibalism, reproductive synchrony, and generation occur on the same time scale. This is, of course, not a realistic depiction of the larid system or even most systems, because behaviors such as cannibalism and reproductive synchrony often occur on a daily time scale, whereas maturation time might

J. M. Cushing et al., *Modeling Behavior and Population Dynamics*, Interdisciplinary Applied Mathematics 57, https://doi.org/10.1007/978-3-031-34283-7_12

be measured in years. In Chap. 13, we will construct a framework for models with two time scales to address the question of how daily behaviors affect population dynamics.

Our modeling framework in this chapter is that used in Sect. 11.3 except that we distinguish two classes of adults, namely, those who are reproductively active and those who are not. The model has the demographic vector

$$\mathbf{x} = \begin{pmatrix} x_1 \\ x_2 \\ x_3 \end{pmatrix} = \begin{pmatrix} \text{juveniles} \\ \text{reproductively active adults} \\ \text{reproductively inactive adults} \end{pmatrix}.$$

Juveniles are produced by reproductively active adults, and, after a unit of time, juveniles mature to become reproductively active adults. In this model, we require that reproductively active adults must, after one unit of time, become reproductively inactive for at least one time unit, after which they return to reproductive activity with a certain probability. This modeling assumption of alternating adult reproductive states is motivated by the 2-day ovulation cycle in glaucous-winged gulls (Chap. 9), in which an egg is laid (reproduction occurs) every other day. Here this idea is generalized in a simple proof-of-concept model in which newborns become adults after one model time step and adults alternate between fecund and non-fecund states.

This model lacks an important feature for application to gulls, however. Namely, it *assumes commensurate time scales for juvenile maturation and adult ovulation cycle.* This issue is addressed in Chap. 13 where these two time scales are separated.

The projection matrix has the form

$$P = \begin{pmatrix} 0 & b & 0 \\ s_1 & 0 & s_3 g \\ 0 & s_2 & s_3(1-g) \end{pmatrix},$$

where b is the adult fertility rate, s_1 is the probability a juvenile survives to become a reproductively active adult, and s_2, s_3 are the reproductively active and inactive adult survival rates, respectively. The fraction g is the probability an inactive adult becomes active (per unit time). If $g < 1$, an adult can remain in the inactive state in order to synchronize reproduction with other adults.

The next step is to incorporate density-dependent factors into the entries of P which account for the biological mechanisms of interest. For example, we can model the negative and positive effects that adult cannibalism of juveniles have on the juvenile and adult survival in the same manner that we did in Sect. 11.3. A key to our model's treatment of reproductive synchrony is the density dependence of the probability g that an inactive adult becomes active. We assume that if there are no active adults (i.e., $x_2 = 0$) at any point in time t, then all inactive adults will become active at time $t+1$. We also assume that if the number of active adults x_2 present at time t increases, then the fraction of inactive adults that become active at time $t+1$ will decrease. Under these assumptions, we have

$$g = g(x_2) \text{ where } g(0) = 1 \text{ and } g'(x_2) < 0.$$

The resulting model has a projection matrix of the general form

$$P(\mathbf{x}) = \begin{pmatrix} 0 & p_{12}(\mathbf{x}) & 0 \\ p_{21}(\mathbf{x}) & 0 & p_{23}(\mathbf{x}) \\ 0 & p_{32}(\mathbf{x}) & p_{33}(\mathbf{x}) \end{pmatrix}.$$

For notational convenience we rewrite the entries as

$$P(\mathbf{x}) = \begin{pmatrix} 0 & b\bar{p}_{12}(\mathbf{x}) & 0 \\ s_1\bar{p}_{21}(\mathbf{x}) & 0 & s_3\bar{p}_{23}(\mathbf{x}) \\ 0 & s_2\bar{p}_{32}(\mathbf{x}) & s_3\bar{p}_{33}(\mathbf{x}) \end{pmatrix} \quad (12.1)$$

where the factors $\bar{p}_{ij}$ satisfy $\bar{p}_{ij}(\mathbf{0}) = 1$ for $i \neq j$ so that the inherent (i.e., density-free) fertility and survival rates

$$b > 0, \quad 0 < s_i < 1$$

appear explicitly in the model. Mathematically we make the following assumptions on the density factors $\bar{p}_{ij}(\mathbf{x})$:

(A4) (a) $\bar{p}_{ij} \in C^2(\Omega^3, \bar{R}^3_+)$ where Ω^3 is an open set in R^3 containing $\bar{R}^3_+$ such that

$$0 \leq s_1\bar{p}_{21}(\mathbf{x}),\ s_3\bar{p}_{23}(\mathbf{x}),\ s_2\bar{p}_{32}(\mathbf{x}),\ s_3\bar{p}_{33}(\mathbf{x}) \leq 1$$

(b) $\bar{p}_{ij}(\mathbf{0}) = 1$ for $i \neq j$ and $\bar{p}_{33}(\mathbf{x}) \equiv 0$ for all $\mathbf{x} \in \bar{R}^3_+$ with $x_2 = 0$.

In the absence of density effects the projection matrix (12.1) reduces to

$$P(0) = \begin{pmatrix} 0 & b & 0 \\ s_1 & 0 & s_3 \\ 0 & s_2 & 0 \end{pmatrix}.$$

This is also the Jacobian matrix of the nonlinear matrix model

$$\mathbf{x}(t+1) = P(\mathbf{x}(t))\,\mathbf{x}(t) \quad (12.2)$$

with projection matrix (12.1) evaluated at the extinction equilibrium $\mathbf{x} = \mathbf{0}$. By the linearization principle [8], the (local) stability of $\mathbf{x} = \mathbf{0}$ can be determined from the eigenvalues of $P(\mathbf{0})$, which are $\pm\sqrt{bs_1 + s_2s_3}$ and 0. Since the dominant (Perron) eigenvalue $\sqrt{bs_1 + s_2s_3}$ is not strictly dominant (the eigenvalue $-\sqrt{bs_1 + s_2s_3}$ has the same absolute value), the matrix $P(\mathbf{0})$ is *imprimitive.* (See Appendix 11.6.) This seemingly minor technical fact has extraordinary consequences, as we will see.

As a prelude to these consequences, we notice that the extinction equilibrium $\mathbf{x} = \mathbf{0}$ of the linear matrix equation with projection matrix $P(\mathbf{0})$, i.e.,

$$\mathbf{x}(t+1) = P(\mathbf{0})\,\mathbf{x}(t), \tag{12.3}$$

is stable if $\sqrt{bs_1 + s_2 s_3} < 1$ and unstable if $\sqrt{bs_1 + s_2 s_3} > 1$. Equivalently, $\mathbf{x} = \mathbf{0}$ is stable if $R_0 < 1$ and unstable if $R_0 > 1$ where

$$R_0 = b\frac{s_1}{1 - s_2 s_3}. \tag{12.4}$$

This quantity is the inherent reproduction number, calculated as the dominant eigenvalue of the next-generation matrix

$$F(\mathbf{0})\,(I - T(\mathbf{0}))^{-1} = \begin{pmatrix} b\frac{s_1}{1-s_2s_3} & \frac{b}{1-s_2s_3} & b\frac{s_3}{1-s_2s_3} \\ 0 & 0 & 0 \\ 0 & 0 & 0 \end{pmatrix}$$

where $P(\mathbf{0}) = F(\mathbf{0}) + T(\mathbf{0})$ and

$$F(\mathbf{0}) = \begin{pmatrix} 0 & b & 0 \\ 0 & 0 & 0 \\ 0 & 0 & 0 \end{pmatrix}, \quad T(\mathbf{0}) = \begin{pmatrix} 0 & 0 & 0 \\ s_1 & 0 & s_3 \\ 0 & s_2 & 0 \end{pmatrix}$$

are the inherent fertility and transition matrices; see [2, 4, 10].

At the critical point $R_0 = 1$ the projection matrix is

$$P(\mathbf{0}) = \begin{pmatrix} 0 & \frac{1-s_2s_3}{s_1} & 0 \\ s_1 & 0 & s_3 \\ 0 & s_2 & 0 \end{pmatrix} \tag{12.5}$$

and the linear matrix equation (12.3) has infinitely many positive equilibria, namely, the positive constant multiples of a positive eigenvector associated with eigenvalue $+1$, specifically,

$$c\begin{pmatrix} 1 - s_2 s_3 \\ s_1 \\ s_1 s_2 \end{pmatrix}, \quad c > 0.$$

In this linear case, as c approaches 0 this set of positive equilibria approaches the extinction equilibrium $\mathbf{x} = \mathbf{0}$ and we say that a "branch" of positive equilibria bifurcates from $\mathbf{x} = \mathbf{0}$ at $R_0 = 1$.

Since -1 is also an eigenvalue of $P(\mathbf{0})$ when $R_0 = 1$, there are also infinitely many nonnegative, periodic solutions of period 2 (called 2-cycles), namely,

$$c\begin{pmatrix} 0 \\ \frac{1}{s_2}\frac{s_3}{1-s_3} \\ 0 \end{pmatrix} \to c\begin{pmatrix} \frac{1-s_2s_3}{s_1s_2}\frac{s_3}{1-s_3} \\ 0 \\ \frac{s_3}{1-s_3} \end{pmatrix} \to c\begin{pmatrix} 0 \\ \frac{1}{s_2}\frac{s_3}{1-s_3} \\ 0 \end{pmatrix} \to c\begin{pmatrix} \frac{1-s_2s_3}{s_1s_2}\frac{s_3}{1-s_3} \\ 0 \\ \frac{s_3}{1-s_3} \end{pmatrix} \to \cdots, \quad c > 0.$$

Notice that these 2-cycles, which mathematically lie on the boundary of the positive cone R^3_+, imply that all adults are alternately reproductively active or reproductively inactive. In other words, they are reproductively synchronized. We will refer to these as *synchronous 2-cycles*. As c approaches 0 this set of synchronous 2-cycles approaches the extinction equilibrium and we say that a branch of synchronous 2-cycles bifurcates from $\mathbf{x} = \mathbf{0}$ at $R_0 = 1$.

The bifurcation of both positive equilibria and synchronous 2-cycles at $R_0 = 1$ in the linear matrix equation (12.3) with projection matrix (12.5) leads us to suspect that this possibility also occurs for the nonlinear equation (12.2) with projection matrix (12.1). Because positive equilibria imply the presence of both types of adults at each time step and because synchronous 2-cycles imply the opposite, namely, that the two adult classes are alternatingly present, this model provides an opportunity to study if and when reproductive synchrony occurs. To do this, we first need to demonstrate mathematically that both bifurcations actually occur and then to provide the means to determine which of the two is stable.

Example 8 *As a motivating example we consider the projection matrix*

$$P(\mathbf{x}) = \begin{pmatrix} 0 & b\frac{1}{1+x_2} & 0 \\ \frac{3}{4} & 0 & \frac{1}{2}\frac{1}{1+x_2} \\ 0 & \frac{1}{2} & \frac{1}{2}\left(1 - \frac{1}{1+x_2}\right) \end{pmatrix} \tag{12.6}$$

in which there is density-dependent adult fertility and a density-dependent reproductive recovery term g. If the initial demographic vector $\mathbf{x}(0)$ contains only active adults

$$\mathbf{x}(0) = \begin{pmatrix} 0 \\ x_2 \\ 0 \end{pmatrix}, \quad x_2 > 0,$$

then it is not difficult to see that the resulting solution of the matrix equation has the sign structure

$$\begin{pmatrix} 0 \\ + \\ 0 \end{pmatrix} \to \begin{pmatrix} + \\ 0 \\ + \end{pmatrix} \to \begin{pmatrix} 0 \\ + \\ 0 \end{pmatrix} \to \begin{pmatrix} + \\ 0 \\ + \end{pmatrix} \to \begin{pmatrix} 0 \\ + \\ 0 \end{pmatrix} \to \cdots \tag{12.7}$$

and we say the solution is synchronous. If this sequence is periodic with period 2, then we refer it as a synchronous 2-cycle. Mathematically, a 2-cycle corresponds to a fixed point of the composite of the map $P(\mathbf{x})\mathbf{x}$ and to find one we calculate

$$\begin{pmatrix} 0 \\ x_2 \\ 0 \end{pmatrix} \to \begin{pmatrix} b\frac{x_2}{1+x_2} \\ 0 \\ \frac{1}{2}x_2 \end{pmatrix} \to \begin{pmatrix} 0 \\ \frac{1}{4}x_2 + \frac{3}{4}b\frac{x_2}{1+x_2} \\ 0 \end{pmatrix}$$

and solve for a positive root x_2 of the algebraic equation

$$x_2 = \frac{1}{4}x_2 + \frac{3}{4}b\frac{x_2}{1+x_2}$$

to obtain $x_2 = b - 1$. From (12.6) we see that

$$s_1 = \frac{3}{4}, \quad s_2 = s_3 = \frac{1}{2}$$

and from Eq. (12.4) that $R_0 = b$ and, therefore,

$$x_2 = R_0 - 1.$$

For each $R_0 > 1$ this example has a unique positive synchronous 2-cycle alternating between the two demographic vectors

$$\begin{pmatrix} 0 \\ R_0 - 1 \\ 0 \end{pmatrix} \leftrightarrow \begin{pmatrix} R_0 - 1 \\ 0 \\ \frac{1}{2}(R_0 - 1) \end{pmatrix}. \tag{12.8}$$

Moreover, for each $R_0 > 1$ there also exists a (unique) positive equilibrium. It requires some tedious algebra to solve the equilibrium equation $\mathbf{x} = P(\mathbf{x})\mathbf{x}$ for

$$\begin{pmatrix} x_1 \\ x_2 \\ x_3 \end{pmatrix} = \begin{pmatrix} R_0 \frac{x_2}{x_2+1} \\ x_2 \\ x_2 \frac{x_2+1}{x_2+2} \end{pmatrix},$$

where

$$x_2 = \frac{3}{8}R_0 + \frac{1}{8}\sqrt{9R_0^2 + 36R_0 + 4} - \frac{5}{4}$$

and to show it is positive if and only if $R_0 > 1$.

The point of Example 8, with its complicated formulas for equilibria and synchronous 2-cycles, is to show that the matrix equation (12.2) with projection matrix (12.1) can exhibit simultaneously two different reproductive dynamics: a synchronous 2-cycle, in which there is reproductive synchrony, and a positive equilibrium, in which there is not.

Notice that the formulas for the positive equilibria and synchronous 2-cycles in Example 8 show that both entities are continuous as functions of R_0 and that both approach the extinction equilibrium $\mathbf{x} = \mathbf{0}$ as R_0 approaches 1, that is to say, both branches bifurcate from the extinction equilibrium pair $(R_0, \mathbf{x}) = (1, \mathbf{0})$. Note, however, that in contrast to the linear case (12.3) with projection matrix (12.5) when $R_0 = 1$ for all positive equilibria and synchronous 2-cycles from the bifurcating branches, the branches for the nonlinear model in Example 8 are such that there exists a (unique) positive equilibrium and a (unique) synchronous 2-cycle for each $R_0 > 1$. One way to visualize the difference between the linear and nonlinear case is to draw a bifurcation diagram, which is a plot of R_0 on the horizontal axis and, in the case of a the positive equilibrium branch, a quantity representing the equilibrium on the vertical axis (e.g., total population size $\|\mathbf{x}\|$). In the linear case, the plot would be a vertical straight line located at $R_0 = 1$, which is why this bifurcation is sometimes called a vertical bifurcation. In the nonlinear case, however, the bifurcation diagram would be a curve emanating from the point $(R_0, \mathbf{x}) = (1, \mathbf{0})$ and extending to the right for all $R_0 > 1$. It is as if the linear bifurcation diagram has been bent or warped by the nonlinearities. See Fig. 12.1. Similar diagrams can be drawn for the synchronous 2-cycle branches.

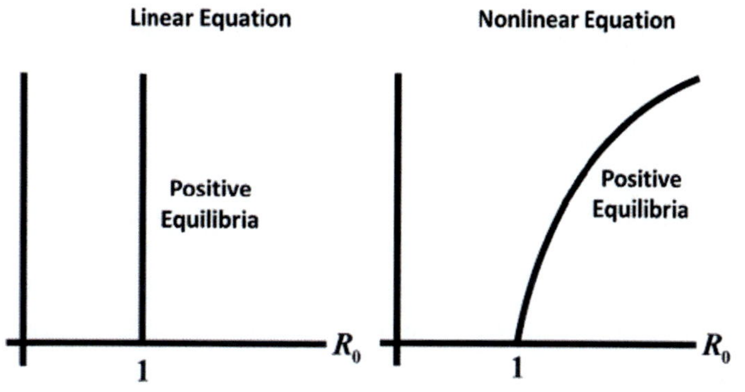

Figure 12.1: Schematic diagrams of bifurcation diagrams for a linear matrix equation (12.3) and a nonlinear matrix equation (12.2). Similar bifurcation diagrams can be drawn for the bifurcating synchronous 2-cycles

With regard to the bifurcating positive equilibria, this is exactly what occurs in the cannibalism model studied in Sect. 11.3. Indeed, it occurs in very general nonlinear matrix models with primitive projection matrices [2, 6]. What is different in Example 8 is the occurrence of bifurcating synchronous 2-cycles that arise from the lack of primitivity of the inherent projection matrix $P(\mathbf{0})$ (in particular, from the eigenvalue -1). The important question yet to be answered concerns stability. Which of the two is stable? Can both be stable? Neither? Is stability related to the direction of bifurcation?

Rather than pursue this special example any further, we take up the question of the existence and stability of both synchronous 2-cycles and positive equilibria in the general model (12.2) with projection matrix (12.1).

12.2 The General Reproductive Synchrony Model

There is a large literature on a special type of imprimitive matrix models called semelparous Leslie models. See [5] and the literature cited therein. These models arose in studies of the famous species of cicadas whose periodic outbreaks have long fascinated biologists. The existence and stability of synchronous cycles in semelparous Leslie models have been used to account for the synchronized cohorts of these cicadas (and other periodical insects). The matrix equation (12.2) with projection matrix (12.1) is not a Leslie model, however, and the analysis in this literature does not apply to it.

The nature of the bifurcation that occurs at $R_0 = 1$ when the projection matrix in a nonlinear matrix model is imprimitive at $\mathbf{x} = \mathbf{0}$ can, in general, be very complex and it is not well understood, especially in higher dimensions. One thing that can generally be said is that positive equilibria do bifurcate at $R_0 = 1$ from $\mathbf{x} = \mathbf{0}$ and that in addition synchronous cycles (of some period) also bifurcate from $\mathbf{x} = \mathbf{0}$ [5]; however, it is not easy to sort out which is an attractor and which is not. Indeed, other more complicated attractors can also bifurcate as well (such as the so-called cycle chains, which consist of heteroclinic orbits connecting the phases of synchronous cycles) [3].

Luckily, our model of interest (12.2) with projection matrix (12.1) is a special type of imprimitive matrix model (said to have the beta property) studied in [11]. That study provides conditions under which both positive equilibria and synchronous 2-cycles bifurcate from $\mathbf{x} = \mathbf{0}$ and $R_0 = 1$ and provides diagnostic quantities that can be used to determine their stability properties. We will not present these general results here, but instead apply them to the matrix equation (12.2) with projection matrix (12.1). We closely follow the analysis given in [13].

Using formula (12.4), we introduce R_0 explicitly into the projection matrix (12.1):

$$P(\mathbf{x}) = \begin{pmatrix} 0 & R_0 \frac{1-s_2 s_3}{s_1} \bar{p}_{12}(\mathbf{x}) & 0 \\ s_1 \bar{p}_{21}(\mathbf{x}) & 0 & s_3 \bar{p}_{23}(\mathbf{x}) \\ 0 & s_2 \bar{p}_{32}(\mathbf{x}) & s_3 \bar{p}_{33}(\mathbf{x}) \end{pmatrix}, \quad R_0 = b \frac{s_1}{1 - s_2 s_3}. \tag{12.9}$$

If $\mathbf{x} \in \bar{R}^3_+$ is an equilibrium (fixed point) of the nonlinear matrix equation (12.2) with this projection matrix, then we call $(R_0, \mathbf{x}) \in R_+ \times \bar{R}^3_+$ an *equilibrium pair*. Note that $(R_0, \mathbf{0})$ is an *extinction equilibrium pair* for all values of R_0. If $(R_0, \mathbf{x}) \in R_+ \times R^3_+$ is an equilibrium pair, we refer to it as a *positive equilibrium pair*. A *synchronous 2-cycle* of this matrix equation is a solution that sequentially alternates between two demographic vectors

$$\mathbf{x}_2 := \begin{pmatrix} 0 \\ x_2 \\ 0 \end{pmatrix} \text{ and } \mathbf{x}_{13} := \begin{pmatrix} x_1 \\ 0 \\ x_3 \end{pmatrix} \text{ with } x_i > 0 \, .$$

We represent the 2-cycle by its two points $[\mathbf{x}_2, \mathbf{x}_{13}]$ and let $(R_0, [\mathbf{x}_2, \mathbf{x}_{13}])$ denote a *synchronous 2-cycle pair*.

As pointed out above, the extinction equilibrium loses stability as R_0 increases through 1. Theorems 9 and 10 below describe, respectively, the bifurcation of positive equilibria and synchronous 2-cycles that result from this destabilization. Unlike the cannibalism model in Sect. 11.3, however, stability is not determined solely by the direction of bifurcation as determined by the sign of a single diagnostic quantity κ. For the matrix equation (12.2) with projection matrix (12.9), the bifurcation is more complicated and requires the use of several diagnostic quantities.

The partial derivatives

$$\partial_k^0 \bar{p}_{ij} := \left.\frac{\partial \bar{p}_{ij}(\mathbf{x})}{\partial x_k}\right|_{(R_0,\mathbf{x})=(1,\mathbf{0})}.$$

measure the sensitivities of the entries in the projection matrix (12.9) to changes in class-specific densities x_k at low densities (i.e., near $\mathbf{x} = \mathbf{0}$). We organize these sensitivities into two categories based on two different groups of individuals: the reproductive individuals in class x_2 and the non-reproductive individuals in classes x_1 and x_3. A derivative such as $\partial_1^0 \bar{p}_{12}$, which measures how the fecundity of reproductive individuals (those in class x_2), is affected by changes in the density x_1 of non-reproductive individuals, we call a *between-group sensitivity*. By contrast, a *within-group sensitivity* is a derivative such as $\partial_2^0 \bar{p}_{12}$, which measures the effect that the density of reproductive individuals x_2 has on their own fertility. To describe the bifurcation at $(R_0, \mathbf{x}) = (1, \mathbf{0})$ we need to define some linear combinations of these sensitivities. To do this, we use the eigenvector

$$\mathbf{v}_+ = \begin{pmatrix} 1 - s_2 s_3 \\ s_1 \\ s_1 s_2 \end{pmatrix}$$

of the inherent projection matrix (12.5) associated with eigenvalue 1 and denote its projections onto the x_2-axis and the (x_1, x_3)-plane on the boundary ∂R_+^3 of the positive cone by

$$\boldsymbol{\pi}_2 := \begin{pmatrix} 0 \\ s_1 \\ 0 \end{pmatrix}, \qquad \boldsymbol{\pi}_{13} := \begin{pmatrix} 1 - s_2 s_3 \\ 0 \\ s_1 s_2 \end{pmatrix}.$$

We also introduce the notation

$$D_{\boldsymbol{\pi}_2}^0 \bar{p}_{ij} := \nabla^0 \bar{p}_{ij} \boldsymbol{\pi}_2, \qquad D_{\boldsymbol{\pi}_{13}}^0 \bar{p}_{ij} := \nabla^0 \bar{p}_{ij} \boldsymbol{\pi}_{13}$$

for the directional derivatives of $\bar{p}_{ij}$, where the gradient $\nabla \bar{p}_{ij}$ is a row vector and the superscript "0" denotes evaluation at $(R_0, \mathbf{x}) = (1, \mathbf{0})$. It follows from A4(b) that

$$D_{\boldsymbol{\pi}_2}^0 \bar{p}_{33} = s_1 \partial_2^0 \bar{p}_{33}, \qquad D_{\boldsymbol{\pi}_{13}}^0 \bar{p}_{33} = 0.$$

We use the quantities

$$\begin{aligned} c_b &:= (1 - s_2 s_3)\, D^0_{\boldsymbol{\pi}_{13}} \bar{p}_{12} + (1 - s_2 s_3)\, D^0_{\boldsymbol{\pi}_2} \bar{p}_{21} \\ &\quad + s_2 s_3 D^0_{\boldsymbol{\pi}_2} \bar{p}_{23} + s_2 s_3 D^0_{\boldsymbol{\pi}_{13}} \bar{p}_{32} + s_1 s_2 s_3^2 \partial^0_2 \bar{p}_{33} \end{aligned} \tag{12.10}$$

$$\begin{aligned} c_w &:= (1 - s_2 s_3)\, D^0_{\boldsymbol{\pi}_2} \bar{p}_{12} + (1 - s_2 s_3)\, D^0_{\boldsymbol{\pi}_{13}} \bar{p}_{21} \\ &\quad + s_2 s_3 D^0_{\boldsymbol{\pi}_{13}} \bar{p}_{23} + s_2 s_3 D^0_{\boldsymbol{\pi}_2} \bar{p}_{32} \end{aligned}, \tag{12.11}$$

which are linear combinations of these directional derivatives, to define

$$a_+ := c_w + c_b, \quad a_- := c_w - c_b. \tag{12.12}$$

These measure, respectively, the total density sensitivities in the population and the difference between within-group and between-group density sensitivity in the population.

The following theorem describes the properties of the bifurcation of positive equilibria from $(R_0, \mathbf{0}) = (1, \mathbf{0})$. Recall that we say a bifurcation is forward (backward) if, in a neighborhood of the bifurcation point $(R_0, \mathbf{0}) = (1, \mathbf{0})$, the positive equilibria exist for $R_0 > 1$ ($R_0 < 1$). We say the bifurcation is stable (unstable) if, in a neighborhood of the bifurcation point $(R_0, \mathbf{0}) = (1, \mathbf{0})$, the positive equilibria are stable (unstable).

Theorem 9 ([13]) *Consider the matrix equation (12.2) with projection matrix (12.9) under assumption A4. If $a_+ > 0$ then a backward and unstable bifurcation of positive equilibrium pairs $(R_0, \mathbf{x})$ occurs at $(1, \mathbf{0})$. If $a_+ < 0$ then a forward bifurcation of positive equilibrium pairs $(R_0, \mathbf{x})$ occurs at $(1, \mathbf{0})$ which is stable if $a_- < 0$ and unstable if $a_- > 0$.*

Theorem 9 is a special case of, and the motivation for, a general bifurcation theorem concerning the dynamic dichotomy that occurs in a general class of nonlinear matrix models as the extinction equilibrium destabilizes [11, 12].

A negative derivative $\partial^0_k \bar{p}_{ij} < 0$ implies a decrease of the vital rate $\bar{p}_{ij}$ with an increase in class x_k at low density. A positive derivative $\partial^0_k \bar{p}_{ij} > 0$ implies increase of the vital rate $\bar{p}_{ij}$ with an increase in class x_k at low density (which is called a component Allee effect in [1]). The quantity

$$a_+ = (1 - s_2 s_3)\, D^0_{\mathbf{v}_+} \bar{p}_{12} + (1 - s_2 s_3)\, D^0_{\mathbf{v}_+} \bar{p}_{21} + s_2 s_3 D^0_{\mathbf{v}_+} \bar{p}_{23} + s_2 s_3 D^0_{\mathbf{v}_+} \bar{p}_{32} + s_1 s_2 s_3^2 \partial^0_2 \bar{p}_{33},$$

whose sign determines the direction of bifurcation in Theorem 9, is a weighted sum of all negative and positive density effects. A forward bifurcation occurs, i.e., $a_+ < 0$, if the negative density sensitivities outweigh the positive density sensitivities (at low population densities). A backward bifurcation requires the occurrence of component Allee effects which outweigh the negative density sensitivities (at low population densities).

As discussed in Sect. 11.2, a backward bifurcation opens the possibility of a strong Allee effect and for (initial condition dependent) survival when $R_0 < 1$.

Suppose, on the other hand, that there are no component Allee effects and there is at least one within-group negative effect. It follows that $c_b \leq 0$ and $c_w < 0$ and, as a result, that $a_+ < 0$ and the bifurcation is forward. From Theorem 9 we find then that the bifurcation is stable if $|c_b| < |c_w|$ and unstable if $|c_b| > |c_w|$. *This means, in this case, that the bifurcation is stable if the magnitude of the within-group density sensitivity is larger than the magnitude of the between-group density sensitivity, and that it is unstable if the opposite is true.*

Note that according to Theorem 9, a forward bifurcation can be unstable, specifically when $a_+ < 0$ and $a_- > 0$. In this case, both the extinction and positive equilibria are unstable for $R_0 \gtrsim 1$.

Theorem 10 ([13]) *Consider the matrix equation (12.2) with projection matrix (12.9) under assumption A4. If* $c_w > 0$ *then a backward and unstable bifurcation of synchronous 2-cycles pairs* $(R_0, [x_2, x_{13}])$ *occurs at* $(1, \mathbf{0})$*. If* $c_w < 0$ *then a forward bifurcation of synchronous 2-cycles pairs* $(R_0, [x_2, x_{13}])$ *occurs at* $(1, \mathbf{0})$ *which is stable if* $a_- > 0$ *and unstable if* $a_- < 0$.

The direction of bifurcation of the synchronous 2-cycles, unlike that of the positive equilibria, is not determined by the sign of a_+, which as pointed out above is a weighted sum of directional derivatives in the direction of $\bar{v}_+$ lying in the positive cone. Instead, the direction of bifurcation of the synchronous 2-cycles is determined by the sign of c_w which, by its definition (12.11), is a weighted sum of directional derivatives in the directions $\boldsymbol{\pi}_2$ and $\boldsymbol{\pi}_{13}$. These directions lie in the invariant components of the boundary ∂R^3_+ wherein the synchronous solutions lie.

Remark 11 *Unlike the (primitive) matrix model for cannibalism in Sect. 11.3, the stability of the bifurcations at* $(R_0, \mathbf{x}) = (1, \mathbf{0})$ *is not determined solely by the direction of bifurcation in the imprimitive matrix equation (12.2) and (12.9). While backward bifurcations of both positive equilibria and synchronous 2-cycles are indeed unstable, we see from Theorems 9 and 10 that when both bifurcations are forward their stability criteria are exactly opposite, that is to say, one is a stable bifurcation and the other is an unstable bifurcation. This* ***dynamic dichotomy*** *means biologically that, upon destabilization of the extinction equilibrium as* R_0 *increases through* 1*, either the population approaches reproductive synchrony or a positive equilibrium with both adult classes present at each time step, depending on the sign of* a_-.

While Theorems 9 and 10 account in general terms for the bifurcations that can occur in the imprimitive model equation (12.2) with projection matrix (12.9), they are restricted to a neighborhood of the bifurcation point $(R_0, \mathbf{0}) = (1, \mathbf{0})$. Other dynamic features and attractors can occur outside a neighborhood of the bifurcation point. For example, in the case of backward and hence unstable bifurcations, these theorems do not give information about what attractors are present. They do not provide information about the possibility of strong Allee effects in this case, which are of interest in our applications (cf. Sect. 11.3). The dynamics that occur outside a neighborhood of the bifurcation point depend in general on the specifics of the nonlinearities appearing in the model equations.

In the following sections we turn next to some matrix models studied in [11] and [13] with specific entries $\bar{p}_{ij}(\mathbf{x})$ in their projection matrices that are designed to capture the mechanisms in which we are interested, namely, adult cannibalism of juveniles and adult reproductive synchrony. We use these example models for two main purposes:

(1) To illustrate the application of Theorems 9 and 10 and the bifurcation phenomena described in them.
(2) To use these results, together with numerical simulations of the models, to investigate several bifurcations scenarios involving the relationship between cannibalism and reproductive synchrony.

These investigations will by no means constitute a thorough examination of these example models, with respect to either their mathematical or biological features; they serve only to illustrate some of their possibilities.

12.3 A Model Study of Reproductive Synchrony

The motivating example with projection matrix (12.6) is a special case of the projection matrix

$$P(\mathbf{x}) = \begin{pmatrix} 0 & b\frac{1}{1+c_2x_2} & 0 \\ s_1 & 0 & s_3\frac{1}{1+c_gx_2} \\ 0 & s_2 & s_3\left(1-\frac{1}{1+c_gx_2}\right) \end{pmatrix}, \qquad R_0 = b\frac{s_1}{1-s_2s_3} \tag{12.13}$$

which contains the density-dependent term

$$\frac{1}{1+c_2x_2}$$

accounting for adult regulation of fertility and the term

$$g(\mathbf{x}) = \frac{1}{1+c_gx_2}$$

for the probability that an inactive adult becomes active (per unit time). No cannibalism is incorporated into this example. The coefficient $c_g > 0$ in this model is a measure of the tendency for the two adult classes to synchronize. If this *synchrony coefficient* c_g is small. then nearly all the reproductively inactive adults x_3 will become active and move to the active adult class x_2 at the next time step. Since all active adults are assumed in this model to move to the inactive class at the next time step, the individuals in these two groups of adults will remain highly segregated. On the other hand, if c_g is large then most reproductively inactive adults will remain inactive at the next time step when the class of active adults will join them in class x_3. How this propensity to reproductively synchronize affects the bifurcation at $(R_0, \mathbf{x}) = (1, \mathbf{0})$ is determined by the diagnostic quantities (12.10), (12.11), and (12.12), which are

$$\begin{aligned} c_b &= -s_1 s_2 s_3 (1 - s_3) c_g < 0 \\ c_w &= -s_1 (1 - s_2 s_3) c_2 < 0 \\ a_+ &= -s_1 (1 - s_2 s_3) c_2 - s_1 s_2 s_3 (1 - s_3) c_g < 0 \\ a_- &= -s_1 (1 - s_2 s_3) c_2 + s_1 s_2 s_3 (1 - s_3) c_g . \end{aligned}$$

Theorems 9 and 10 imply that the bifurcations at $(R_0, \mathbf{x}) = (1, \mathbf{0})$ of both the positive equilibria and the synchronous 2-cycles are forward and that these bifurcations exhibit the dynamic dichotomy described in Remark 11.

Specifically, the bifurcating positive equilibria are stable if $a_- < 0$ (and the bifurcating synchronous 2-cycles are unstable), while the reverse is true if $a_- > 0$. If we consider a_- as a function of the synchrony coefficient c_g we conclude that if c_g is small, specifically if

$$c_g < \frac{1 - s_2 s_3}{s_2 s_3 (1 - s_3)} c_2 , \tag{12.14}$$

then the positive equilibria are stable and the synchronous 2-cycles are unstable, but that if c_g is large, i.e.,

$$c_g > \frac{1 - s_2 s_3}{s_2 s_3 (1 - s_3)} c_2$$

then the synchronous 2-cycles are stable and the positive equilibria are unstable. *Thus, reproductive synchrony occurs if the coefficient c_g is sufficiently large.*

Figure 12.2 shows sample solutions illustrating each of these two cases.

Numerical explorations of the example in Fig. 12.2 reveal some dynamic phenomena not covered by Theorems 9 and 10. In case (B) of reproductive synchrony, larger values of $R_0 > 1$ show a destabilization of the bifurcating synchronous 2-cycles, which results in another bifurcation of a branch of stable, but now *positive* 2-cycles. With even further increases in R_0 these positive 2-cycles disappear (in a backward period doubling bifurcation) that results in stable positive equilibria. This sequence of bifurcations as R_0 increases, shown in Fig. 12.3, emphasizes that the stability results of Theorems 9 and 10 are, in general, valid in only a neighborhood of the bifurcation at $(R_0, \mathbf{x}) = (1, \mathbf{0})$. *Thus, the reproductive synchrony guaranteed by the bifurcation at $R_0 = 1$ might, as it is in this sample simulation, be lost for larger values of R_0.*

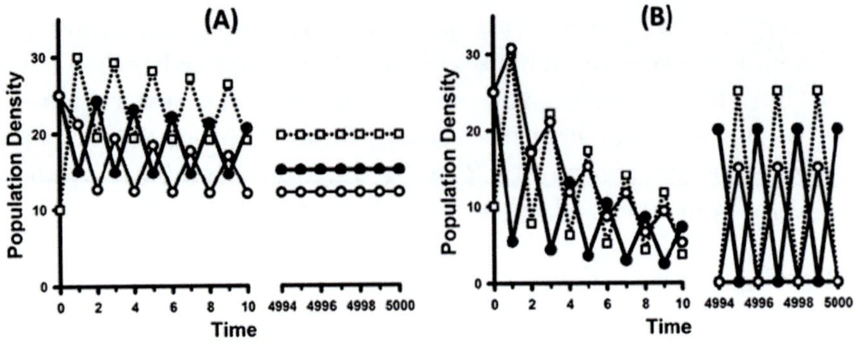

Figure 12.2: Sample solutions of the matrix equation (12.2) with projection matrix (12.13) illustrate the two alternatives from the bifurcating dynamic dichotomy that occur with two different values of the synchrony coefficient c_g. The model parameters are $b = 1.5$, $s_1 = 0.5$, $s_2 = 0.75$, $s_3 = 0.5$, and $c_2 = 0.01$ and the initial condition is $(x_1(0), x_2(0), x_3(0)) = (10, 25, 25)$. The inherent net reproduction number $R_0 = 1.2$ is larger than 1. The open squares are the juvenile class densities. The solid (open) circles are the reproductively active (inactive) adult class densities. **(a) Small synchrony coefficient.** With $c_g = 0.01$ we have $c_b = -9.4 \times 10^{-4}$, $c_w = -3.1 \times 10^{-3}$, $a_+ = -4.1 \times 10^{-3}$, and $a_- = -2.2 \times 10^{-3} < 0$. As predicted by Theorem 9 we see that, after some oscillatory transients, the population approaches an equilibrium with both adult classes present at all times. **(b) Large synchrony coefficient.** With $c_g = 1$ we have $c_b = -9.4 \times 10^{-2}$, $c_w = -3.1 \times 10^{-3}$, $a_+ = -9.7 \times 10^{-}$, and $a_- = 9.1 \times 10^{-2} > 0$. As predicted by Theorem 10 we see that, after some oscillatory transients, the population approaches a synchronous 2-cycle with only one adult class present at all times (reproductive synchrony). Figure reproduced from [13] with permission.

12.4 A Cannibalism and Reproductive Synchrony Model

We can include adult cannibalism of juveniles in the projection matrix (12.13) of reproductive synchrony model in Sect. 12.3 by means similar to those we used for the cannibalism model in Sect. 11.3. The difference is that there are two classes of adults for which we must account.

As in the projection matrix (11.10) we incorporate adult-on-juvenile cannibalism into the projection matrix in two ways: an adverse effect on juvenile survival and a beneficial effect on adult survival. These are described, respectively, by the factors $\bar{p}_{21}(\mathbf{x})$ and the three factors $\bar{p}_{32}(\mathbf{x})$, $\bar{p}_{23}(\mathbf{x})$, and $\bar{p}_{33}(\mathbf{x})$ in (12.9), which we treat individually as follows.

The expression

$$\bar{p}_{21}(\mathbf{x}) = \prod_{i=2}^{3} \left(1 - \frac{1}{1 + c_1 x_1} \frac{v_i x_i}{1 + v_i x_i}\right), \quad c_1, v_i \geq 0 \tag{12.15}$$

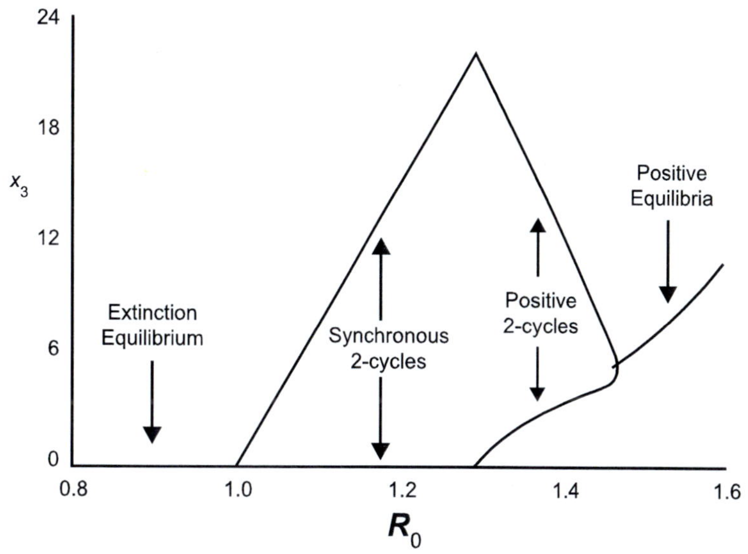

Figure 12.3: The x_3 (reproductively active adult) component of the attractor is plotted as a function of R_0 for the parameter values used in plot (b) of Fig. 12.2.

$$v_i \geq 0$$

for juvenile survival is constructed under the following modeling assumptions. The probability that an individual juvenile is cannibalized in the presence of x_i adults is modeled by a familiar predator-prey Holling II functional of the form

$$\frac{v_i x_i}{1 + v_i x_i}. \tag{12.16}$$

This probability of cannibalism is further modified by the *prey saturation assumption* that an individual juvenile's probability of being a victim decreases in the presence of more juveniles, which accounts for the factor $1/(1 + w_1 x_1)$. The parenthetical term in (12.15) is the probability a juvenile will *not* be cannibalized (per unit time) and the product arises from assuming that cannibalism by the two adult classes are independent events.

With regard to the positive effect that cannibalism has on x_2 adult survival, we set

$$\bar{p}_{32}(\mathbf{x}) = \sigma_2 \left(\frac{1}{1 + c_1 x_1} \frac{v_2}{1 + v_2 x_2} x_1 \right), \tag{12.17}$$

where $\sigma_2(z)$ is an increasing function of its argument z, which in $\bar{p}_{32}(\mathbf{x})$ is the total number of juveniles cannibalized by a single x_2 adult. We introduce a similar factor into the survival probability of x_3 adults and obtain

$$\bar{p}_{23}(\mathbf{x}) = \frac{1}{1 + c_g x_2} \sigma_3 \left(\frac{1}{1 + c_1 x_1} \frac{v_3}{1 + v_3 x_3} x_1 \right) \tag{12.18}$$

$$\bar{p}_{33}(\mathbf{x}) = \left(1 - \frac{1}{1 + c_g x_2}\right) \sigma_3 \left(\frac{1}{1 + c_1 x_1} \frac{v_3}{1 + v_3 x_3} x_1\right). \tag{12.19}$$

In order to meet the requirements A4 needed in Theorems 9 and 10, terms $\sigma_i(z)$ must satisfy (in addition to being twice continuously differentiable) the constraints

$$\sigma_i(0) = 1, \quad \partial_z \sigma_i(z) \geq 0, \quad 0 < s_i \sigma_i(z) < 1 \text{ for all } z \geq 0.$$

Following [11, 13] we write

$$\begin{aligned} \sigma_i(z) &= 1 + \frac{s_{im} - s_i}{s_i} \frac{c_{\sigma i} z}{1 + c_{\sigma i} z} \\ 0 &< s_i \leq s_{im} < 1, \quad c_{\sigma i} \geq 0. \end{aligned}$$

Adult survival probability increases from $s_i \sigma_i(0) = s_i$ in the absence of cannibalism to a maximum of $\lim_{z \to \infty} s_i \sigma_i(z) = s_{im}$ when cannibalism occurs.

When $v_i = 0$ cannibalism is absent and the model reduces to the synchrony model in Sect. 12.3. When one or both $v_i > 0$, then this model incorporates the same features of cannibalism as in the model investigated in Sect. 11.3: the negative and positive trade-off effects on juvenile and adult survival, respectively, and a victim saturation effect (if $c_1 > 0$).

In summary, all these assumptions and submodels lead to a matrix equation (12.2) with the projection matrix (12.9) and density factors

$$\begin{aligned} \bar{p}_{12}(\mathbf{x}) &= \tfrac{1}{1+c_2 x_2} \\ \bar{p}_{21}(\mathbf{x}) &= \textstyle\prod_{i=2}^{3} \left(1 - \tfrac{1}{1+c_1 x_1} \tfrac{v_i x_i}{1+v_i x_i}\right) \\ \bar{p}_{32}(\mathbf{x}) &= \sigma_2 \left(\tfrac{1}{1+c_1 x_1} \tfrac{v_2}{1+v_2 x_2} x_1\right) \\ \bar{p}_{23}(\mathbf{x}) &= \tfrac{1}{1+c_g x_2} \sigma_3 \left(\tfrac{1}{1+c_1 x_1} \tfrac{v_3}{1+v_3 x_3} x_1\right) \\ \bar{p}_{33}(\mathbf{x}) &= \left(1 - \tfrac{1}{1+c_g x_2}\right) \sigma_3 \left(\tfrac{1}{1+c_1 x_1} \tfrac{v_3}{1+v_3 x_3} x_1\right) \end{aligned} \tag{12.20}$$

to which the bifurcation Theorems 9 and 10 apply.

Calculations show that the diagnostic quantities, whose signs determine the direction of bifurcation and the stability properties of the bifurcating positive equilibria and synchronous 2-cycles, are

$$\begin{aligned} c_b &= -s_1 s_2 s_3 (1 - s_3) c_g + v_2 (1 - s_2 s_3) \left(s_2 s_3 \partial_z^0 \sigma_2 - s_1\right) \\ c_w &= -s_1 (1 - s_2 s_3) c_2 + v_3 (1 - s_2 s_3) \left(\partial_z^0 \sigma_3 - s_1 s_2\right), \end{aligned} \tag{12.21}$$

where

$$\partial_z^0 \sigma_i = \frac{s_{im} - s_i}{s_i} c_{i\sigma} > 0.$$

While these formulas allow one to ascertain which coefficients contribute to the various bifurcation possibilities in Theorems 9 and 10, it is clear that there are many scenarios possible with regard to the role that cannibalism plays in determining population survival and its dynamic characteristics. Moreover, beyond the local bifurcations described by these general theorems, there is the question of what attractors and dynamics are possible outside the neighborhood of the bifurcation point $(R_0, \mathbf{x}) = (1, \mathbf{0})$ (e.g., can there be multiple attractors, strong Allee effects, etc.).

In what follows, we will focus our attention on only a few scenarios of interest that are motivated by the observations in gull colonies discussed in Chaps. 7–9 in Part II.

Cannibalism-Induced Reproductive Synchrony. When cannibalism is absent ($v_2 = v_3 = 0$), the projection matrix (12.9) reduces to the projection matrix (12.13) for the reproductive synchrony model. In this case, we saw in Sect. 12.3 that both positive equilibria and synchronous 2-cycle bifurcations are forward and that only one of the two is stable. In particular, the positive equilibria are stable (and the synchronous 2-cycles are unstable) if

$$a_- = -s_1 (1 - s_2 s_3) c_2 + s_1 s_2 s_3 (1 - s_3) c_g < 0 ,$$

i.e., if the synchrony coefficient c_g is less than the threshold (12.14):

$$c_g < \frac{1 - s_2 s_3}{s_2 s_3 (1 - s_3)} c_2 .$$

This is illustrated in Fig. 12.2a. In this case, we ask whether or not it is possible that the stability properties in the dynamic dichotomy can be reversed by introducing cannibalism into the model, that is to say, *is it possible for the positive equilibria to be stable and the synchronous 2-cycles to be unstable when cannibalism is absent (i.e.,* $v_2 = v_3 = 0$*), but for the reverse to be true when cannibalism is introduced into the model (i.e., when at least one* v_i *is positive)?*

We can use Theorems 9 and 10 to address this question by considering the nature of the bifurcation at $R_0 = 1$ as determined by the diagnostic quantities c_b and c_w given by (12.21) together with $a_+ = c_w + c_b$ and $a_- = c_w - c_b$. If c_b and c_w, and hence a_+, are negative, then both positive equilibrium and the synchronous 2-cycle bifurcations are forward. From the formulas (12.21) we see that this occurs if the cannibalism coefficients v_2 and v_3 are not too large. In this case, which bifurcating branch is stable is determined by the sign of a_-. The answer to our question is "yes" if $a_- < 0$ in the absence of cannibalism and $a_- > 0$ when cannibalism is present. The first case $a_- < 0$ when $v_2 = v_3 = 0$ occurs (as noted above) if the synchrony coefficient c_g is less than the threshold (12.14). By inspection we see that the second case $a_- > 0$ occurs if, for example, $v_3 > 0$ and $c_{3\sigma}$ are sufficiently large, i.e., if cannibalism is sufficiently beneficial for the survival of the non-reproducing adults x_3. Figure 12.4 illustrates this scenario for which *the population dynamic exhibits reproductive synchrony when cannibalism is present, but not when it is absent.* The basic mechanism responsible is the prey saturation effect that reduces an individual juvenile's probability of being cannibalized.

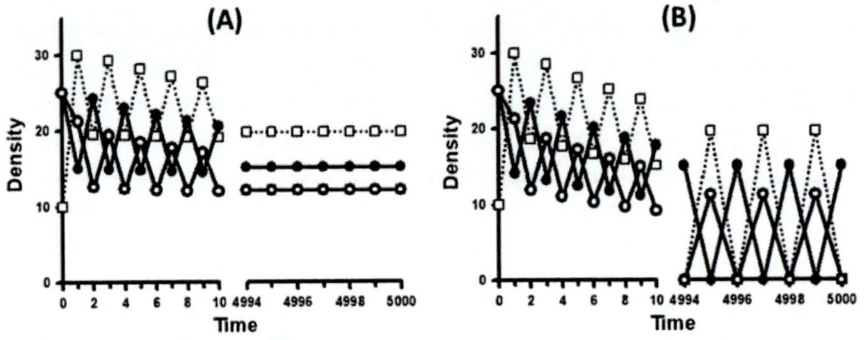

Figure 12.4: The solution of the matrix equation (12.2) with initial conditions $(x_1, x_2, x_3) = (10, 25, 25)$ illustrates the two alternatives in the bifurcating dynamic dichotomy with coefficients $b = 1.5$, $s_1 = 0.5$, $s_2 = 0.75$, $s_3 = 0.5$, $c_2 = 0.01$, and $c_g = 0.01$ in the projection matrix (12.9) with entries (12.20). The inherent net reproduction number $R_0 = 1.2$ is larger than 1. The open squares are the juvenile class densities. The solid (open) circles are the reproductively active (inactive) adult class densities. **(a) No cannibalism and no reproductive synchrony**. With cannibalism being absent ($v_2 = v_3 = 0$), the population approaches an equilibrium with both adult classes present at all times. This is the same simulation as in Fig. 12.2a. **(b) Cannibalism and reproductive synchrony**. With cannibalism being present ($v_2 = v_3 = 1$, $s_{2m} = s_{3m} = 0.9$, $c_{2\sigma} = c_{3\sigma} = 0.1$, $c_1 = 1$), the population approaches a synchronous 2-cycle. This is predicted by Theorem 10 since the quantities $c_b = -0.31 < 0$, $c_w = -0.19 < 0$, and $a_+ = -0.50 < 0$ are negative, but now $a_- = 0.12 > 0$ is positive, which implies the synchronous 2-cycle bifurcation is stable

A Non-cannibalistic Population in a Deteriorated Environment. We saw above that the matrix model with projection matrix (12.9) and (12.20) predicts that cannibalism of adults on juveniles can produce reproductive synchrony. Motivated by the observations described in Chap. 7 that egg cannibalism by gulls correlates with environmental conditions, in particular with environmental food resource availability, we extend the model to include a parameter that measures resource availability $\rho > 0$.

We assume the inherent fertility and survival probabilities are proportional to the (per capita per unit time) resource uptake rate

$$\eta(\rho) := \frac{c_4 \rho}{1 + c_4 \rho}, \qquad c_4 > 0$$

and write

$$\begin{aligned} b(\rho) &= \beta \eta(\rho), \qquad s_i(\rho) = s_i^0 \eta(\rho) \\ \beta &> 0, \qquad 0 < s_i^0 < 1 \,. \end{aligned} \tag{12.22}$$

Here β and s_i^0 are the maximal possible fertility and survival rates, which are approached as the available resource increases without bound. The inherent reproduction number

$$R_0(\rho) = b(\rho)\frac{s_1(\rho)}{1 - s_2(\rho)\, s_3(\rho)}$$

is an increasing function of the resource availability ρ, ranging over the interval

$$0 \leq R_0(\rho) < \beta\frac{s_1^0}{1 - s_2^0 s_3^0}$$

for $\rho \geq 0$. There exists a (unique) critical resource level

$$\rho^* := \frac{1}{c_4}\frac{1}{\sqrt{\beta s_1^0 + s_2^0 s_3^0} - 1} \tag{12.23}$$

for which $R_0(\rho) = 1$ if and only if

$$\lim_{\rho\to\infty} R_0(\rho) = \beta\frac{s_1^0}{1 - s_2^0 s_3^0} > 1, \tag{12.24}$$

which we assume holds. Then $R_0(\rho) < 1$ for low resource levels $\rho < \rho^*$ and $R_0(\rho) > 1$ for resources $\rho > \rho^*$ exceeding the threshold ρ^*.

With $v_1 = v_2 = 0$ the population is non-cannibalistic and the model (12.2) with the projection matrix (12.9) and density factors (12.20) reduces to the model in Sect. 12.3 in which, we have seen, the bifurcations at $(R_0(\rho), \mathbf{x}) = (1, \mathbf{0})$ are forward and display a dynamic dichotomy between positive equilibria and synchronous 2-cycles. For the bifurcations to occur as the environmental resource level ρ increases, it is necessary to assume inequality (12.24) holds so that there is a unique value of ρ at which $R_0(\rho) = 1$, namely, ρ^* defined by (12.23). For $\rho < \rho^*$ we have $R_0(\rho) < 1$ which implies extinction, as shown in the example plots in Figs. 12.5a and 12.6a. *As resource availability ρ increases through the threshold ρ^* the extinction equilibrium destabilizes and the resulting bifurcation produces a dynamic dichotomy: the non-cannibalistic model population survives with $\rho > \rho^*$ without or with reproductive synchrony if c_g is small or large, respectively.*

The plots in Figs. 12.5b and 12.6b illustrate these alternatives. These results for $\rho \gtrsim \rho^*$ are predicted by Theorems 9 and 10. *The dynamic properties that occur near the bifurcation point might or might not persist away from it, however.* This is illustrated by the plot in Fig. 12.6c which shows the disappearance of the synchronous 2-cycles and a re-equilibration for larger resource levels ρ.

Cannibalism and Reproductive Synchrony in a Deteriorated Environment. To capture the trade-off between resource gathering and cannibalism, we assume the cannibalism coefficient is negatively correlated with $\eta(\rho)$ and replace v_i by $v_i(1-\eta(\rho))$ in the projection matrix entries (12.20). The matrix model has the projection matrix

$$P(\mathbf{x}) = \begin{pmatrix} 0 & R_0(\rho)\frac{1-s_2(\rho)s_3(\rho)}{s_1(\rho)}\bar{p}_{12}(\mathbf{x}) & 0 \\ s_1(\rho)\bar{p}_{21}(\mathbf{x}) & 0 & s_3(\rho)\bar{p}_{23}(\mathbf{x}) \\ 0 & s_2(\rho)\bar{p}_{32}(\mathbf{x}) & s_3(\rho)\bar{p}_{33}(\mathbf{x}) \end{pmatrix}$$

$$R_0(\rho) = b(\rho)\frac{s_1(\rho)}{1-s_2(\rho)s_3(\rho)}$$

with entries

$$\begin{aligned}
\bar{p}_{12}(\mathbf{x}) &= \frac{1}{1+c_2x_2} \\
\bar{p}_{21}(\mathbf{x}) &= \prod_{i=2}^{3}\left(1-\frac{1}{1+c_1x_1}\frac{v_i(1-\eta(\rho))x_i}{1+v_i(1-\eta(\rho))x_i}\right) \\
\bar{p}_{32}(\mathbf{x}) &= \sigma_2\left(\frac{1}{1+c_1x_1}\frac{v_2(1-\eta(\rho))}{1+v_2(1-\eta(\rho))x_2}x_1\right) \\
\bar{p}_{23}(\mathbf{x}) &= \frac{1}{1+c_gx_2}\sigma_3\left(\frac{1}{1+c_1x_1}\frac{v_3(1-\eta(\rho))}{1+v_3(1-\eta(\rho))x_3}x_1\right) \\
\bar{p}_{33}(\mathbf{x}) &= \left(1-\frac{1}{1+c_gx_2}\right)\sigma_3\left(\frac{1}{1+c_1x_1}\frac{v_3(1-\eta(\rho))}{1+v_3(1-\eta(\rho))x_3}x_1\right).
\end{aligned}$$

In this model a change in the environment is measured by a change in the environmental resource availability ρ. As we have seen, the matrix equation (12.2) with projection (12.9) and constituents (12.20) and (12.22) can, when cannibalism is present, have a forward bifurcation and exhibit a dynamic dichotomy between stable positive equilibrium and synchronous 2-cycles. However, when the benefit to adult survival of cannibalistic resources is significant enough, a backward bifurcation can result. In that case, as discussed in Sect. 11.2, there is the potential of a strong Allee effect for values of $R_0 < 1$. *For a cannibalistic population this would mean that survival is possible in an environment so severely degraded that food resources ρ drop below the threshold ρ^* and $R_0(\rho) < 1$, in which case the population would go extinct in the absence of cannibalism.* We saw this scenario with the cannibalism model in Sect. 11.3 in Chap. 11.

The added possibility with projection matrix (12.9) and constituents (12.22) and (12.20) is now that a strong Allee effect can occur that involves synchronous 2-cycles rather than positive equilibria. This is illustrated in Fig. 12.7 where three plots of sample time series solutions are shown at decreasing values of the environmental resource availability ρ (and hence $R_0(\rho)$), in simulation of a degrading environment. For the largest value of ρ the reproduction number $R_0(\rho)$ is greater than 1 and the population shown in Fig. 12.7a survives and approaches a positive equilibrium. In Fig. 12.7b the resource ρ has decreased so that $R(\rho)$ is close to but still greater than 1 with the result that the population survives, but now approaches a synchronous 2-cycle. The lower resource level has raised the intensity of cannibalism and invoked the prey saturation effect so as to lead to reproductive synchrony. The bifurcation diagnostic quantities are positive, however, and this implies a backward bifurcation of synchronous 2-cycles at $R_0(\rho) = 1$ (which explains why the population densities are not close to $\mathbf{0}$). In Fig. 12.7c the population still approaches a synchronous 2-cycle even though ρ has been reduced so low that $R_0(\rho)$ is less than 1 and the extinction equilibrium is also stable. This illustrates that a strong Allee effect occurs at this environmental resource level as a result of cannibalism. *The model studied here predicts that adult cannibalism of juveniles allows for long-term survival in a severely degraded environment and that reproductive synchrony can be a response to that cannibalism. This survival possibility occurs with the two caveats caveats C_1 and C_2 discussed in Sect. 11.3 of Chap. 11, namely, that survival is initial condition dependent (i.e., population densities must remain sufficiently high so as to stay out of the Allee basin of attraction) and that there exists a "tipping point" threshold, below which a degraded environment results in a sudden catastrophic extinction event.*

12.5 Conclusion

In the matrix equation (12.2) with projection matrix (12.9), the components (12.20) and (12.22) account for three basic mechanisms:

> (1) Adult-on-juvenile cannibalism is inversely correlated with non-cannibalistic food resource availability and has a positive effect on adult survival.
> (2) A victim saturation effect in which increased prey density decreases the probability that an individual juvenile is cannibalized.
> (3) A reproductive refractory period in which a female cannot reproduce during two successive time units, a situation that allows for the possibility that females can synchronize their reproductive activity.

We found analytically that there is a dynamic dichotomy between two possible attracting states: equilibria and synchronous 2-cycles, which biologically represent asynchronous and synchronized reproduction, respectively.

As the environment degrades (ρ decreases) several scenarios are possible in this model population. One of these is particularly relevant, and nicely corresponds, to the observa-

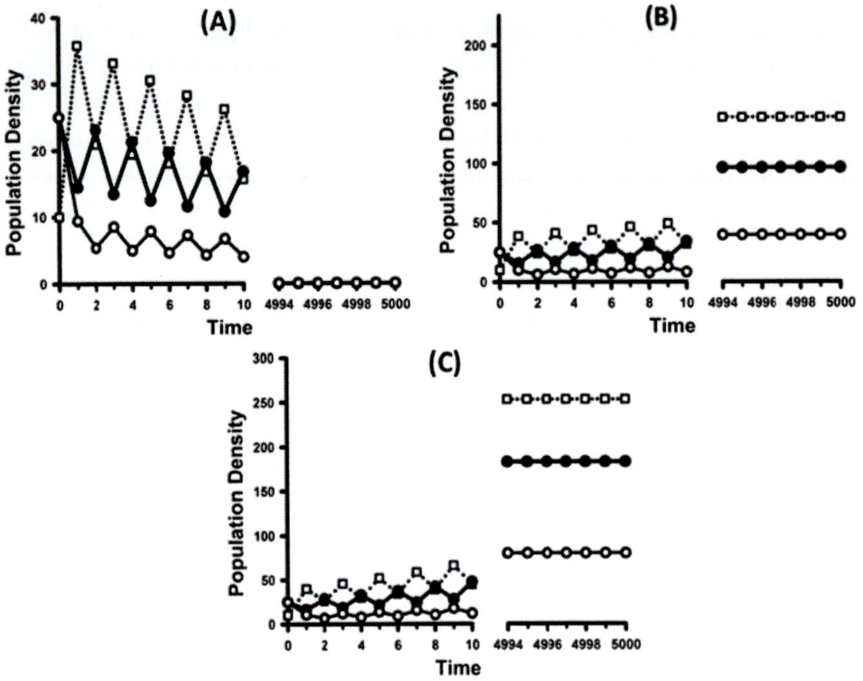

Figure 12.5: Shown are the time series solutions starting from initial condition $(x_1, x_2, x_3) = (10, 25, 25)$ of the matrix equation (12.2) with projection matrix (12.9) and constituents (12.20) and (12.22) **without cannibalism** ($v_2 = v_3 = 0$). Other coefficients are $\beta = 2$, $s_1^0 = 0.75$, $s_2^0 = s_3^0 = 0.50$, $c_4 = 0.05$, $c_2 = c_g = 0.001$, and $c_1 = 20$. The critical resource threshold is $\rho^* = 61.94$. The open squares are the juvenile class densities. The solid (open) circles are the reproductively active (inactive) adult class densities. (**a**) $\rho = 55$: In this case $R_0 = 0.9320$ and $c_w = -0.4761 \times 10^{-3}$, $c_b = -0.4683 \times 10^{-4}$, $a_+ = -0.5229 \times 10^{-3}$, $a_- = 0.4292 \times 10^{-3}$. (**b**) $\rho = 75$: In this case $R_0 = 1.107$ and $c_w = -0.4998 \times 10^{-3}$, $c_b = -0.5584 \times 10^{-4}$, $a_+ = -0.5557 \times 10^{-3}$, $a_- = 0.4440 \times 10^{-3}$. (**c**) $\rho = 90$: In this case $R_0 = 1.206$ and $c_w = -0.5109 \times 10^{-3}$, $c_b = -0.6068 \times 10^{-4}$, $a_+ = -0.5716 \times 10^{-4}$, $a_- = 0.4503 \times 10^{-3}$. Figure reproduced from [13] with permission.

tions in glaucous-winged gull colonies described in Chaps. 7–9. A decrease in environmental food availability stimulates a higher intensity of egg cannibalism that in turn induces reproductive synchrony in the population which, prior to the environmental change, was asynchronous. Moreover, cannibalism can save the population from extinction in that the population would go extinct in its absence. This mode of avoiding extinction comes with the two caveats C_1 and C_2 discussed in Sect. 11.3 of Chap. 11, namely, initial condition-dependent survival and the existence of a "tipping point" threshold, below which a degraded environment results in a sudden catastrophic extinction event.

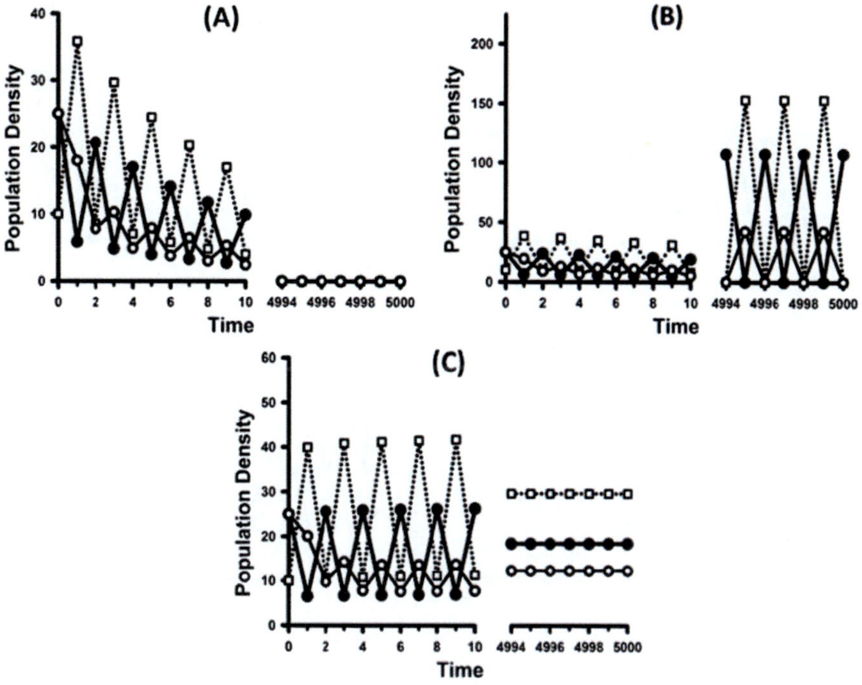

Figure 12.6: Shown are the time series solutions starting from initial condition $(x_1, x_2, x_3) = (10, 25, 25)$ of the matrix equation (12.2) with projection matrix (12.9) and constituents (12.20) and (12.22) **without cannibalism** ($v_2 = v_3 = 0$). Other coefficients are the same as in Fig. 12.5 except **the synchrony coefficient is increased** to $c_g = 1$. The open squares are the juvenile class densities. The solid (open) circles are the reproductively active (inactive) adult class densities. (**a**) $\rho = 55$: In this case $R_0 = 0.9320$ and $c_w = -0.4761 \times 10^{-3}$, $c_b = -0.4683 \times 10^{-1}$, $a_+ = -0.4731 \times 10^{-1}$, $a_- = 0.4636 \times 10^{-1}$. (**b**) $\rho = 75$: In this case $R_0 = 1.107$ and $c_w = -0.4998 \times 10^{-3}$, $c_b = -0.5584 \times 10^{-1}$, $a_+ = -0.5634 \times 10^{-1}$, $a_- = 0.5534 \times 10^{-1}$. (**c**) $\rho = 90$: In this case $R_0 = 1.206$ and $c_w = -0.5109 \times 10^{-3}$, $c_b = -0.6068 \times 10^{-1}$, $a_+ = -0.6119 \times 10^{-1}$, $a_- = 0.6017 \times 10^{-1}$. Figure reproduced from [13] with permission.

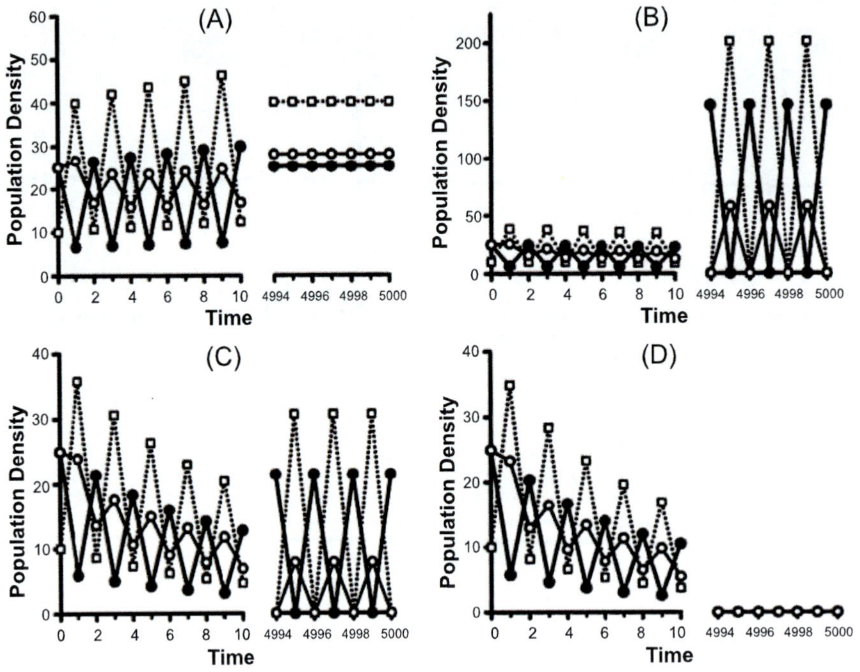

Figure 12.7: Shown are the time series solutions starting from initial condition $(x_1, x_2, x_3) = (10, 25, 25)$ of the matrix matrix equation (12.2) with projection matrix (12.9) and with constituents (12.20) and (12.22) **with cannibalism present** $(v_2 = v_3 = 5)$. Other coefficients are $\beta = 2$, $s_1^0 = 0.75$, $s_2^0 = s_3^0 = 0.50$, $c_4 = 1/20$, $c_2 = 0.001$, $c_g = 1$, $c_1 = 20$, $s_{2m} = s_{3m} = 0.95$, and $c_{\sigma 2} = c_{\sigma 3} = 300$. The open squares are the juvenile class densities. The solid (open) circles are the reproductively active (inactive) adult class densities. (**a**) $\rho = 90$: In this case $R_0 = 1.206$ and $c_w = 34.01$, $c_b = 33.68$, $a_+ = 67.69$, $a_- = 0.3346$. (**b**) $\rho = 75$: In this case $R_0 = 1.107$ and $c_w = 37.28$, $c_b = 36.80$, $a_+ = 73.98$, $a_- = 0.3738$. (**c**) $\rho = 55$: In this case $R_0 = 0.9320$ and $c_w = 41.66$, $c_b = 41.21$, $a_+ = 82.87$, $a_- = 0.4484$. (**d**) $\rho = 50$: In this case $R_0 = 0.8772$ and $c_w = 42.68$, $c_b = 42.21$, $a_+ = 84.90$, $a_- = 0.4727$. Simulations show that a tipping point occurs at approximately $\rho = 50.8$. Figures reproduced from [13] with permission.

References

1. Courchamp F, Berec L, Gascoigne J (2008) Allee Effects in Ecology and Conservation. Oxford University Press, Oxford

2. Cushing JM (1998) An Introduction to Structured Population Dynamics. CBMS-NSF Regional Conference Series in Applied Mathematics, vol 71, SIAM, Philadelphia

3. Cushing JM (2009) Three stage semelparous Leslie models. J Math Biol 59:75–104

4. Cushing JM, Zhou Y (1994) The net reproductive value and stability in structured population models. Nat Resour Model 8:1–37

5. Cushing JM, Henson SM (2012) Stable bifurcations in nonlinear semelparous Leslie models. J Biol Dyn 6:80–102

6. Cushing JM, Farrell AP (2019) A bifurcation theorem for nonlinear matrix models of population dynamics. J Differ Equ Appl 26(1):25–44. http://dx.doi.org/10.1080/10236198.2019.1699916

7. Darling F (1938) Bird Flocks and the Breeding Cycle. Cambridge University Press, Cambridge

8. Elaydi SN (2005) An Introduction to Difference Equations, 3rd edn. Springer, New York

9. Henson SM, Cushing JM, Hayward JL (2011) Socially induced ovulation synchrony and its effect on seabird populations. J Biol Dyn 5:495–516

10. Li C-K, Schneider H (2002) Applications of Perron-Frobenius theory to population dynamics. J Math Biol 44:450–462

11. Veprauskas A (2016) On the dynamic dichotomy between positive equilibria and synchronous 2-cycles in matrix population models, PhD dissertation. University of Arizona

12. Veprauskas A (2018) Synchrony and the dynamic dichotomy in a class of matrix population models. SIAM J Appl Math 178(5):2491–2510

13. Veprauskas A, Cushing JM (2017) A juvenile-adult population model: climate change, cannibalism, reproductive synchrony, and strong Allee effects. J Biol Dyn 11(1):1–24. https://doi.org/10.1080/17513758.2015.1131853

Chapter 13

An Across-Season Model

Our purpose in this chapter is to address a significant time-scale discrepancy between the models studied in Chaps. 11 and 12 and the gull populations that motivated them. This discrepancy arises because, unlike in the models, reproductive synchrony and juvenile maturation in gulls occur on different time scales. This modeling project was initially reported in [3] and [5].

13.1 Introduction

A significant discrepancy exists between the cannibalism and reproductive synchrony models studied in Chaps. 11 and 12 and the population dynamics of the gull populations that motivated those models. This discrepancy arises because in those models juvenile maturation and adult reproductive synchrony are on commensurate time scales. However, in gull populations reproductive synchrony happens within a breeding season, on a daily time scale, while juveniles take several seasons (years) to reach maturity. Our modeling approach, which aims to alleviate this discrepancy, is to apply a within-breeding season projection matrix followed by the application of an across-season projection matrix to account for overwinter mortality and maturation. We follow the modeling methodology and analysis given in [3] and [5].

The across-season model is derived in Sect. 13.2. The model is an example taken from a general class of periodically forced matrix models studied in [3], for which some analysis is given in Sect. 13.3 and then applied to the across-season model in Sect. 13.4. The main goal is to investigate the population dynamic consequences of within-season cannibalism and reproductive synchrony as they depend on environmental resource availability.

J. M. Cushing et al., *Modeling Behavior and Population Dynamics*, Interdisciplinary Applied Mathematics 57, https://doi.org/10.1007/978-3-031-34283-7_13

13.2 A Seasonal Cannibalism and Reproductive Synchrony Model

We specify four classes: $x_1(t)$ denotes the number of new eggs, that is, eggs laid on day t; $x_2(t)$ denotes the number of juveniles, by which we mean older eggs and chicks; $x_3(t)$ is the number of reproductively active adults; and $x_4(t)$ is the number of reproductively inactive adults. The time unit is 1 day. Motivated by the gull biology in Chap. 8, we assume that the ovulation cycle for an adult female is 2 days long and that each female lays an egg approximately every other day. The number of reproductively active adults $x_3(t)$ is the number of female adults that are in the first day of the hormone cycle on day t. These are the females that lay an egg on day t. The number of reproductively inactive adults $x_4(t)$ is the number of female adults in their second day of their hormone cycle; these adults do not therefore lay an egg on day t. After the breeding season of length $k-1$ days, individuals are exposed to between-season mortality. All surviving juveniles are assumed to mature during this between-season interval, after which the females are placed in either class x_3 or class x_4 at the start of the following season.

According to this modeling scheme, the demographic state vector

$$\mathbf{x} = \begin{pmatrix} x_1 \\ x_2 \\ x_3 \\ x_4 \end{pmatrix} = \begin{pmatrix} \text{eggs} \\ \text{juveniles} \\ \text{reproductively active adults} \\ \text{reproductively inactive adults} \end{pmatrix}$$

is advanced through a breeding season by applying a within-season projection matrix $k-1$ times, after which it is advanced to the start of the next breeding season by the application of an across-season projection matrix. This process is then repeated periodically with period k. This leads to a matrix equation

$$\mathbf{x}(t+1) = P(t, \mathbf{x}(t))\, \mathbf{x}(t), \tag{13.1}$$

where the projection matrix $P(t, \mathbf{x})$ is k-periodic in t and is defined as follows.

The within-season projection matrix W. The within-season projection matrix associated with the compartmental diagram in Fig. 13.1 has the form

$$W = \begin{pmatrix} 0 & 0 & w_{13} & 0 \\ w_{21} & w_{22} & 0 & 0 \\ 0 & 0 & 0 & w_{34} \\ 0 & 0 & w_{43} & w_{44} \end{pmatrix}. \tag{13.2}$$

The terms $w_{ij} = w_{ij}(\mathbf{x})$ describing vital per capita rates (birth, survival, and transition rates among compartments) are assumed to be density dependent. The term w_{13} is the number of eggs produced per reproductively active female (per day) which we describe by

$$w_{13}(\mathbf{x}) = b\beta(\mathbf{x}),$$

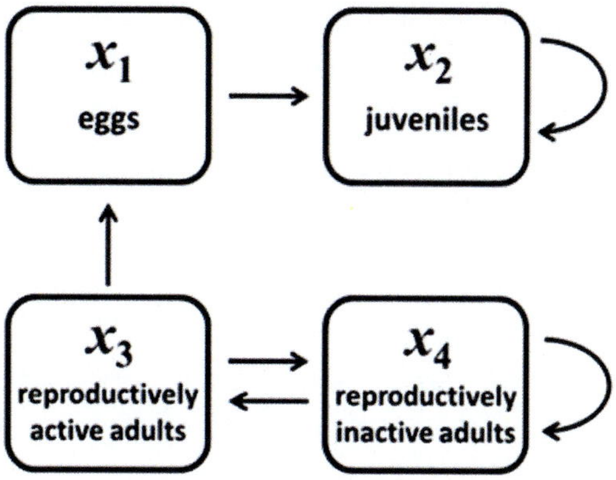

Figure 13.1: A compartmental diagram for the within-season dynamics described by the projection matrix W. The time step is 1 day. Reproduced from [3] with permission.

where $\beta(\mathbf{0}) = 1$ so that b is the inherent or intrinsic egg-laying rate (in the absence of density effects) and the factor $\beta(\mathbf{x})$ describes how fertility is affected by population density.

Let $s_i\sigma_i(\mathbf{x})$ denote the fractions of i-class individuals that survive one time unit (day), with $\sigma_i(\mathbf{0}) = 1$ so that s_i is the inherent or intrinsic (i.e., density-free) survival rate. Note that W implies a surviving class $i = 1$ individual (new egg) always moves to the juvenile class $i = 2$ at the next time step and a surviving class $i = 3$ (reproductively active) individual always moves to the reproductively inactive class $i = 4$ at the next time step (because of the reproductive refractory period). Thus,

$$w_{21}(\mathbf{x}) = s_1\sigma_1(\mathbf{x}), \quad w_{22}(\mathbf{x}) = s_2\sigma_2(\mathbf{x}), \quad w_{43}(\mathbf{x}) = s_3\sigma_3(\mathbf{x}) .$$

We assume a surviving class $i = 4$ (non-reproductively active) individual moves to class $i = 3$, i.e., becomes reproductively active, with probability g and otherwise remains in class $i = 4$. Thus, we have

$$w_{34}(\mathbf{x}) = s_4\sigma_4(\mathbf{x})\,g(x_3), \quad w_{44}(\mathbf{x}) = s_4\sigma_4(\mathbf{x})(1 - g(x_3)) .$$

Note that we have assumed the fraction $g = g(x_3)$ depends only on the number of reproductively active females x_3.

Mathematically, we assume all density-dependent terms are twice continuously differentiable on an open set Ω^4 containing the nonnegative cone R^4_+ and satisfy

$$b\beta(\mathbf{x}) \geq 0, \quad 0 \leq s_i\sigma_i(\mathbf{x}) < 1, \quad 0 < g(x_3) \leq 1$$
$$\beta(\mathbf{0}) = 1, \quad \sigma_i(\mathbf{0}) = 1, \quad g(0) = 1$$

for all $\mathbf{x} \in \Omega^4$.

Next, we incorporate the negative and positive feedback mechanisms of adult on egg cannibalism into the model by further specifying the density-dependent survival factors

$\sigma_i(\mathbf{x})$ in a manner similar to that used in Chaps. 11 and 12 (and [3]). Let $\pi_i(x_1, x_i)\, x_i$ denote the probability an egg is cannibalized in the presence of x_i adults of class $i = 3, 4$ so that

$$\sigma_1(\mathbf{x}) = \prod_{i=3}^{4} \left(1 - \pi_i(x_1, x_i)\, x_i\right)$$

where

$$0 \leq \pi_i(x_1, x_i)\, x_i \leq 1$$

for all $\mathbf{x} \in \Omega^4$. The dependence of π_i on x_1 allows for a victim saturation effect on cannibals in analogy with the familiar predator saturation effect of prey on predators:

$$\partial_1 \pi_i := \frac{\partial \pi_i}{\partial x_1} \leq 0, \quad \partial_i \pi_i := \frac{\partial \pi_i}{\partial x_i} \geq 0 .$$

In this model we assume that adult survival is increased by cannibalism resources so that

$$\sigma_i(\mathbf{x}) = \sigma_i(\pi_i(x_1, x_i)\, x_1), \quad i = 3 \text{ and } 4$$

where σ_i is an increasing function of the number of eggs cannibalized $\pi_i(x_1, x_i)\, x_1$ by an individual i-class adult, with

$$\sigma_i'(z) := \frac{d\sigma_i(z)}{dz} > 0,$$

but that juvenile survival is density independent: $\sigma_2(\mathbf{x}) \equiv 1$. This is motivated by the empirical result in Chap. 9 that gull cannibals focus on new eggs, taking them within the first 24 hours after they are laid. Cannibals are much less likely to cannibalize older eggs or chicks. Finally, we require

$$g'(x_3) := \frac{dg}{dx_3} < 0$$

so that a larger active adult density on any given day decreases the probability that inactive adults become active the next day. For this reason, we refer to g as the synchrony propensity factor.

Lastly, we assume adult fertility is regulated only by adult population density, specifically that it is a decreasing function

$$\beta(\mathbf{x}) = \varphi(c_{\varphi 3} x_3 + c_{\varphi 4} x_4)$$

of a weighted total adult population size $c_{\varphi 3} x_3 + c_{\varphi 4} x_4$ ($c_{\varphi i} \geq 0$ but not both equal to 0):

$$\varphi'(z) := \frac{d\varphi}{dz} < 0.$$

In addition to all these density effects on the vital rates in the seasonal projection matrix $W(\mathbf{x})$, we want to account for a trade-off between environmental resource availability and cannibalism activity, as motivated by Chap. 7.

We let $\rho > 0$ denote the amount of available environmental resource and assume $b = b(\rho)$ and $\pi_j = \pi_j(\rho, x_1, x_j)$ are functions of ρ satisfying

$$b'(\rho) > 0, \quad b(0) = 0, \quad \partial_\rho \pi_j(\rho, x_1, x_j) < 0 \ ,$$

the middle condition implying that egg-laying ceases in the absence of the environmental resource ρ. In a more general model, the inherent survivorship s_i could also depend on ρ, but here we will investigate only the case in which they are independent of ρ.

In summary, under all the preceding assumptions, the within-season projection matrix (13.2) becomes

$$W(\rho, \mathbf{x}) = \begin{pmatrix} 0 & 0 & w_{13}(\rho, \mathbf{x}) & 0 \\ w_{21}(\rho, \mathbf{x}) & w_{22}(\rho, \mathbf{x}) & 0 & 0 \\ 0 & 0 & 0 & w_{34}(\rho, \mathbf{x}) \\ 0 & 0 & w_{43}(\rho, \mathbf{x}) & w_{44}(\rho, \mathbf{x}) \end{pmatrix} \tag{13.3}$$

with

$$w_{13}(\rho, \mathbf{x}) = b(\rho)\, \varphi\left(c_{\varphi_3} x_3 + c_{\varphi_4} x_4\right) \tag{13.4a}$$

$$w_{21}(\rho, \mathbf{x}) = s_1 \left(1 - \pi_3(\rho, x_1, x_3)\, x_3\right)\left(1 - \pi_4(\rho, x_1, x_4)\, x_4\right), \ w_{22}(\rho, \mathbf{x}) = s_2 \tag{13.4b}$$

$$w_{34}(\rho, \mathbf{x}) = s_4 \sigma_4\left(\pi_4(\rho, x_1, x_4)\, x_1\right) g(x_3) \tag{13.4c}$$

$$w_{43}(\rho, \mathbf{x}) = s_3 \sigma_3\left(\pi_3(\rho, x_1, x_3)\, x_1\right), \ w_{44}(\rho, \mathbf{x}) = s_4 \sigma_4\left(\pi_4(\rho, x_1, x_4)\, x_1\right)(1 - g(x_3)) \tag{13.4d}$$

The Across-Season Projection Matrix A. Our goal in this model is to place the within-season effects of reproduction, reproductive synchrony, cannibalism, and resource availability on a time scale different from the maturation or generation time. For gulls, reproduction activities occur during mating season on a daily time scale, and several years are needed for juveniles to reach maturity. In our proof-of-concept modeling exercise here, we will depart from this maturation time scale and assume juvenile maturation in the model population occurs on a yearly time scale. It will be clear how the model could be extended to describe longer maturation periods at the cost of considerable more complexity.

We assume the census takes place at the end of each day of the breeding season. On the last day of the breeding season, there is a demographic vector $\mathbf{x}$ with individuals present in possibly each of the four classes. During the following across-season time interval, lasting to the start of the next breeding season, all surviving non-adult individuals mature. Let $v_i \geq 0$ denote the (possibly density independent) probability of survival of an i-class individual during this time interval. On the first day of the next breeding season, the population consists only of adults, and we assume a fraction v of these are reproductively active on that first day. This gives us the across-season projection matrix

$$A = \begin{pmatrix} 0 & 0 & 0 & 0 \\ 0 & 0 & 0 & 0 \\ v_1 v & v_2 v & v_3 v & v_4 v \\ v_1(1-v) & v_2(1-v) & v_3(1-v) & v_4(1-v) \end{pmatrix}. \tag{13.5}$$

The resulting projection matrix in Eq. (13.1) is

$$P(t,\mathbf{x}) = \begin{cases} W(\mathbf{x}) & \text{for } t = 0, 1, \cdots, k-2 \\ A & \text{for } t = k-1 \end{cases}$$

repeated periodically with period k, i.e., for fixed $\mathbf{x}$, $P(t+k,\mathbf{x}) = P(t,\mathbf{x})$ for all $t \geq 0$.

While this periodically forced, nonlinear matrix model is complex and might seem analytically intractable, it is possible to subject it to a considerable amount of analysis with regard to the stability and destabilization of the extinction equilibrium and the properties of the bifurcation to stable survival dynamics. In Sect. 13.3 we describe the analysis of a class of periodically forced matrix equations, which we apply to this across-season model in Sect. 13.4.

13.3 A Class of Periodic Matrix Models

Consider the $m \times m$ matrix equation

$$\mathbf{x}(t+1) = P(t, \mathbf{x}(t))\,\mathbf{x}(t) \tag{13.6}$$

with the t periodic projection matrix

$$P(t,\mathbf{x}) = \begin{cases} W(\mathbf{x}) & \text{for } t = 0, 1, \cdots, k-2 \\ A & \text{for } t = k-1 \end{cases} \tag{13.7}$$

extended periodically in t with period $k \geq 2$ so that

$$P(t+k, \mathbf{x}) = P(t, \mathbf{x})$$

for all $t \in Z := \{0, 1, 2, 3, \cdots\}$. We assume

> (A5) The $m \times m$ matrix $W(\mathbf{x}) = (w_{ij}(\mathbf{x}))$ has twice continuously differentiable entries $w_{ij} : \Omega \to R^{+}$, where Ω is an open set containing $\bar{R}_{+}$, and the $m \times m$ matrix $A = (a_{ij})$, $a_{ij} \in \bar{R}_{+}$, has rank 1.

Our goal in this section is to establish some theorems concerning the asymptotic dynamics of Eq. (13.6) by reducing it to a scalar difference equation. The key to this reduction is the assumption in A5 that the matrix A has rank 1. This mathematical assumption arises from two features of the scenario we have in mind. First, all individuals present at the start of a breeding season, regardless of their demographic class at the end of the previous breeding season, are mature adults. Thus, any non-adult individuals at the end of a breeding season will, if they survive, mature to become adult individuals at the beginning of the next breeding season. Second, if there are multiple adult classes in the model, then the adults present at the beginning of a breeding season are distributed throughout these classes with the same proportions every breeding season. An example is the model in Sect. 13.2 in which there are two adult classes, namely, reproductively active adults and reproductively inactive adults.

Clearly the *extinction equilibrium*, $\mathbf{x}(t) = \mathbf{0}$, $t \in Z$, is a solution of the nonautonomous, periodically forced matrix equation (13.6). By a *nontrivial k-cycle* of (13.6), we mean a periodic solution of period k (i.e., $\mathbf{x}(t+k) = \mathbf{x}(t)$ for all t) that is not the extinction equilibrium $\mathbf{x} = \mathbf{0}$. By a *non-extinction k-cycle* , we mean a nontrivial, nonnegative k-cycle $\mathbf{x}(t) \in \bar{R}_+^m \backslash \{\mathbf{0}\}$ for each t. If, on the other hand, a k-cycle $\mathbf{x}(t)$ is not in $\bar{R}_+^m$ for some t, then we refer to $\mathbf{x}(t)$ as an *infeasible k-cycle.*

One can analyze the asymptotic dynamics of any k-periodically forced difference equation by analyzing the asymptotic dynamics of the $(k-1)^{st}$ composite equation (which is a time-autonomous equation) [4]. An equilibrium of the composite equation is the initial condition for a k-periodic cycle of the original periodic equation (and vice versa). Moreover, the stability/instability properties of the k-cycle are the same as the stability/instability properties of the fixed point of the composite.

Our goal in this section is to show, under assumption A5, that the composite equation associated with the periodic matrix equation (13.6) and projection matrix (13.7) reduces to a scalar (one-dimensional) difference equation and to draw some conclusions about the existence and stability of k-cycles.

At the beginning of each season, the population's demographic vector is $\mathbf{x}(nk)$ for $n \in Z$. The range of the matrix A is one dimensional (assumption A5), and, as a result, the columns of A are all nonnegative multiples $v_i\mathbf{a}$ of a vector which we denote by

$$\mathbf{a} = \text{col}(a_i) \in \bar{R}_+^m \backslash \{\mathbf{0}\}.$$

Without loss in generality (by rescaling the v_i if necessary), we can assume

$$\sum_{i=1}^{m} a_i = 1. \tag{13.8}$$

Thus, the across-season matrix has the form

$$A = \begin{pmatrix} v_1\mathbf{a} & v_2\mathbf{a} & \cdots & v_m\mathbf{a} \end{pmatrix}.$$

Define the vector

$$\mathbf{v} := \text{col}(v_i) \in \bar{R}_+^m \backslash \{\mathbf{0}\}.$$

We see from the projection matrix (13.7) that $\mathbf{x}(nk) \in R_+^m$ lies in the range of A for $n \in Z\backslash\{\mathbf{0}\}$, and therefore, we can write

$$\mathbf{x}(nk) = \alpha(n)\,\mathbf{a}, \quad n \in Z\backslash\{\mathbf{0}\} \tag{13.9}$$

for a scalar $\alpha(n) \geq 0$. Independently of the initial condition $\mathbf{x}(0)$, the vectors in the sequence $\mathbf{x}(k), \mathbf{x}(2k), \mathbf{x}(3k), \ldots$ (which are the demographic population vectors at the beginning of each season after the first) are all multiples of $\mathbf{a}$. Therefore, in studying the dynamics of the model, there is no loss in generality if we restrict the initial condition $\mathbf{x}(0)$ to be a multiple of $\mathbf{a}$; specifically we assume

$$\mathbf{x}(0) = \alpha \mathbf{a}, \tag{13.10}$$

where $\alpha \geq 0$. From (13.8) it follows that

$$\alpha = \sum_{i=1}^{m} x_i(0)$$

is the initial *total population size.*

Applying the matrix difference equation (13.6) for $t = 0, 1, 2, \ldots, k-1$, we obtain the sequence

$$\begin{aligned}
\mathbf{x}(0) &= \alpha \mathbf{a} \\
\mathbf{x}(1) &= W(\mathbf{x}(0))\, \alpha \mathbf{a} \\
\mathbf{x}(2) &= W(\mathbf{x}(1))\, W(\mathbf{x}(0))\, \alpha \mathbf{a} \\
\mathbf{x}(3) &= W(\mathbf{x}(2))\, W(\mathbf{x}(1))\, W(\mathbf{x}(0))\, \alpha \mathbf{a} \\
&\vdots \\
\mathbf{x}(i) &= \left(\prod_{t=i-1}^{t=0} W(\mathbf{x}(t)) \right) \alpha \mathbf{a} \text{ for } i = 1, \cdots, k-1.
\end{aligned}$$

where

$$\prod_{t=i-1}^{t=0} W(\mathbf{x}(t)) := W(\mathbf{x}(i-1)) \cdots W(\mathbf{x}(1))\, W(\mathbf{x}(0)).$$

Define the matrix

$$Q(\alpha) := \prod_{t=k-2}^{t=0} W(\mathbf{x}(t)). \tag{13.11}$$

At the end of the first season, the population vector is $\mathbf{x}(k-1) = Q(\alpha)\, \alpha \mathbf{a}$, and at the beginning of the next season, it is

$$\mathbf{x}(k) = AQ(\alpha)\, \alpha \mathbf{a}.$$

The vector $AQ(\alpha)\, \mathbf{a}$ is a multiple of $\mathbf{a}$, so we write

$$AQ(\alpha)\, \mathbf{a} = r(\alpha)\, \mathbf{a}. \tag{13.12}$$

In conclusion we have

$$\mathbf{x}(k) = r(\alpha)\, \alpha \mathbf{a}.$$

The matrix equation (13.6) with the projection matrix (13.7) maps an initial demographic vector $\alpha\mathbf{a}$ at the beginning of a season to another multiple of $\mathbf{a}$, namely, $r(\alpha)\alpha\mathbf{a}$, at the beginning of the next season (k time steps later).

The sequence of demographic vectors at the beginning of the nth season $\mathbf{x}(nk) = \alpha(n)\mathbf{a}$ satisfies the difference equation

$$\mathbf{x}((n+1)k) = r(\alpha(n))\mathbf{x}(nk) \ ,$$

where the sequence $\alpha(n)$ is given by the scalar recursion formula (difference equation)

$$\alpha(n+1) = r(\alpha(n))\alpha(n), \quad n \in Z. \tag{13.13}$$

Thus, $r(\alpha)$ is the per-season population growth rate. Note that (13.8) implies

$$\alpha(n) = \text{ total population size at the beginning of season } n.$$

We can study the existence and stability of k-cycles of the original periodically forced matrix equation (13.6) with the projection matrix (13.7) by studying the existence and stability of equilibria of the autonomous, scalar difference equation (13.13). The fixed point $\alpha = 0$ of (13.13) corresponds to the extinction equilibrium, and a positive fixed point of (13.13) corresponds to a non-extinction k-periodic cycle of the matrix equation (13.6) with the projection matrix (13.7).

To study the scalar equation (13.13), we need to analyze the scalar function $r(\alpha)$ that defines that equation. Multiplying both sides of (13.12) by the transpose $\mathbf{a}^T$ of $\mathbf{a}$ and dividing both sides of the result by $\mathbf{a}^T\mathbf{a}$, we get

$$r(\alpha) = \frac{\mathbf{a}^T AQ(\alpha)\mathbf{a}}{\mathbf{a}^T\mathbf{a}}.$$

Since

$$\mathbf{a}^T A = \begin{pmatrix} v_1\mathbf{a}^T\mathbf{a} & v_2\mathbf{a}^T\mathbf{a} & \cdots & v_m\mathbf{a}^T\mathbf{a} \end{pmatrix} = \mathbf{a}^T\mathbf{a}\,\mathbf{v}^T$$

we obtain the formula

$$r(\alpha) = \mathbf{v}^T Q(\alpha)\mathbf{a} \tag{13.14}$$

for the scalar difference equation (13.13). It follows from assumption A5 that $r(\alpha)$ is twice continuously differentiable on an open interval containing R^1_+ and satisfies $r(\alpha) \geq 0$ for $\alpha \geq 0$.

The equilibrium $\alpha = 0$ of the difference equation (13.13) corresponds to the extinction equilibrium $\mathbf{x} = \mathbf{0}$ of Eq. (13.6). A straightforward application of the linearization principle to the difference equation (13.13) shows that $\alpha = 0$ is locally asymptotically stable if $r(0) < 1$ and is unstable if $r(0) > 1$, where

$$r(0) = \mathbf{v}^T Q(0)\mathbf{a} = \mathbf{v}^T W^{k-1}(\mathbf{0})\mathbf{a} \tag{13.15}$$

is the *inherent (i.e., density-free) population growth rate* (at the beginning of the breeding season).

It is not straightforward to relate $r(0)$ to the model parameters appearing in the periodic matrix equation (13.6). Nonetheless, by choosing a model parameter appearing in (13.6) and studying its effect on $r(0)$, we can derive some general facts about the existence and stability of non-extinction k-cycles of the across-season model as they depend on the chosen parameter.

Let μ be a parameter in the within-season matrix W and hence in the projection matrix (13.7) of the matrix equation (13.6). We denote the appearance of μ in these matrices by writing $W = W(\mu, \mathbf{x})$ and $P = P(\mu, \mathbf{x})$. The components of the composite matrix $Q = Q(\mu, \mathbf{x})$ also depend on μ which, in turn, implies the population growth rate $r = r(\mu, \alpha)$ does as well. We rewrite (13.13) to indicate the dependence of $r(\mu, \alpha)$ on μ :

$$\alpha(n+1) = r(\mu, \alpha(n))\,\alpha(n), \quad n \in Z \tag{13.16}$$

$$r(\mu, \alpha) = \mathbf{v}^T Q(\mu, \alpha)\,\mathbf{a}. \tag{13.17}$$

The equation for a nontrivial equilibrium (fixed point) $\alpha \neq 0$ of (13.16) is

$$r(\mu, \alpha) = 1. \tag{13.18}$$

If α is a solution of this equation, i.e., if it is an equilibrium of the scalar equation (13.16), then we say the pair $(\mu, \alpha) \in \bar{R}^2$ is an *equilibrium pair.* Clearly, for all values of μ, the *extinction equilibrium pair* $(\mu, 0)$ of the scalar equation (13.16) corresponds to the extinction equilibrium pair $(\mu, \mathbf{0})$ of the across-season matrix equation (13.6) with the projection matrix (13.7). A nontrivial equilibrium pair (μ, α), $\alpha \neq 0$, corresponds to a nontrivial k-cycle $(\mu, \mathbf{x}(t))$ of the across-season matrix equation (13.6) with the projection matrix (13.7), and vice versa. *Positive (or negative) equilibrium pairs* (μ, α) of the scalar equation (13.16), i.e., equilibrium pairs with $\alpha > 0$ or $\alpha < 0$, respectively, correspond to *non-extinction* or *infeasible* k-cycle pairs $(\mu, \mathbf{x}(t))$ of the across-season matrix equation (13.6) with the projection matrix (13.7), respectively.

The question concerning extinction versus survival focuses on the stability properties of the extinction equilibrium pair $(\mu, 0)$ of the scalar equation (13.16) and, therefore (by the linearization principle), on the sign of $r(\mu, 0) - 1$. We focus on the destabilization of the extinction equilibrium by assuming that there is a value μ_0 of the parameter μ at which $r(\mu, 0) - 1$ changes sign.

(A6) Let μ denote a parameter that appears in the entries $w_{ij} = w_{ij}(\mu, x)$ of the matrix $W = W(\mu, x)$ in A5. Assume the entries $w_{ij}(\mu, x)$ are twice continuously differentiable with respect to μ on an open interval containing a parameter value μ_0 for which $r(\mu_0, 0) = 1$ and

$$\partial_\mu^0 r := \left.\frac{\partial r(\mu, \alpha)}{\partial \mu}\right|_{(\mu,\alpha)=(\mu_0,0)} \neq 0.$$

Furthermore, assume

$$\partial_\alpha^0 r := \left.\frac{\partial r(\mu,\alpha)}{\partial\alpha}\right|_{(\mu,\alpha)=(\mu_0,0)} \neq 0.$$

The following Theorem 12 describes the bifurcation alternatives for non-extinction k-cycle pairs $(\mu, \mathbf{x}(n))$ of the matrix equation (13.6) with the projection matrix (13.7) that occur at the parameter value $\mu = \mu_0$ in A6 where the extinction equilibrium destabilizes. A proof, which can be found in [3], is based on the general bifurcation theorems for the scalar difference equations applied to Eq. (13.16) [1].

Theorem 12 ([3]) *Assume that A5 and A6 hold and that $\partial_\mu^0 r > 0$. The extinction equilibrium pairs $(\mu, \mathbf{0})$ of the periodically forced matrix equation (13.6) with the projection matrix (13.7) are locally asymptotically stable for $\mu \lesssim \mu_0$ and unstable for $\mu \gtrsim \mu_0$. There exists a continuum of nontrivial k-cycle solution pairs $(\mu, \mathbf{x}(t))$ of the matrix equation (13.6) with the projection matrix (13.7) that (transcritically) bifurcates from the extinction pair $(\mu, \mathbf{0})$ at $\mu = \mu_0$.*

(a) Suppose $\partial_\alpha^0 r < 0$. For $\mu \gtrsim \mu_0$ the pairs $(\mu, \mathbf{x}(t))$ are (locally asymptotically) stable non-extinction k-cycle pairs. For $\mu \lesssim \mu_0$ the pairs $(\mu, \mathbf{x}(t))$ are unstable infeasible k-cycle pairs. In this case, we say a ***forward, stable bifurcation*** *occurs at $\mu = \mu_0$.*

(b) Suppose $\partial_\alpha^0 r > 0$. For $\mu \lesssim \mu_0$ the pairs $(\mu, \mathbf{x}(t))$ are unstable non-extinction k-cycle pairs. For $\mu \gtrsim \mu_0$ the pairs $(\mu, \mathbf{x}(t))$ are stable infeasible k-cycle pair. In this case, we say a ***backward, unstable bifurcation*** *occurs at $\mu = \mu_0$.*

Remark 13 *Near the bifurcation point $(\mu_0, \mathbf{0})$, the bifurcating non-extinction k-cycles guaranteed by Theorem 12 are stable if $\partial_\alpha^0 r < 0$ and unstable if $\partial_\alpha^0 r > 0$ (regardless of the sign of $\partial_\mu^0 r$). Can we relate this crucial quantity directly to the entries w_{ij} of the matrix W? A formula for $\partial_\alpha^0 r$ is derived in [3]. It is, perhaps not unexpectedly, quite complicated, and so, for our purposes here, suffice it to say that $\partial_\alpha^0 r$ is a linear combination, with nonnegative coefficients, of the derivatives $\partial_l^0 w_{ij}$. Since it is the sign of $\partial_\alpha^0 r$ that is of primary interest in the bifurcation Theorem 12, we conclude that the bifurcating non-extinction k-cycles are stable if the negative derivatives $\partial_l^0 w_{ij}$ dominate any positive derivatives. In other words, a forward, stable bifurcation of non-extinction k-cycles occurs if, at low population densities, negative density effects outweigh positive density effects (component Allee effects). (And a backward, unstable bifurcation occurs if the opposite is true.)*

Remark 14 *If $\partial_\mu^0 r < 0$, then the results described in Theorem 12 remain valid, but with inequalities reversed.*

Theorem 12 describes the nature of the bifurcation that occurs in a neighborhood of the bifurcation point $(\mu_0, \mathbf{0})$. What stability properties the bifurcating k-cycles have outside this neighborhood, however, is dependent on the model-specific characteristics. It is a well-known fact about the scalar difference equations, such as (13.16), that stable bifurcating equilibria can, as μ is manipulated further and further from μ_0, destabilize and result in the bifurcation of new attractors, which in turn can destabilize; the result can be a complicated cascade of bifurcations and complicated attractors, including strange and chaotic attractors. Thus, these possibilities exist for the k-cycles of the matrix equation (13.6) with the projection matrix (13.7) as well. They depend crucially on the properties of the nonlinearities used in the model components.

Nonetheless, some general properties concerning equilibrium stability of a scalar difference equation like (13.16), and hence the k-cycles of (13.6) with the projection matrix (13.7), outside the neighborhood of the bifurcation point can be ascertained from geometric properties of the bifurcation diagram of the scalar equation [1–3]. The bifurcation diagram is the graph of the Eq. (13.18) in the μ,α-plane. Any positive equilibrium pair (μ,α) located on a decreasing segment of this graph is unstable. Although those located on an increasing segment are not necessarily stable, those located on an increasing segment sufficiently close to a bifurcation point are stable. See Fig. 13.2. These facts about the equilibrium pairs of the scalar difference equation (13.16) allow for the determination of forward and backward bifurcations at $\mu=\mu_0$, multiple attractor scenarios (such as strong Allee effects and hysteresis), and tangent (or blue-sky or saddle-node) bifurcations that define tipping points . These phenomena concerning the equilibria of the difference equation translate to the same phenomena concerning k-cycles of the seasonal matrix models (13.6) and (13.7).

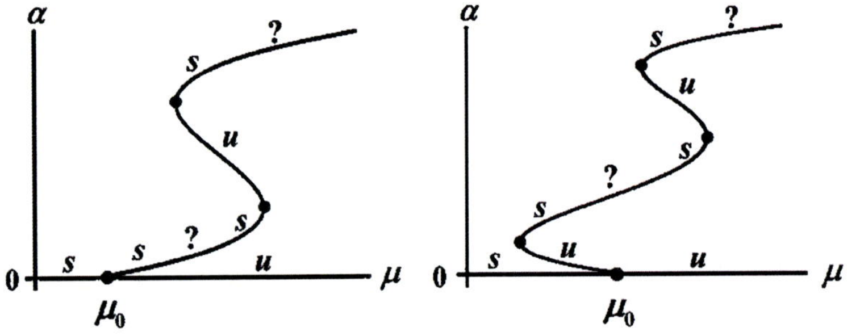

Figure 13.2: These plots illustrate the relationship between the geometry of the graph of positive equilibrium pairs of the scalar difference equation (13.16), obtained from (13.18), and their stability and instability when $\partial_\mu r > 0$ for all (μ,α). An s indicates a stable equilibrium and u indicates an unstable equilibrium. The question marks indicate that stability on increasing segments of the graph is only guaranteed in general near bifurcation points, indicated by sold circles. Figure reproduced from [3] with permission.

Example 15 *Consider a* 2×2 *matrix equation of the form (13.6) and (13.7) with the two-dimensional demographic vector*

$$\mathbf{x} = \begin{pmatrix} x_1 \\ x_2 \end{pmatrix} = \begin{pmatrix} \textit{juveniles} \\ \textit{adults} \end{pmatrix}$$

and within-season and across-season matrices

$$W(x_1, x_2) = \begin{pmatrix} s_1 \sigma(x_2) & b\varphi(x_2) \\ 0 & s_2 \end{pmatrix}, \qquad A = \begin{pmatrix} 0 & 0 \\ v_1 & v_2 \end{pmatrix}.$$

Here the parameters satisfy

$$0 < s_1, s_2 < 1, \; b > 0$$

$$0 \leq v_i < 1 \text{ and } v_1 > 0 \; .$$

We assume the density factors σ and φ are twice continuously differentiable on an open half line containing $\bar{R}_+$ and satisfy

$$\sigma(0) = \varphi(0) = 1, \; \varphi(x) > 0 \text{ and } 0 < s_1 \sigma(x) < 1 \text{ on } \bar{R}_+.$$

In this model, $b\varphi(x_2)$ juveniles are produced per adult per time unit where b is the inherent (density-free) birth rate of juveniles. The parameters s_1 and s_2 are the inherent per unit time survival probabilities of juveniles and adults, respectively, and $\sigma(x_2)$ is the density affecting juvenile survival. Note that no juvenile matures to an adult during the reproductive season. Finally, from the across-season matrix A, we see that the probabilities that a juvenile survives and that an adult survives from the end of one season to the start of the next season are, respectively, v_1 and v_2. A fraction v_1 of juveniles at the end of a season mature into adults at the beginning of the next season.

In this example it is assumed that adult fertility is regulated by adult population density in a negative way, i.e.,

$$\varphi'(x) < 0, \; x \geq 0.$$

We also assume that an increased adult density affects juvenile survival in a positive way (a component Allee effect). This could be attributed, for example, to protection of juveniles by adults. Thus, we assume

$$\sigma'(x) > 0, \; x \geq 0.$$

We take $\mu = b$ as the bifurcation parameter in assumption A6 and Theorem 12.

From the across-season matrix A we have that

$$\mathbf{a} = \begin{pmatrix} 0 \\ 1 \end{pmatrix}.$$

By induction we calculate $Q(\alpha)$ from (13.11) and obtain

$$Q(\mu, \alpha) = \begin{pmatrix} s_1 \sigma(\alpha) & b\varphi(\alpha) \\ 0 & s_2 \end{pmatrix}$$

if $k = 2$ *and*

$$Q(\mu,\alpha) = \begin{pmatrix} s_1^{k-1}\prod_{i=0}^{k-2}\sigma\left(s_2^i\alpha\right) & \mu\left(\sum_{i=0}^{k-3} s_1^{k-2-i}s_2^i\varphi\left(s_2^i\alpha\right)\prod_{j=i+1}^{k-2}\sigma\left(s_2^j\alpha\right) + s_2^{k-2}\varphi\left(s_2^{k-2}\alpha\right)\right) \\ 0 & s_2^{k-1} \end{pmatrix}$$

if $k \geq 3$. *When used in (13.14), these formulas give*

$$r(\mu,\alpha) = \begin{cases} v_1\mu\varphi(\alpha) + v_2 s_2 & \text{if } k = 2 \\ v_1\mu\psi(\alpha) + v_2 s_2^{k-1} & \text{if } k \geq 3 \end{cases},$$

where

$$\psi(\alpha) := \sum_{i=0}^{k-3}\left(s_1^{k-2-i}s_2^i\varphi\left(s_2^i\alpha\right)\prod_{j=i+1}^{k-2}\sigma\left(s_2^j\alpha\right)\right) + s_2^{k-2}\varphi\left(s_2^{k-2}\alpha\right) > 0. \tag{13.19}$$

From

$$r(\mu,0) = \begin{cases} v_1\mu + v_2 s_2 & \text{if } k = 2 \\ v_1\mu\psi(0) + v_2 s_2^{k-1} & \text{if } k \geq 3 \end{cases}$$

$$0 < \psi(0) = \sum_{i=0}^{k-2} s_1^{k-2-i}s_2^i = \begin{cases} \frac{s_2^{k-1} - s_1^{k-1}}{s_2 - s_1} & \text{if } s_2 \neq s_1 \\ (k-1)\, s_1^{k-2} & \text{if } s_2 = s_1 \end{cases}$$

we see that the bifurcation point $\mu_0 = b_0$ *where* $r(\mu,0) = 1$ *(and the extinction equilibrium destabilizes) is*

$$\mu_0 = \begin{cases} (1 - v_2 s_2)/v_1 & \text{if } k = 2 \\ \left(1 - v_2 s_2^{k-1}\right)/v_1\psi(0) & \text{if } k \geq 3 \end{cases}.$$

Note that

$$0 < \partial_\mu r(\mu,\alpha) = \begin{cases} v_1\varphi(\alpha) & \text{if } k = 2 \\ v_1\psi(\alpha) & \text{if } k \geq 3 \end{cases}$$

and consequently by Theorem 12 the direction of bifurcation (and the stability of bifurcating k*-cycles at* $\mu = \mu_0$*) is determined by the sign of*

$$\partial_\alpha^0 r = \begin{cases} v_1\mu_0\varphi'(0) & \text{if } k = 2 \\ v_1\mu_0\psi'(0) & \text{if } k \geq 3 \end{cases}.$$

Since $\partial_\alpha^0 r < 0$ *when the reproductive season has length* $k = 2$, *we conclude that a forward and stable bifurcation occurs. Thus, for* $\mu = b < \mu_0$ *the extinction equilibrium is stable. For at least* $b > \mu_0$ *close to* μ_0, *the population has a non-extinction 2-cycle (or, a stable equilibrium if the population is viewed only at the beginning of each reproductive season).*

For longer reproductive seasons $k \geq 3$, we see that the sign of $\partial_{\alpha}^{0} r$ is the same as the sign of $\psi'(0)$ whose sign therefore determines the direction of bifurcation at $b = \mu_0$ and the stability of the non-extinction k-cycles. From (13.19) a calculation using the product and chain rules shows $\psi'(0)$ is a linear combination of $\varphi'(0)$ and $\sigma'(0)$, specifically

$$\psi'(0) = \left(\sum_{i=0}^{k-2} s_1^{k-2-i} s_2^{2i}\right) \varphi'(0) + \left(\sum_{i=0}^{k-3} s_1^{k-2-i} s_2^{2i} \sum_{j=i+1}^{k-2} s_2^{j-i}\right) \sigma'(0).$$

If $\psi'(0) < 0$, then a forward and stable bifurcation occurs, and if $\psi'(0) > 0$, then a backward and unstable bifurcation occurs. The first case certainly occurs if the density effects on juvenile survival are negative $\sigma'(0) < 0$.

However, if there is an Allee component in juvenile survival, i.e., if $\sigma'(0) > 0$, then the formula for $\psi'(0)$ quantifies the magnitude of this effect (relative to the negative density effect on adult fertility $\varphi'(0)$) that will cause a backward and unstable bifurcation to occur. The dynamics of the population are determined by the scalar difference equation (13.16), namely,

$$\alpha(n+1) = \left(v_1 \mu \psi(\alpha(n)) + v_2 s_2^{k-1}\right) \alpha(n), \quad n \in Z$$

which is subject to the geometric analysis in Fig. 13.2. It follows that a backward bifurcation implies a strong Allee effect for an interval of $\mu = b$ values less than μ_0. A large value of $\sigma'(0) > 0$ means that adult density has a significant positive effect on juvenile survival (e.g., by parental care, protection from predators, etc.)

A strong Allee effect permits a population to survive when its inherent birth rate b is so low that extinction would occur in the absent of the Allee component (as might be caused, for example, by a deteriorated environment). This survival scenario is subject to the two caveats C_1 and C_2 discussed in Sect. 11.3 of Chap. 11, namely, that survival is initial condition dependent (i.e., population densities must remain sufficiently high so as to stay out of the Allee basin of attraction) and that there exists a "tipping point" threshold below which low value of b results in a sudden catastrophic extinct event.

13.4 Analysis of the Seasonal Cannibalism and Reproductive Synchrony Model

In this section we return to the cannibalism and reproductive synchrony model with across-season maturation in Sect. 13.2. We apply Theorem 12 to the matrix equation (13.1) with the within-season projection matrix (13.3)–(13.4) and across-season projection matrix (13.5) using the environmental resource availability as the bifurcation parameter $\mu = \rho$.

To locate the resource threshold above which the extinction equilibrium destabilizes, we need to find a value of $\rho = \rho_0$ such that $r(\rho_0, 0) = 1$ where

$$r(\rho, 0) = \mathbf{v}^{\tau} W^{k-1}(\rho, \mathbf{0}) \mathbf{v}$$

with matrix

$$W(\rho,\mathbf{0})=\begin{pmatrix}0&0&b(\rho)&0\\ s_1&s_2&0&0\\ 0&0&0&s_4\\ 0&0&s_3&0\end{pmatrix}.$$

A calculation shows

$$W^2(\rho,\mathbf{0})=\begin{pmatrix}0&0&0&b(\rho)s_4\\ s_1s_2&s_2^2&b(\rho)s_1&0\\ 0&0&s_3s_4&0\\ 0&0&0&s_3s_4\end{pmatrix}$$

and an induction shows that the even and odd powers of $W(\rho,\mathbf{0})$ are

$$W^{2n}(\rho,\mathbf{0})=\begin{pmatrix}0&0&0&b(\rho)s_4(s_3s_4)^{n-1}\\ s_1s_2^{2n-1}&s_2^{2n}&b(\rho)A_{2n}&b(\rho)B_{2n}\\ 0&0&(s_3s_4)^n&0\\ 0&0&0&(s_3s_4)^n\end{pmatrix}$$

$$0<A_{2n}:=s_1\sum_{i=1}^{n}(s_3s_4)^{n-i}\left(s_2^2\right)^{i-1}=\begin{cases}s_1\frac{(s_3s_4)^n-\left(s_2^2\right)^n}{s_3s_4-s_2^2} & \text{if } s_3s_4\neq s_2^2\\ ns_1\left(s_2^2\right)^{n-1} & \text{if } s_3s_4=s_2^2\end{cases}$$

$$0<B_{2n}:=s_1s_2s_4\sum_{i=1}^{n-1}(s_3s_4)^{n-1-i}\left(s_2^2\right)^{i-1}=\begin{cases}s_1s_2s_4\frac{(s_3s_4)^{n-1}-\left(s_2^2\right)^{n-1}}{s_3s_4-s_2^2} & \text{if } s_3s_4\neq s_2^2\\ (n-1)s_1s_2s_4\left(s_2^2\right)^{n-2} & \text{if } s_3s_4=s_2^2\end{cases}$$

and

$$W^{2n-1}(\rho,\mathbf{0})=\begin{pmatrix}0&0&b(\rho)(s_3s_4)^{n-1}&0\\ s_1s_2^{2n-2}&s_2^{2n-1}&b(\rho)A_{2n-1}&b(\rho)B_{2n-1}\\ 0&0&0&s_4(s_3s_4)^{n-1}\\ 0&0&s_3(s_3s_4)^{n-1}&0\end{pmatrix}$$

$$0<A_{2n-1}:=s_1s_2\sum_{i=1}^{n-1}(s_3s_4)^{n-1-i}\left(s_2^2\right)^{i-1}=\begin{cases}s_1s_2\frac{(s_3s_4)^{n-1}-\left(s_2^2\right)^{n-1}}{s_3s_4-s_2^2} & \text{if } s_3s_4\neq s_2^2\\ s_1s_2(n-1)\left(s_2^2\right)^{n-2} & \text{if } s_3s_4=s_2^2\end{cases}$$

$$0<B_{2n-1}:=s_1s_4\sum_{i=1}^{n-1}(s_3s_4)^{n-1-i}\left(s_2^2\right)^{i-1}=\begin{cases}s_1s_4\frac{(s_3s_4)^{n-1}-\left(s_2^2\right)^{n-1}}{s_3s_4-s_2^2} & \text{if } s_3s_4\neq s_2^2\\ s_1s_2(n-1)\left(s_2^2\right)^{n-2} & \text{if } s_3s_4=s_2^2\end{cases}$$

From these formulas we find that

$$r(\rho,0)=\begin{cases} b(\rho)\left[v_1(1-v)s_4(s_3s_4)^{n-1}+v_2(vA_{2n}+(1-v)B_{2n})\right] & \text{if } k=2n+1 \\ \quad +(v_3v+v_4(1-v))(s_3s_4)^n & \text{for } n=1,2,3,\dots \\ b(\rho)\left[v_1v(s_3s_4)^{n-1}+v_2(vA_{2n-1}+(1-v)B_{2n-1})\right] & \text{if } k=2n \\ \quad +[v_3(1-v)s_4+v_4vs_3](s_3s_4)^{n-1} & \text{for } n=2,3,4,\dots \end{cases} \tag{13.20}$$

Since we have assumed that $b'(\rho)>0$, $b(0)=0$, it follows that

$$r(0,0)<1, \quad \partial_\rho r(\rho,0)>0.$$

If $r(\rho,0)<1$ holds for all $\rho>0$, then the extinction equilibrium is stable for all resource levels ρ. Suppose, on the other hand, that $r(\rho,0)>1$ for sufficiently high resource levels ρ. Specifically, suppose

$$1<\lim_{\rho\to+\infty} r(\rho,0)\le+\infty. \tag{13.21}$$

In this case, since $r(\rho,0)$ is a strictly increasing function of ρ, there exists a unique value of $\rho=\rho_0>0$ at which $r(\rho_0,0)=1$. The local bifurcation at $\rho=\rho_0$ described in Theorem 12 and the global bifurcation alternatives shown in Fig. 13.2 apply to this model.

We are interested in circumstances when a backward bifurcation occurs because this typically leads to (initial condition dependent) survival and dangerous tipping points in severely degraded environments (as discussed in Sect. 13.3 and in Chaps. 11 and 12). By a severely degraded environment, we mean one in which resource levels ρ are so low that $r(\rho,0)<1$ and hence the population is threatened with extinction because the extinct equilibrium is stable. We are particularly interested in the role that reproductive synchrony plays.

As pointed out in Remark 13, a backward bifurcation occurs when there are, at low densities, dominant positive density effects (component Allee effects) present in the model. There are, in fact, both negative and positive density effects in the entries (13.4) of the projection matrix W. Specifically, negative low density effects occur in the entry w_{12} as a result of the assumed adult density (x_3 and x_4) regulation of fertility ($\varphi'(z)<0$) and in the entry w_{21} from the effect of cannibalism on egg survival ($\partial_i\pi_i\ge0$). Positive low density effects occur in the entries w_{34}, w_{43}, w_{44} with respect to increases in x_3 and x_4 from the assumed positive effect of cannibalism on adult survival ($\sigma_i'(z)>0$).

In addition, the adult reproductive synchrony term $g(x_3)$ contributes a negative feedback to w_{34}, but a positive feedback on w_{44} ($g'<0$). As a result, the effect of the reproductive synchrony on the direction of bifurcation is equivocal in the general model and, as we see from Examples 16 and 17, depends on model's inherent parameter values, namely, the survivorships s_i and v_i and the season length $k-1$.

Example 16 *As a specific example of the across-season model, we take*

$$\varphi(z) = \frac{1}{1 + c_{\varphi 3}x_3 + c_{\varphi 4}x_4}, \quad b(\rho) = b_0\rho \tag{13.22a}$$

$$\pi_j(x_1, x_j) = \frac{1}{1 + w_1 x_1}\frac{\omega_j(\rho)}{1 + \omega_j(\rho)x_j}, \quad \omega_j(\rho) = \frac{w_j}{1 + c_\rho \rho} \tag{13.22b}$$

$$\sigma_j(x_1, x_j) = 1 + \frac{s_{jm} - s_j}{s_j}\frac{c_{\sigma j}z}{1 + c_{\sigma j}z}, \quad g(x_3) = \frac{1}{1 + c_g x_3} \tag{13.22c}$$

for the nonlinear components in the projection matrix W given by (13.5). All of the parameters in these expressions are nonnegative.

Note that by setting a coefficient equal to zero, the corresponding mechanism is removed from the model. For example, $c_{\sigma j} = 0$ removes the positive benefit of cannibalism to the survival probability of adult x_j. Similarly, $c_g = 0$ removes the synchrony propensity, $w_j = 0$ removes cannibalism by x_j, and $w_1 = 0$ removes the victim saturation effect, etc. It is assumed that $0 < s_j < s_{jm} < 1$ so that the survival component $s_j\sigma_j$ increases from s_j to a maximum s_{jm} as the number of cannibalized victims z increases from 0 to $+\infty$. An increase in a parameter increases the effect (or intensity) of the corresponding nonlinear mechanism. These nonlinearities satisfy all of the requirements laid out above and, in particular, satisfy (13.21).

With the environmental resource availability $\mu = \rho$ as a bifurcation parameter, we calculate the bifurcation value

$$\rho_0 = \begin{cases} \frac{1}{b_0}\frac{1-(v_3 v+v_4(1-v))(s_3 s_4)^n}{v_1(1-v)s_4(s_3 s_4)^{n-1}+v_2(vA_{2n}+(1-v)B_{2n})} & \text{if } k = 2n+1 \text{ for } n = 1,2,3,\ldots \\ \frac{1}{b_0}\frac{1-[v_3(1-v)s_4+v_4 v s_3](s_3 s_4)^{n-1}}{v_1 v(s_3 s_4)^{n-1}+v_2(vA_{2n-1}+(1-v)B_{2n-1})} & \text{if } k = 2n \text{ for } n = 2,3,4,\ldots \end{cases}.$$

from (13.20) at which the local forward and backward bifurcation alternatives and global equilibrium properties shown in Fig. 13.2 hold.

For the parameter values displayed in the caption, Fig. 13.3 shows two bifurcation diagrams. In these diagrams the total population size α of the attractor at the beginning of each season is plotted as a function of the environmental resource availability ρ. Cannibalism is present and the bifurcations are backward, creating a strong Allee effect in a degraded environment when $\rho < \rho_0$. By this we mean (initial condition dependent) survival for an interval of environmental resource levels ρ below the threshold ρ_0 where the inherent population growth rate $r(0)$ is less than 1 and the extinction equilibrium is stable.

The bifurcation diagram with $c_g = 0$ (when reproductive synchrony is not possible) illustrates a cannibalism-induced backward bifurcation. If cannibalism were also absent, then the bifurcation would be forward. We know this because there would then be only negative density effects (Remark 13). In this sense, cannibalism is beneficial in this example because it allows survival in a degraded environment when in the absence of cannibalism extinction would result (although in the presence of tipping points below which extinction occurs).

The bifurcation diagram with $c_g > 0$ indicates added benefits toward the survival of the population due to reproductive synchrony. It does this in two ways: the presence of the synchrony propensity term results in a higher total population size (at the start of each season) and also a reduced environmental resource "tipping point" where the saddle-node bifurcation leads to catastrophic extinction.

Figure 13.4 shows some sample time series plots at the indicated locations on the bifurcation diagrams. Note the within-season adult class synchronized oscillations when $c_g > 0$ and their absence when $c_g = 0$.

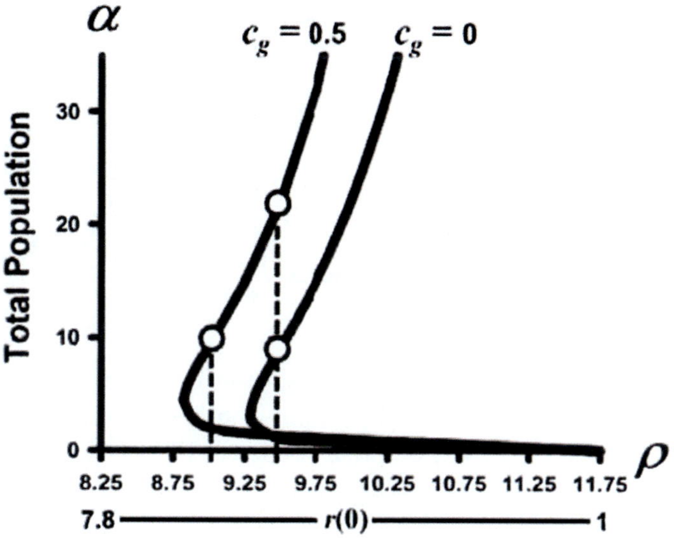

Figure 13.3: Shown are two backward bifurcation diagrams plotting total population size a at the beginning of the season of the attractor as a function of the environmental resource availability ρ using the nonlinearities (13.22) in (13.1) with the projection matrices (13.3) and (13.5). In one plot ($c_g = 0$), the synchrony propensity term is absent and in the other plot ($c_g = 0.5$) it is present. Parameter values are $b_0 = 0.7$, $c_{\varphi 3} = c_{\varphi 4} = 0.001$, $c_\rho = 1$, $c_{\sigma 3} = c_{\sigma 4} = 1$, $w_1 = 1$, $w_3 = w_4 = 2$, $s_1 = s_2 = 0.3$, $s_3 = s_4 = 0.9$, $s_{3m} = s_{4m} = 0.99$, and $\nu_1 = \nu_2 = \nu_4 = \nu_5 = 0.95$. The bifurcation value is $\rho^* \approx 11.75$. Sample time series plots at the open circles are shown in Fig. 13.4. Figure reproduced from [3] with permission.

Example 17 *The sample simulations in Example 16 show the benefit of reproductive synchrony that is possible in a cannibalistic population. This is not, however, always predicted by the model. In fact, the opposite can be the case. The two sample bifurcation diagrams in Fig. 13.5 show a case in which reproductive synchrony reduces population size and increases the tipping point for the environmental resource availability. These diagrams use the same within-season parameter values as in Fig. 13.4, but change the across-season survival probabilities as indicated in the captions.*

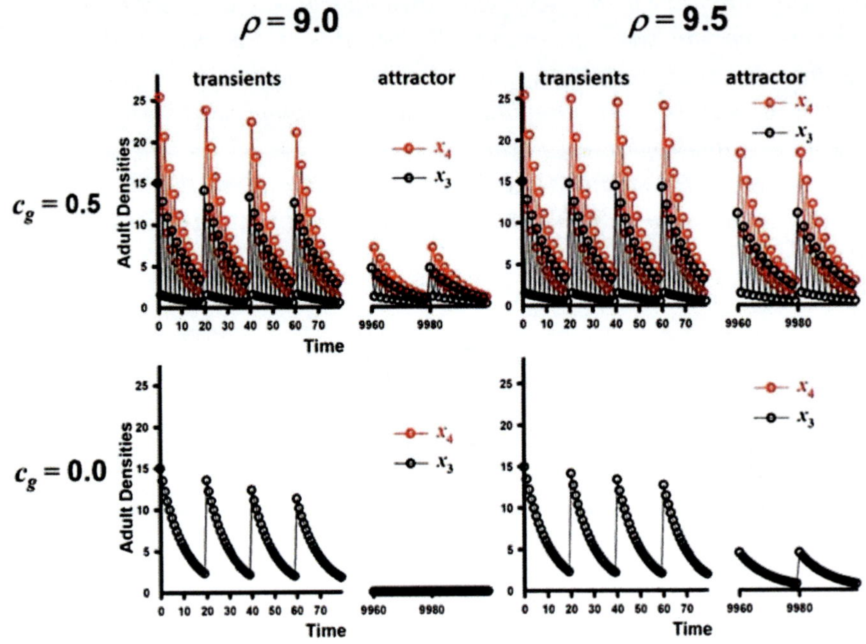

Figure 13.4: Sample adult class time series are shown for initial conditions $x_1 = x_2 = 0$, $x_3 = \nu\alpha$, $x_4 = \alpha(1-\nu)$ with $\nu = 0.5$ and $\alpha = 30$. The first column of plots is for $\rho = 9.0$ and the second column is for $\rho = 9.5$ (open circles). Figure reproduced from [3] with permission.

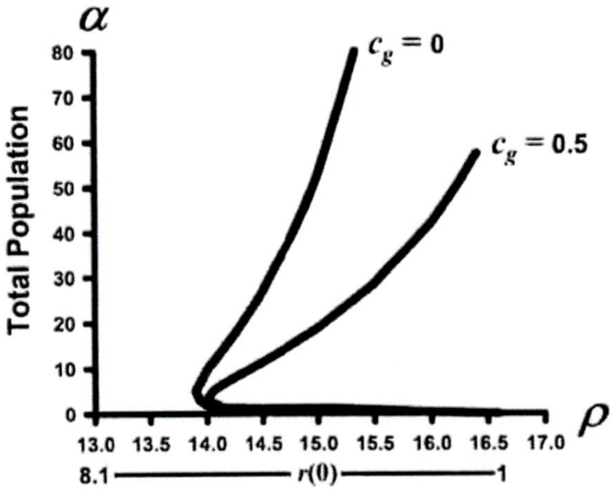

Figure 13.5: Shown are the same two bifurcation plots that appear in Fig. 13.3, with the same parameter values except for a change in the across-season survival probabilities to $\nu_1 = 0.5$, $\upsilon_2 = \nu_3 = \nu_4 = 0.99$. Figure reproduced from [3] with permission.

13.5 Conclusion

The model in this chapter brings our modeling efforts a step closer to the life history strategies of the gull populations that motivated them by separating the breeding season time scale from the maturation or generation time scale. The density effects, including cannibalism, and reproductive timing occur during a breeding season, but maturation occurs between seasons. The model leads to a periodically forced, matrix equation whose mathematical analysis is more involved than that of the autonomous matrix equations studied in Chaps. 11 and 12. Nonetheless, our analysis provides similar bifurcation scenarios to those seen in the simpler autonomous models. Namely, it describes the bifurcation that occurs as the extinction state destabilizes, and it provides diagnostic quantities that determine the direction of bifurcation and hence stability of the resulting survival states.

Using specific examples we show how *adult-on-juvenile cannibalism can, under the right circumstances, result in a long-term survival benefit in a degraded environment that would otherwise lead to extinction and can also lead to reproductive synchrony. Moreover, these examples also show that a high propensity to synchronize can have other benefits as well, such as increased population levels at the beginning of each new season.*

Although the model in this chapter captures more detail relevant to the glaucous-winged gull populations that motivated the study, there are several significant simplifications in the model that still separate it from the biology of gulls. First, the model assumes that adult females can lay eggs throughout the breeding season, whereas in reality each female gull only lays (approximately) three eggs during the season. Moreover, the model does not track the order in which eggs are laid, and so any egg can be cannibalized on the day it is laid, whereas in gull populations it is the first egg laid in the nest that is the most likely to be cannibalized. Second, the model assumes that all surviving juveniles mature to reproducing adults at the beginning of the next breeding season, whereas juvenile glaucous-winged gulls actually require 4 years to mature. Another feature that could be of importance is density-dependent across-season survival, which is ignored in this model. The modeling methodology used in this chapter can include these additional features, but it would result in a higher dimensional model that includes several different age classes of juveniles. Although there would be a dimensional reduction for the dynamics of the composite map from the start of the season to the start of the next season, it would no longer necessarily be a reduction to dimension one and hence to a tractable one-dimensional difference equation. The resulting periodically forced model would therefore be more challenging to analyze.

References

1. Cushing JM (2016) One dimensional maps as population and evolutionary dynamic models. In: Cushing JM, Saleem M, Srivastava HM, Khan MA, Merajuddin M (eds) Applied Analysis in Biological and Physical Sciences, Springer Proceedings in Mathematics & Statistics, vol 186. Springer, India, pp 41–62

2. Cushing JM (2019) Difference equations as models of evolutionary population dynamics. J Biol Dyn 13:103–127. https://doi.org/10.1080/17513758.2019.1574034

3. Cushing JM, Henson SM (2018) Periodic matrix models for seasonal dynamics of structured populations with application to a seabird population. J Math Biol 77:1689–1720. https://doi.org/10.1007/s00285-018-1211-4

4. Elaydi SN (2005) An Introduction to Difference Equations, 3rd edn. Springer, New York

5. Henson SM, Cushing JM, Hayward JL (2011) Socially induced ovulation synchrony and its effect on seabird populations. J Biol Dyn 5:495–516

Index

J. M. Cushing et al., *Modeling Behavior and Population Dynamics*, Interdisciplinary Applied Mathematics 57, https://doi.org/10.1007/978-3-031-34283-7